Probabilistic Graphical Models for Computer Vision

Computer Vision and Pattern Recognition Series

Series Editors

Also in the Series:

Lin and Zhang, Low-Rank Models in Visual Analysis: Theories, Algorithms and Applications, 2017, ISBN: 9780128127315
Zheng et al., Statistical Shape and Deformation Analysis: Methods, Implementation and Applications, 2017, ISBN: 9780128104934
De Marsico et al., Human Recognition in Unconstrained Environments: Using Computer Vision, Pattern Recognition and Machine Learning Methods for Biometrics, 2017, ISBN: 9780081007051
Saha et al., Skeletonization: Theory, Methods and Applications, 2017, ISBN: 9780081012918

Probabilistic Graphical Models for Computer Vision

Qiang Ji

Academic Press is an imprint of Elsevier
125 London Wall, London EC2Y 5AS, United Kingdom
525 B Street, Suite 1650, San Diego, CA 92101, United States
50 Hampshire Street, 5th Floor, Cambridge, MA 02139, United States
The Boulevard, Langford Lane, Kidlington, Oxford OX5 1GB, United Kingdom

Notices

Knowledge and best practice in this field are constantly changing. As new research and experience broaden our understanding, changes in research methods, professional practices, or medical treatment may become necessary.

Practitioners and researchers must always rely on their own experience and knowledge in evaluating and using any information, methods, compounds, or experiments described herein. In using such information or methods they should be mindful of their own safety and the safety of others, including parties for whom they have a professional responsibility.

To the fullest extent of the law, neither the Publisher nor the authors, contributors, or editors, assume any liability for any injury and/or damage to persons or property as a matter of products liability, negligence or otherwise, or from any use or operation of any methods, products, instructions, or ideas contained in the material herein.

Library of Congress Cataloging-in-Publication Data
A catalog record for this book is available from the Library of Congress

British Library Cataloguing-in-Publication Data
A catalogue record for this book is available from the British Library

ISBN: 978-0-12-803467-5

For information on all Academic Press publications
visit our website at https://www.elsevier.com/books-and-journals

Publisher: Mara Conner
Acquisition Editor: Tim Pitts
Editorial Project Manager: Mariana L. Kuhl
Production Project Manager: Kamesh Ramajogi
Designer: Matthew Limbert

Typeset by VTeX

Contents

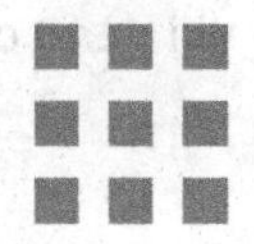

List of Figures

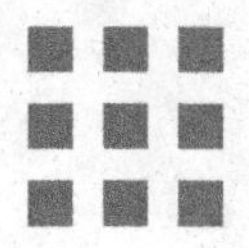

List of Tables

Background and motivation

1.1 Introduction

Probabilistic graphical models (PGMs) are a diagrammatic representation of the joint probability distribution of a set of random variables. PGMs bring graph theory and probability theory together for the purpose of multivariate statistical modeling. The power of PGMs lies in their integration of graphs' intuitive and powerful data structure with the rigor of the probability. The graph representation is intuitive and has semantic meanings that reflects the humans' understanding of the domain. With their built-in conditional independencies, PGMs allow the often intractable joint probability distribution to be compactly factorized as a product of local functions for each node. This factorization significantly reduces the number of parameters needed to specify a joint probability distribution. Moreover, hierarchical representation of PGMs allows for the capturing of knowledge at different levels of abstraction and the systematic encoding of domain-specific knowledge.

In addition to providing powerful representations, PGMs provide mechanisms for automatically learning a model from data and performing inference within the model in a principled manner. The graph structure allows efficient manipulation of the joint probability for both learning and inference. Many years of research on PGMs have produced a body of well-established algorithms for PGM learning and inference. The research on PGMs remains active, and many new developments are being made continuously. Finally, as a general mathematical model, PGMs generalize over many existing well-established multivariate models, including mixture models, factor analysis, hidden Markov models, Kalman filters, and Ising models. The flexibility of PGMs means that they perform well in various applications. As a result, PGMs provide a unified theoretical framework for systemically formulating and solving many real-world problems, problems in bioinformatics, medical diagnoses, natural language processing, computer vision (CV), pattern recognition, computer networks, telecommunications, and control theory.

The idea of using graphs to represent the interactions among variables can be traced back to the 1900s in several different fields. In statistical physics, Gibbs [1] used undirected graphs to characterize system with multiple particles. Each particle was treated as a random variable. In the area of genetics, Wright [2,3] used directed graphs to model the inheritance relationships of natural species. In the area of statistics, Bartlett [4] first studied the relationships between variables in log-linear models. This idea was further developed in [5,6] through the introduction of a Markov field to model multivariate Gaussian distributions. In the early 1970s, Grenander et al. [7] introduced the pattern theory as the key mathematical model tool for modeling and solving vision problems. Through the pattern theory, Grenander proposed a flexible probabilistic graph be employed to model the un-

Probabilistic Graphical Models for Computer Vision. https://doi.org/10.1016/B978-0-12-803467-5.00006-X

derlying structural dependencies in the data and their uncertainties, and that the graphical model then be used to perform pattern recognition via Bayesian inference from the observed data.

PGMs became widely accepted in computer science in the late 1980s due to major theoretical breakthroughs. The book *Probabilistic Reasoning in Intelligent Systems* [8] by Judea Pearl laid the foundation for later research on Bayesian networks. In the meantime, Lauritzen and Spiegelhalter [9] proposed algorithms for efficient reasoning in PGMs, making fundamental contributions to probabilistic inference. Another reason that PGMs became widely accepted is the development of more advanced expert systems. Several early attempts involved building these expert systems based on the naive Bayes model [10,11]. The Pathfinder system [12] is an example of a system that uses a Bayesian network to assist general pathologists with diagnoses in hematopathology. Since then the PGM framework has gained explosive attentions from a wide range of communities, including communication, medical diagnosis, gene analysis, financial prediction, risk analysis, and speech recognition fields. Several highly influential books came along with rapid developments [13–16].

An exciting development in CV over the last decade has been the gradual widespread adoption of PGMs in many areas of CV and pattern recognition. In fact, PGMs have become ubiquitous for addressing a wide variety of CV problems, ranging from low-level CV tasks, such as image labeling and segmentation, to middle-level computer vision tasks, such as object recognition and 3D reconstruction, to high level computer vision tasks, such as semantic scene understanding and human activity recognition. Specifically, for image segmentation and image labeling, Markov random fields (MRFs) and conditional random fields (CRFs) have become de facto state-of-the-art modeling frameworks. The same class of models has also been effectively applied to stereo reconstruction, demonstrating their wide applicability. Bayesian networks have been used for representing causal relationships for a variety of vision tasks, including facial expression recognition, active vision, and visual surveillance. Similarly, dynamic PGMs such as hidden Markov models, dynamic Bayesian networks, and their variants have been routinely used for object tracking, human action, and activity recognition.

The history of PGMs in CV closely follows that of graphical models in general. Research by Judea Pearl and Steffen Lauritzen in the late 1980s played a seminal role in introducing this formalism to areas of artificial intelligence and statistical learning. Not long after, the formalism spread to fields including statistics, systems engineering, information theory, pattern recognition, and CV. One of the earliest occurrences of graphical models in the vision literature was a paper by Binford, Levitt, and Mann [17]. It described the use of a Bayesian network in a hierarchical probability model to match 3D object models to groupings of curves in a single image. The following year marked the publication of Pearl's influential book on graphical models [8]. In 1993, IEEE Transactions on Pattern Analysis and Machine Intelligence (TPAMI), the leading CV journal, published its first special issue (SI) on graphical models [18]. The SI focused on methods for the construction, learning, and inference of directed graphical models (e.g., Bayesian networks) and their applications to spatial structural relationships modeling for perceptual grouping in CV. Ten years later, in 2003, TPAMI published its second SI on graphical models in computer vision [19], demonstrating the progress of PGM research in computer vision. Since then, many technical papers have been published that address different aspects and continuing applications of PGMs in computer vision. In 2009, TPAMI published its third special issue on applications of PGMs in CV [20]. Compared to the first and second TPAMI SIs, the third SI expanded in both theoretical and application scopes. It included both directed and undirected graphical models and applications in both CV and pattern recognition. Since then, to meet CV researchers' increasing demands for PGMs, a series of well-attended workshops and tutorials on PGMs and their applications in CV have been held continuously in major CV conferences and journals, including 2011 CVPR Workshop on Inference in Graphical Models with Structured Potentials [21], 2011 CVPR Tutorial on Structured Prediction and Learning in Computer Vision [22], 2011 ICCV Tutorial on Learning with Inference for Discrete Graphical Models [23], 2013 ICCV workshop on inference in PGMs [24], 2014 ECCV Workshop on Graphical Models in Computer Vision [25], 2014 CVPR Tutorial on Learning and Inference in Discrete Graphical Models [26], and 2015 ICCV Tutorial on Inference in Discrete Graphical Models [27]. The most recent TPAMI SI on Higher Order Graphical Models in Computer Vision: Modelling, Inference and Learning [28] specifically addresses the inference and learning of high-order PGMs for various CV tasks. These series of workshops, tutorials, and special issues further demonstrate the importance and significance of PGMs for CV and increasing demands for and interest in PGMs by CV researchers. The latest developments in deep learning have further demonstrated the importance of graphical models since the building blocks of some major deep learning architectures, such as the deep Boltzmann machine (DBM) and the deep belief network are all special kinds of PGMs.

Several factors have contributed to the widespread use of PGMs in CV. First, many CV tasks can be modeled as structured learning and prediction problems that involve a set of input variables $\mathbf{X}$ and a set of output variables $\mathbf{Y}$. The goal of structured learning and prediction is to learn a model $\mathcal{G}$ that relates $\mathbf{X}$ to $\mathbf{Y}$ such that during testing we can predict $\mathbf{Y}$ given $\mathbf{X}$ using $\mathcal{G}$. Whereas input variables typically capture images or their derivatives (various image features), the output variables are the target variables we want to estimate. For example, for image segmentation, $\mathbf{X}$ represent pixels or their features, whereas output $\mathbf{Y}$ represent their labels. For 3D reconstruction, $\mathbf{X}$ represent the image features of each pixel, whereas $\mathbf{Y}$ represent their 3D coordinates or 3D normals. For body pose estimation and tracking, $\mathbf{X}$ represent body image features, whereas $\mathbf{Y}$ represent the 3D joint angles or joint positions.

The key in structured learning and prediction is to capture structural relationships among elements of $\mathbf{X}$ and $\mathbf{Y}$ and between $\mathbf{X}$ and $\mathbf{Y}$. PGMs are well suited for such structured learning and prediction problems due to their powerful and effective capability in modeling various types of relationships between the random variables and the availability of principled statistical theories and algorithms for inference and learning. Using probabilistic models such as Bayesian networks (BNs) and Markov networks (MNs), we can systemically capture the spatio-temporal relationships between the input and target variables. Specifically, PGMs can be used to capture either the joint probability distribution of $\mathbf{X}$ and $\mathbf{Y}$, that is, $p(\mathbf{X}, \mathbf{Y})$ with either BN or MN or the conditional probability $p(\mathbf{Y}|\mathbf{X})$ with a conditional random field (CRF). PGM learning can be used to automatically learn the parameters for these models. Given the learned models that capture either the joint or conditional distributions of the input and output, the prediction problem can be solved as a maximum a posterior probability (MAP) inference problem, that is, finding $\mathbf{y}$ that maximizes $p(\mathbf{y}|\mathbf{X} = \mathbf{x})$, $\mathbf{y}^* = \arg\max_{\mathbf{y}} p(\mathbf{y}|\mathbf{x})$. Well-established PGM learning and inference algorithms are available to learn the models and to perform inference.

Second, uncertainties are abundant in CV problems due to signal noise and to ambiguity or incomplete knowledge about the target variables. They exist in the images, in their features, in the outputs, and in the relationships between inputs and outputs of the algorithms. Any solutions to CV problems should systematically account for uncertainties, propagate them, and evaluate their effects on the estimated target variables. Through the probability theories, PGMs provide a powerful means to capture uncertainties and to incorporate and propagate them into the estimation of the target variables.

Third, as a Bayesian framework, PGMs can encode high-level domain knowledge, such as the physical laws that govern the properties and behaviors of the target variables or CV theories such as the projection/illumination models that relate the 2D images to their corresponding 3D models. High-level knowledge is important in constraining and regularizing ill-posed CV problems. Hence PGMs offer a unified model to systematically encode high-level knowledge and to combine it with data.

In summary, PGMs provide a unified framework for rigorously and compactly representing the image observations, target variables, their relationships, their uncertainties, and high-level domain knowledge and for performing recognition and classification through rigorous probabilistic inference. As a result, PGMs are being increasingly applied to addressing a wide range of CV problems, and these applications have led to significant performance improvement.

1.2 Objectives and key features of this book

To meet CV researchers' increasing needs for and interest in PGMs and to fill the void created by the lack of a book that specifically discusses PGMs in the context of CV, this book provides a comprehensive and systematic treatment of the topics of PGMs and their applications in CV. To this end, we first provide an in-depth discussion of the basic concepts and well-established theories of PGM models and theories with a focus on PGM models that have been widely used in CV and in the latest developments in PGM models such as the PGM-based deep models, which could potentially benefit CV problems. We will discuss PGM models and theories at a level suitable for CV researchers to understand. In addition, our discussion will be accompanied by corresponding pseudocode, which is easy for CV researchers to understand and can be quickly and easily implemented.

We then move on to demonstrate applications of PGM models to a wide range of CV problems from low-level CV tasks, such as image segmentation and labeling, middle-level computer vision tasks, such as object detection and recognition, object tracking, and 3D reconstruction, to high-level CV tasks, such as facial expression recognition and human activity recognition. For each CV problem, we will show how a PGM can be used to model the problem, how to learn the PGM model from training data, how to encode the high-level knowledge into the PGM models, and how to use the learned PGM models to infer the unknown target variables given their image measurements. We will also contrast PGM formulations and solutions against those of conventional machine learning models for the same CV task. Through these application examples, we intend to demonstrate that PGMs provide a unified framework that allows CV problems to be solved in a principled and rigorous manner and the advantages of PGMs over other learning models.

Finally, to provide a self-contained and stand-alone book, we will offer the necessary background materials on such topics as probability calculus, basic estimation methods, optimization methods, and sampling methods. To fully take advantage of this book, readers must have basic mathematical knowledge of calculus, probability, linear algebra, and optimization and mastery of one high-level programming language such as C/C++ or Python.

In summary, resulted from the author's many years of research and teaching in CV and PGMs, this book offers a comprehensive and self-contained introduction to the field of probabilistic graphical models for CV. It represents the first such book that introduces PGMs specifically for the purpose of CV.

1.3 PGM introduction

As a graphical representation of probabilities, a PGM consists of nodes and links, where nodes represent RVs (RVs), whereas links represent their probabilistic dependencies. Depending on the types of links, graphical models can be classified into directed or undirected PGMs. The directed PGMs consist of directed links, whereas the undirected PGMs consist of undirected links. The commonly used directed PGMs in CV include Bayesian networks (BNs) and hidden Markov models (HMMs), whereas the commonly used undirected graphs for CV include Markov networks (MNs) (also called Markov random fields (MRFs)) and conditional random fields (CRFs). For undirected PGMs, the links typically capture the mutual dependencies or correlations among the RVs, whereas links for directed graphical models often capture the causal relationships among random variables. For CV, PGMs can be used to represent the elements of an image or a video and their relationships. Figure 1.1 shows an example of a directed and an undirected PGM for image segmentation. The directed PGM in Fig. 1.1A is a BN. Its nodes represent respectively the image regions/superpixels (Rs), edges of image regions (Es), and vertexes (Vs), and its directed links capture the natural causal relationships between image regions, edges, and vertexes. The causal relationships include neighboring regions' intersections, which produce edges, whose interactions, in turn, produce vertexes. A BN is parameterized by the conditional probability p for each node. A BN hence captures the joint probability distribution among random variables that represent image regions, edges, and vertexes. The

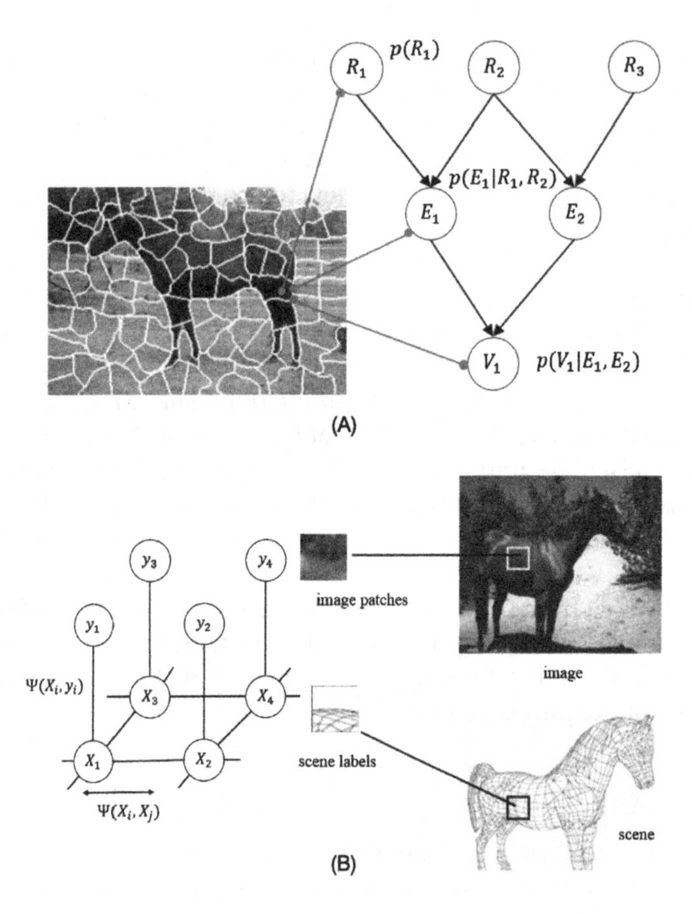

FIGURE 1.1 Examples of PGMs for image segmentation: (A) Bayesian Network and (B) Markov Network.

undirected PGM in Fig. 1.1B is an MN. Its nodes represent image regions (Ys) and their labels (Xs). Its undirected links capture the mutual dependencies between image regions and their labels. For the MN in Fig. 1.1B, the links among the label nodes capture the mutual influence between the neighboring label nodes, whereas the link between a label node and the corresponding image node captures the relation between the image label and its image features. The model is parameterized by the potential function ψ for each node. Like a BN, a MN captures the joint probability distribution of the random variables that represent image regions and their labels. Compared to BNs, MNs are more natural in modeling many CV problems since they do not impose directed edges or require causal relationships among random variables and their parameterizations are more general. Bayesian networks, in contrast, encode more conditional independencies among its variables, and their parameterization is limited to probability. However, BNs are computationally more tractable than MNs in terms of both learning and inference.

Given its topology, a PGM embeds certain independencies among its nodes due to the local Markov condition assumption. For example, in Fig. 1.1A, R_1 and R_2 are marginally independent but become dependent given E_1. Similarly, for Fig. 1.1B, X_1 is independent of X_4 given X_2 and X_3. These built-in independence relationships lead to an important property of PGM, namely, the joint probability distributions of its nodes can be factorized into products of the local functions for each node. For example, for Fig. 1.1A, the joint probability of all nodes can be written as

$$p(R_1, R_2, R_3, E_1, E_2, V_1) = \\ p(R_1)p(R_2)p(R_3)p(E_1|R_1, R_2)p(E_2|R_2, R_3)p(V_1|E_1, E_2). \tag{1.1}$$

With the factorization, a PGM can efficiently and compactly represent the joint probability distribution in high-dimensional space with a small number of parameters. Furthermore, with the built-in conditional independencies, it is possible to perform efficient learning and inference.

1.3.1 PGM issues

Two major issues in PGMs are learning and inference. Learning involves automatically estimating from the training data the PGM's structure or parameters, or both. A learning problem is typically formulated as a continuous optimization problem either through maximum likelihood estimation or maximum a posteriori estimation. Like learning for other models such as support vector machine (SVM) and neural networks (NNs), learning for PGMs is to learn the model or the mapping function that relates the input to the output. Similarly, learning can be done in either a fully supervised or semisupervised manner. With respect to CV, PGM learning is to learn the PGM model that captures the joint probability distribution of the input and output variables. For example, for the BN in Fig. 1.1A, parameter learning is to learn the conditional probability for each node such as $p(R_1)$, $p(E_1|R_1, R_2)$, and $p(V_1|E_1, E_2)$ for nodes R_1, E_1, and V_1 respectively. For the MN model in Fig. 1.1B, parameter learning is to learn the parameters for the unary potential function such as $\Phi(x_i, y_i)$ and those of the pairwise potential function such as $\Psi(x_i, x_j)$.

Inference is concerned with estimating the probability of a set of target variables given the observations of other variables, through which we can infer the most likely states/values for the target variables. For CV, the inference is using the learned PGM model to predict the most likely values of the target variables given the image observations of the input variables. It is often formulated as a discrete minimization problem. For example, for image segmentation, the inference is to infer the most likely labels for each pixel given their image measurements. For the BN in Fig. 1.1A, a possible inference is computing $p(R_1, R_2, R_3|O_R)$ to identify the labels for each image region, that is, $R_1^*, R_2^*, R_3^* = \arg\max_{R_1,R_2,R_3} p(R_1, R_2, R_3|O_R)$, where O_R are image observations of the regions. Similarly, inference for the MN in Fig. 1.1B is computing $p(\mathbf{x}|\mathbf{y})$ to estimate the most likely image patch labels $\mathbf{x}$, that is, $\mathbf{x}^* = \arg\max_{\mathbf{x}} p(\mathbf{x}|\mathbf{y})$.

PGM learning and inference are NP-hard in general. By exploiting the conditional independencies embedded in the PGM, efficient exact and approximate inference and learning methods have been introduced to scale up to large models. In this book, we will first introduce directed and undirected PGMs and then discuss their learning and inference methods, including both exact and approximated methods.

1.4 Book outline

The book is divided into two main parts. The first part introduces the definitions, concepts, and fundamental theories and algorithms for PGMs, whereas the second part focuses on applications of PGMs to different CV problems. Specifically, the book consists of five chapters, and they are organized as follows. In Chapter 1 (this chapter), we introduced PGMs and discussed their significance and motivation for CV and the unique features and focuses of this book. In Chapter 2, we include a review of the necessary background knowledge to make the book self-contained, including probability calculus, basic estimation and optimization methods, and basic sampling techniques. In Chapter 3, we first introduce the basic concepts, definitions, and properties of directed PGMs including BNs, dynamic BNs, and their variants. This will then be followed by a comprehensive introduction to well-established theories for learning and inference with directed PGMs. Among the main theoretical issues to be covered are structure and parameter learning under both complete and incomplete data and exact and approximated inference methods for directed PGMs. Similarly, in Chapter 4, we introduce undirected PGMs and their learning and inference methods. In Chapter 5, we discuss applications of PGMs to a wide range of CV problems from low-level CV tasks, such as image denoising and image segmentation, middle-level CV tasks, such as object detection, recognition, tracking, and 3D reconstruction, to high-level CV tasks, such as facial expression recognition and human activity recognition.

References

[1] J.W. Gibbs, Elementary Principles in Statistical Mechanics, Courier Corporation, 2014.
[2] S. Wright, Correlation and causation, Journal of Agricultural Research 20 (7) (1921) 557–585.
[3] S. Wright, The method of path coefficients, The Annals of Mathematical Statistics 5 (3) (1934) 161–215.
[4] M.S. Bartlett, Contingency table interactions, Supplement to the Journal of the Royal Statistical Society 2 (2) (1935) 248–252.
[5] N. Wermuth, Analogies between multiplicative models in contingency tables and covariance selection, Biometrics (1976) 95–108.
[6] J.N. Darroch, S.L. Lauritzen, T.P. Speed, Markov fields and log-linear interaction models for contingency tables, The Annals of Statistics (1980) 522–539.
[7] U. Grenander, M.I. Miller, M. Miller, et al., Pattern Theory: From Representation to Inference, Oxford University Press, 2007.
[8] J. Pearl, Probabilistic Reasoning in Intelligent Systems: Networks of Plausible Reasoning, 1988.
[9] S.L. Lauritzen, D.J. Spiegelhalter, Local computations with probabilities on graphical structures and their application to expert systems, Journal of the Royal Statistical Society. Series B Methodological (1988) 157–224.
[10] H.R. Warner, A.F. Toronto, L.G. Veasey, R. Stephenson, A mathematical approach to medical diagnosis: application to congenital heart disease, JAMA 177 (3) (1961) 177–183.

[11] F. De Dombal, D. Leaper, J.R. Staniland, A. McCann, J.C. Horrocks, Computer-aided diagnosis of acute abdominal pain, British Medical Journal 2 (5804) (1972) 9–13.
[12] D.E. Heckerman, E.J. Horvitz, B.N. Nathwani, Toward normative expert systems: the pathfinder project, Methods of Information in Medicine 31 (1991) 90–105.
[13] D. Koller, N. Friedman, Probabilistic Graphical Models: Principles and Techniques, MIT Press, 2009.
[14] M.I. Jordan, Learning in Graphical Models, vol. 89, Springer Science & Business Media, 1998.
[15] S.L. Lauritzen, Graphical Models, vol. 17, Clarendon Press, 1996.
[16] R.E. Neapolitan, et al., Learning Bayesian Networks, 2004.
[17] T. Binford, T. Levitt, W. Mann, Bayesian inference in model-based machine vision, in: Proceedings of the 3rd Annual Conference on Uncertainty in Artificial Intelligence, 1987, pp. 73–96.
[18] IEEE Transactions on Pattern Analysis and Machine Intelligence 15 (3) (1993).
[19] J. Rehg, V. Pavlovic, T. Huang, W. Freeman, Guest editors' introduction to the special section on graphical models in computer vision, IEEE Transactions on Pattern Analysis and Machine Intelligence 25 (7) (2003) 785–786.
[20] Q. Ji, J. Luo, D. Metaxas, A. Torralba, T.S. Huang, E.B. Sudderth, Special issue on probabilistic graphical models in computer vision, IEEE Transactions on Pattern Analysis and Machine Intelligence (2009).
[21] CVPR 2011 Workshop on Inference in Graphical Models With Structured Potentials, 2011 [online], available: http://users.cecs.anu.edu.au/~julianm/cvpr2011.html.
[22] CVPR 2011 Tutorial Structured Prediction and Learning in Computer Vision, 2011 [online], available: http://www.nowozin.net/sebastian/cvpr2011tutorial/.
[23] ICCV 2011 Tutorial on Learning With Inference for Discrete Graphical Models, 2011 [online], available: http://www.csd.uoc.gr/~komod/ICCV2011_tutorial.
[24] ICCV 2013 Workshop: Inference for Probabilistic Graphical Models (PGMs), 2013 [online], available: http://cs.adelaide.edu.au/~chhshen/iccv2013_workshop/.
[25] International Workshop on Graphical Models in Computer Vision, 2014, in conjunction with European Conference on Computer Vision [online], available: http://www.tnt.uni-hannover.de/gmcv/index.html.
[26] CVPR 2014 Tutorial on Learning and Inference in Discrete Graphical Models, 2014 [online], available: http://users.cecs.anu.edu.au/~julianm/cvpr2011.html.
[27] ICCV 15 Tutorial – Inference in Discrete Graphical Models, 2015 [online], available: http://cvlab-dresden.de/iccv-15-tutorial-inference-in-discrete-graphical-models/.
[28] K. Alahari, D. Batra, S. Ramalingam, N. Paragios, R. Zemel, Special issue on higher order graphical models in computer vision: modelling, inference and learning, IEEE Transactions on Pattern Analysis and Machine Intelligence (2015) [online], available: http://www.computer.org/csdl/trans/tp/2015/07/07116679.pdf.

Foundation and basics

2.1 Introduction

To make the book self-contained, in this chapter, we introduce the relevant probability calculus, basic estimation and optimization methods, and basic sampling techniques. Understanding these topics is necessary for the subsequent discussion of graphical model theories. Note that this chapter provides only the basic and minimum information necessary to understand the core concepts and theories of PGMs. For an in-depth review and systematic learning of these topics, readers are advised to consult related reference books, such as [1].

2.2 Random variables and probabilities

Before we start the discussion, we first need to define some notations. We use uppercase letters to denote random variables (e.g., X) and the corresponding lowercase letters of the variables to represent their realizations or values (e.g., x). We use a bold-face uppercase letter (e.g., $\mathbf{X}$) to represent a random vector (column vector) that consists of a set of random variables such as $\mathbf{X} = (X_1, X_2, \ldots, X_N)^\top$. Similarly, we use a bold lowercase $\mathbf{x}$ to represent a value of $\mathbf{X}$.

Since PGMs are used to capture the joint probability distribution of a set of random variables, we will first define random variables, the basic probability rules, and the important probability distributions that are often used in PGMs.

2.2.1 Random variable and probability

A random variable (RV) is a variable whose value is uncertain and depends on a chance. The specific value of an RV is produced by a random process that maps an RV into a specific value. The random process may correspond to an experiment, and the outcomes of the experiment correspond to different values of the RV. Let capital X represent an RV, and lowercase $x \in \mathcal{X}$ represent a particular value of X, where $\mathcal{X}$ defines the value space for the RV X. In general, RVs can be discrete or continuous. A discrete RV can be further divided into categorical and integer RVs. For a categorical RV, its value space $\mathcal{X}$ is a finite set of categories $\mathcal{C} = \{c_1, c_2, \ldots, c_K\}$. For an integer RV, its value space $\mathcal{N}$ consists of all possible integer values, including zero. The value space $\mathcal{R}$ for continuous RVs, on the other hand, assumes continuous real values within a certain range.

For a discrete RV, its chance to assume a particular value is numerically quantified by its probability. Mathematically, probability is a measure of uncertainty that lies between 0 and 1. We use $p(X = x)$ (or $p(x)$ for short) to represent the probability that X assumes

Probabilistic Graphical Models for Computer Vision. https://doi.org/10.1016/B978-0-12-803467-5.00007-1

the value of x. According to probability theory, $0 \leq p(x) \leq 1$, where $p(x) = 0$ means that $X = x$ is definitely not true, whereas $p(x) = 1$ means $X = x$ is definitely true without any uncertainty. Furthermore, probability theory states that $\sum_{x \in \mathcal{X}} p(x) = 1$.

For a continuous RV X, instead of computing the probability of X taking on a specific value $x \in \mathcal{X}$ (which is always 0), we are interested in $p(X \in A)$, that is, the probability that X lies in an interval $A \subset \mathcal{R}$. By definition, we have

$$p(X \in A) = \int_A f_x(x)dx, \tag{2.1}$$

where $f_x(x)$, a continuous function $f : \mathcal{R} \mapsto [0, +\infty)$, is the probability density function (pdf) of X, and $\int_{\mathcal{X}} f_x(x)dx = 1$. For a discrete RV, its pdf can be defined as $f_x(x) = \sum_{x_k \in \mathcal{X}} p(x_k)\delta(x - x_k)$, where x_k is the kth value of X, and $\delta()$ is the delta function. Note that whereas $p(x)$, the probability of X, must lie between 0 and 1, $f_x(x)$, the pdf of X, can be any nonnegative number. In addition, whereas $p(x)$ for a specific x is always zero, $f_x(x)$ can be any nonnegative number.

2.2.2 Basic probability rules

In this section, we review some important probability rules and theorems. Given two random variables X and Y, the conditional probability of X given Y is defined as

$$p(X|Y) = \frac{p(X,Y)}{p(Y)}, \tag{2.2}$$

where $p(X, Y)$ represents the joint probability of X and Y. From this definition of the conditional probability we can also derive the **product rule**

$$p(X,Y) = p(X|Y)p(Y). \tag{2.3}$$

The product rule states that the joint probability can be represented as the product of conditional probability and marginal probability. A generalization of the product rule is the **chain rule**, which extends the product rule to N RVs. Let $X_1, X_2, \ldots, X_N$ be N random variables. The chain rule states that the joint probability of the N variables can be expressed as follows:

$$p(X_1, X_2, ..., X_N) = p(X_1)p(X_2|X_1)p(X_3|X_1, X_2)...p(X_N|X_1, X_2, ..., X_{N-1}) \tag{2.4}$$

Like the product rule, the chain rule allows the joint probability of N RVs to be factorized as the product of conditional probabilities for each variable. Please note that the chain rule does not require any assumptions on the RVs and their distributions. The chain rule can be extended to conditional chain rule as follows:

$$\begin{aligned} p(X_1, X_2, \ldots, X_N|Y, Z) &= p(X_1|Y, Z)p(X_2|X_1, Y, Z)p(X_3|X_1, X_2, Y, Z)\ldots \\ &\quad p(X_N|X_1, X_2, \ldots, X_{N-1}, Y, Z) \end{aligned} \tag{2.5}$$

The rule is very useful in situations, where we are interested in the factorization of the conditional joint probabilities. An approximation to the chain rule is the pseudolikelihood **chain rule**, which states that

$$p(X_1, X_2, ..., X_N) \approx \prod_{n=1}^{N} p(X_n|X_{-n}) \tag{2.6}$$

The rule plays an important in developing efficient PGM inference methods.

Assuming that X and Y are discrete RVs and using the **sum rule**, we can derive the marginal distribution of X or Y from their joint distribution by marginalizing over Y:

$$p(X) = \sum_{y} p(X, Y = y). \tag{2.7}$$

The sum rule also applies to continuous RVs, where the summation is replaced by an integral. The sum rule allows one to compute the marginal probability from the joint probability distribution. It can also be used to compute the marginal conditional probability $p(X|Z) = \sum_y p(X, y|Z)$.

Combining the sum rule and product rule yields the **conditional probability rule**, that is, the marginal probability of an RV can be computed from the marginalization of the conditional probability. Let X and Y be two RVs. Then according to the conditional probability rule,

$$p(X) = \sum_{y} p(X|y)p(y). \tag{2.8}$$

This is an important rule since in many real-world problems, the marginal probability of X may be difficult to estimate, but its conditional probability may be easier to estimate. It can be further extended to the marginal conditional probability:

$$p(X|Y) = \sum_{z} p(X|Y, z)p(z|Y). \tag{2.9}$$

A further extension of the conditional probability rule leads to the **Bayes' rule**, which states that

$$p(X|Y) = \frac{p(X)p(Y|X)}{p(Y)}, \tag{2.10}$$

where $p(X)$ is often referred to as the prior probability of X, $p(Y|X)$ is called the likelihood of X, and $p(Y)$ is the probability of the evidence; $p(Y) = \sum_x p(Y|x)p(x)$ is the normalization constant to ensure $p(X|Y)$ sums to 1.

2.2.3 Independencies and conditional independencies

Let X and Y be two RVs. We use $X \perp Y$ to denote that they are marginally independent of each other. If they are marginally independent, then their joint probability can be factorized into a product of their marginal probabilities, that is, $p(X, Y) = p(X)p(Y)$. Given this factorization and following the definition of conditional probability, we also have $p(X|Y) = p(X)$ if $X \perp Y$. This means that knowing Y does not affect the probability of X. We can extend this to N RVs, that is, for N RVs $X_1, X_2, \ldots, X_N$: if they are independent of each other, then we have

$$p(X_1, X_2, \ldots, X_N) = \prod_{n=1}^{N} p(X_n).$$

Besides marginal independencies, RVs may be conditionally independent. In fact, conditional independence is more prevalent in practice. Given three RVs X, Y, and Z, we denote the conditional independence between X and Y given Z as $X \perp Y \mid Z$. Following the marginal independence, we can easily derive that $p(X, Y|Z) = p(X|Z)p(Y|Z)$ if $X \perp Y \mid Z$. Furthermore, if $X \perp Y|Z$, then we can also derive that $p(X|Y, Z) = p(X|Z)$, that is, knowing Y will not contribute our knowledge to X if Z is given. Compared to marginal independence, conditional independence is weaker and relaxed. Note that conditional independence and marginal independence are not equivalent, that is, $X \perp Y \nRightarrow X \perp Y \mid Z$. In fact, two variables that are marginally independent can be conditionally dependent, for example, the V-structure (defined in Section 3.2.2.5) in a BN. Conversely, two marginally dependent variables can become conditionally independent given a third variable, for example, the three variables that are connected in a chain.

Finally, we want to clarify the differences between independence and mutual exclusiveness. Two variables X and Y are mutually exclusive if the presence of one implies the absence of the other. By definition, mutual exclusiveness means that $p(X, Y) = 0$, whereas independence means $p(X, Y) = p(X)p(Y)$. They hence represent different concepts. In fact, mutual exclusiveness implies dependence since the two variables are negatively correlated with each other.

2.2.4 Mean, covariance, correlation, and independence

For a random variable X, its mean (or expectation) is defined as its expected value. We use $E_{p(x)}(x)$ (or μ_X) to represent the expected value of X with respect to its pdf $p(x)$.[1] If X is discrete, its mean is

$$\mu_X = E_{p(x)}(X) = \sum_{x \in \mathcal{X}} x p_x(x).$$

For continuous X, its mean is

$$\mu_X = E_{p(x)}(X) = \int_{x \in \mathcal{X}} x f_x(x) dx.$$

For notational simplicity, the subscript $p(x)$ is often dropped unless otherwise specified. In practice, either $p(x)$ is not available, or the exact computation of the mean may be in-

[1] Note that the mean of X can be taken with respect to any probability density function of X such as $q(x)$.

tractable since it requires integration (or sum) over all possible values of X. As a result, the mean is often approximated by the sample average as an unbiased estimate of the mean.

The variance of an RV measures the expected squared deviation of its values from its mean value. Mathematically, for an RV X, its variance $Var(X)$ (often denoted as σ_X^2) is defined as

$$\begin{aligned}\sigma_X^2 &= E[(X - E(X))^2] \\ &= E(X^2) - E^2(X).\end{aligned}$$

For two random variables X and Y, their covariance Var(X, Y) (denoted as σ_{XY}) can be defined as

$$\begin{aligned}\sigma_{XY} &= E_{p(x,y)}[(X - E(X))(Y - E(Y))] \\ &= E_{p(x,y)}(XY) - E(X)E(Y),\end{aligned}$$

where $p(x, y)$ represents the joint probability density function of X and Y.

The correlation $Cor(X, Y)$ (denoted as ρ_{XY}) between X and Y is defined as

$$\begin{aligned}\rho_{XY} &= \frac{E_{p(x,y)}[(X - E(X))(Y - E(Y)]}{\sqrt{Var(X)Var(Y)}} \\ &= \frac{\sigma_{XY}}{\sigma_X \sigma_Y}.\end{aligned}$$

If X and Y are uncorrelated with each other, that is, $\rho_{XY} = 0$, then $\sigma_{XY} = 0$, and hence $E(XY) = E(X)E(Y)$. If two RVs are independent of each other, then $E(XY) = E(X)E(Y)$, which means $\rho_{XY} = 0$, and hence they are uncorrelated. However, two uncorrelated variables are not necessarily independent since $E(XY) = E(X)E(Y)$ does not mean that X and Y are independent. However, two jointly normally distributed random variables are independent if they are uncorrelated.

The conditional mean of a discrete X given Y is defined as

$$E_{p(x|y)}(X|y) = \sum_x x p(x|y),$$

where $p(x|y)$ is the conditional pdf of X given Y. Similarly, the conditional variance of X given Y is

$$Var(X|y) = E_{p(x|y)}[(X - E(X|y))^2].$$

Note that both $E(X|y)$ and $Var(X|y)$ are functions of y.

We can extend the definitions of mean and variance to random vectors. For a random vector $\mathbf{X}^{N\times 1}$, its mean $E^{N\times 1}(\mathbf{X})$ is a vector of mean values of each element of $\mathbf{X}$, that is, $E(\mathbf{X}) = (E(X_1), E(X_2), \ldots, E(X_N))^\top$. Its variance is defined by its covariance matrix

$$\begin{aligned}\Sigma_{\mathbf{X}}^{N\times N} &= E[(\mathbf{X} - E(\mathbf{X}))(\mathbf{X} - E(\mathbf{X}))^\top] \\ &= E(\mathbf{X}\mathbf{X}^\top) - E(\mathbf{X})E^\top(\mathbf{X}).\end{aligned}$$

The diagonal elements of $\Sigma_{\mathbf{X}}$ measure the variances of elements of **X**, and the off-diagonal elements of $\Sigma_{\mathbf{X}}$ capture the covariances between pairs of elements of **X**.

2.2.5 Probability inequalities

PGM learning and inference theories often involve optimization by maximization or minimization of an objective function. Certain objective functions may be hard to directly optimize. Probability inequalities may be used to construct bounds for these functions. Instead of optimizing the original functions, optimization can then be done for their bounds, which can be often performed more easily. For example, directly maximizing a function may be difficult, but maximizing its lower bound can be much easier. Similarly, instead of directly minimizing a function, we can approximately minimize its upper bound. In this section, we briefly introduce a few of the most widely used probability inequalities. The probability inequalities can be divided into expectation inequality and probability inequality. One of the most frequently used expectation inequalities is **Jensen's inequality**. Let X be an RV, and let ϕ be a concave function, Jensen's inequality can be written as

$$\phi(E(X)) \geq E(\phi(X)). \tag{2.11}$$

Jensen's inequality states that the concave function of the mean of an RV X is greater than or equal to the mean of the function of X. Jensen's inequality gives a lower bound of the function of the expectation. A widely used concave function for PGM is the logarithm. Hence a lower bound can be constructed for the logarithm of the expectation of an RV. Instead of maximizing the original logarithm, we can maximize its lower bound. If the function $\phi(X)$ is convex, then

$$\phi(E(X)) \leq E(\phi(X)). \tag{2.12}$$

In this case, Jensen's inequality provides an upper bound of the function of the mean.

Another expectation inequality is the Cauchy–Schwarz inequality. Given two RVs X and Y that have finite variances, we have

$$E(|XY|) \leq \sqrt{E(X^2)E(Y^2)}. \tag{2.13}$$

Cauchy–Schwarz's inequality can be used to relate covariance with variances. In the particular case where X and Y have zero means, Eq. (2.13) means that their covariance is less than the product of their respective standard deviations, that is,

$$\sigma_{XY} \leq \sigma_X \sigma_Y,$$

that is, their correlation $\frac{\sigma_{XY}}{\sigma_X \sigma_Y}$ is less than or equal to one, and the equality holds when $X = Y$.

Probability inequalities include Markov's inequality, Chebyshev's inequality, and Hoeffding's inequality. Let X be a nonnegative RV with finite $E(X)$. Then, for any $t > 0$,

Markov's inequality states that

$$p(X \geq t) \leq \frac{E(X)}{t}. \tag{2.14}$$

Markov's inequality gives an upper bound for the probability of a random variable greater than or equal to some positive constant.

Chebyshev's inequality is about the range of standard deviations around the mean statistics. It is derived from Markov's inequality. Specifically, it states that for an RV X with finite mean μ and finite nonzero variance σ^2, we have

$$p(|X - \mu| \geq k\sigma) \leq \frac{1}{k^2}, \tag{2.15}$$

where k is a positive real number. Chebyshev's inequality states for an RV X, $1 - \frac{1}{k^2}$ percent of its values are located within k standard deviations of its mean. For example, for a normally distributed RV X, 75% of its values are within two standard deviations of its mean and 89% within three standard deviations. The inequality can be applied to arbitrary distributions and can be used to prove the weak law of large numbers.

Hoeffding's inequality provides an upper bound of the probability that the empirical mean of a set of RVs deviates from its mean. Let $X_1, X_2, \ldots, X_N$ be independent identically distributed (i.i.d.) RVs such that $E(X_n) = \mu$ and $a \leq X_n \leq b$, then for any $\epsilon > 0$,

$$p(|\bar{X} - \mu| \geq \epsilon) \leq 2e^{\frac{-2N\epsilon^2}{(b-a)^2}}, \tag{2.16}$$

where $\bar{X}$ is the empirical mean of the N RVs (i.e. sample average). Hoeffding's inequality states that the probability of the empirical mean to be up to ϵ away from the true mean is larger than $1 - 2e^{\frac{-2N\epsilon^2}{(b-a)^2}}$. It can determine the minimum number of samples N needed to ensure that $\bar{X}$ is up to ϵ away from the true mean (often referred to as the confidence interval) with probability $\alpha = 1 - 2e^{\frac{-2N\epsilon^2}{(b-a)^2}}$. Here, α is called the confidence interval probability and is typically set to 95%.

2.2.6 Probability distributions

In this section, we discuss several important families of distributions commonly used for PGMs. As discussed before, probability distributions capture the probabilities of an RV taking different values in its value space. The distribution can be for a single RV or multiple RVs. The former is referred to as a univariate distribution, whereas the latter is referred to as a multivariate distribution. Depending on the types of RVs, probability distributions can be either discrete or continuous. For a discrete RV, its probability distribution can be represented by a discrete list of the probabilities corresponding to its values (also known as a probability mass function). For a continuous RV, its probability distribution is encoded by the probability density function (pdf). Discrete distributions can be further divided into categorical distributions and integer distributions.

2.2.6.1 Discrete probability distributions

Discrete RVs can be either integer RVs or categorical RVs. We further separately discuss their distributions.

2.2.6.1.1 Categorical probability distributions

For a categorical RV X assume one of K possible categorical values, that is, $x \in \{c_1, c_2, \ldots, c_K\}$, we can denote the probability distribution of X as $X \sim Cat(x|\boldsymbol{\alpha}, K)$ of the form

$$p(X = k) = \alpha_k,$$

where $\boldsymbol{\alpha} = (\alpha_1, \alpha_2, \ldots, \alpha_K)$. In general, we can write a categorical RV distribution as

$$p(X) = \prod_{k=1}^{K-1} \alpha_k^{I(X=k)} (1 - \sum_{k=1}^{K-1} \alpha_k)^{I(X=K)},$$

where the indicator function $I()$ equals 1 if its argument is true or zero. In machine learning, the probability $\alpha_k = p(X = k)$ is often parameterized by the multiclass sigmoid (or softmax) function, that is,

$$\alpha_k = \frac{\exp(w_k)}{\sum_{k'=1}^{K} \exp(w_{k'})},$$

where the probability parameters w_k are learned from data.

The simplest categorical distribution is the **Bernoulli distribution**. It is used to represent the probability distribution of a binary random variable X that can have two possible categories ($K = 2$). Without loss of generality, we can numerically represent the two categorical values of X as either 1 or 0. Denoted as $X \sim$ Ber $(x|\alpha)$, the Bernoulli distribution is defined as

$$\begin{aligned} p(X = 1) &= \alpha, \\ p(X = 0) &= 1 - \alpha, \end{aligned}$$

where α is a real number between 0 and 1. It can also be generically written as

$$p(X) = \alpha^X (1 - \alpha)^{1-X}.$$

The sigmoid (or probit) function is often used to parameterize α:

$$\alpha = \sigma(w) = \frac{1}{1 + \exp(-w)},$$

where the probability parameter w is either manually specified or learned from data.

A special kind of categorical distribution is the **uniform** distribution, where $\alpha_k = \frac{1}{K}$, that is, X has the same chance to assume any one of the K values. For example, if a coin is fair, then the experiment of tossing a coin follows a uniform distribution with the probability of obtaining a head or a tail being equal to 0.5.

2.2.6.1.2 Integer probability distributions

One of the most commonly used integer distributions is the **binomial distribution**. Let X be an integer RV, and let $x \in \mathcal{N}$ be its value, where $\mathcal{N}$ is the integer space. The binomial distribution can be generated through Bernoulli trials (Bernoulli experiment), which involve repeated and independent trials with a Bernoulli variable Y, with Y taking the value 1 or 0 at each trial. Let N be the number of trials, let α be the probability of $Y=1$, and let X represent the number of times that $Y=1$ out of N trails. Then

$$p(X=x) = Bin(x|N,\alpha) = \begin{pmatrix} N \\ x \end{pmatrix} \alpha^x (1-\alpha)^{N-x}, \tag{2.17}$$

where $\binom{N}{x} = \frac{N!}{x!(N-x)!}$ is the binomial coefficient. One well-known Bernoulli trial is the coin toss experiment, where a coin is tossed many times, and each toss results in either a "head" or a "tail". The Bernoulli distribution can be used to describe the probability of obtaining a "head" or a "tail" for each toss, whereas the binomial distribution can be used to characterize the probability distribution of the number of times that "head" appears out of N tosses.

Another widely used integer distribution is the Poisson distribution. Let X be an integer RV, and let $x \in \mathcal{N}$. Then X follows the Poisson distribution if it represents the number of times an event occurs in an interval of time. More precisely, let λ be the event occurrence rate, that is, the average number of times an event occurs in the interval. Then X follows the Poisson distribution, denoted as $X \sim Poisson(x|\lambda)$, if

$$p(X=x) = Poisson(x|\lambda) = e^{\lambda}\frac{\lambda^x}{x!},$$

where x is the number of times the event happens in the interval. Note that the Poisson distribution assumes that each event happens independently, and that the event rate λ is constant across different intervals.

2.2.6.1.3 Multivariate integer probability distributions

Let $\mathbf{X}=(X_1, X_2, \ldots, X_K)^\top$ be a random integer vector with K elements, where X_k are integer RVs with values $x_k \in \mathcal{N}$. Like the binomial distribution, the multinomial distribution can be constructed by performing repeated and independent trials on a categorical random variable Y with K values. Let N be the number of trials and let X_k be the number of times that $Y=k$ appears in the trials. Then $\mathbf{X}$ follows the **multinomial** distribution, denoted as $\mathbf{X} \sim Mul(x_1, x_2, \ldots, x_k|N, \alpha_1, \alpha_2, \ldots, \alpha_K)$, which can be specified as follows:

$$\begin{aligned} p(X_1=x_1, X_2=x_2, \ldots, X_K=x_K) &= Mul(x_1, x_2, \ldots, x_k|N, \alpha_1, \alpha_2, \ldots, \alpha_K) \\ &= \frac{N!}{x_1!x_2!, \ldots, x_K!}\alpha_1^{x_1}\alpha_2^{x_2}\ldots\alpha_K^{x_K}. \end{aligned} \tag{2.18}$$

The multinomial distribution gives the probability of any particular combination of numbers of successes for each possible value of Y. The binomial distribution represents a particular case of multinomial distributions where $K=2$. Corresponding to the coin toss for the binomial distribution, dice rolling is often used as an example of a multinomial distribution. Instead of having two outcomes, tossing a dice typically has six possible outcomes from 1 to 6, that is, $K = 6$. Given N rolls, the number of times each of the six numbers appears follows a multinomial distribution.

2.2.6.2 Continuous probability distributions

Among the continuous probability distributions, the **Gaussian** distribution (or normal distribution) is the most widely used distribution due to its simplicity, special properties, and its ability of approximately modeling the distributions of many real-world random events. In statistics, an RV X follows a Gaussian distribution, denoted as $X \sim \mathcal{N}(x|\mu, \sigma^2)$, where μ is the mean of X, and σ^2 is the variance of X, if its probability density distribution can be written in the form

$$f_X(x) = \mathcal{N}(x|\mu, \sigma^2) = \frac{1}{\sqrt{2\pi}\sigma} \exp[-\frac{(x-\mu)^2}{2\sigma^2}]. \tag{2.19}$$

For a random vector $\mathbf{X}=(X_1, X_2, \ldots, X_N)^\top$, $\mathbf{X}$ follows a multivariate Gaussian distribution, denoted as $\mathbf{X} \sim \mathcal{N}(\mathbf{x}|\boldsymbol{\mu}, \Sigma)$, where $\boldsymbol{\mu}$ is the $N \times 1$ mean vector, and Σ is the $N \times N$ covariance matrix. Mathematically, a multivariate Gaussian distribution can be defined by the multivariate pdf

$$f_{\mathbf{X}}(\mathbf{x}) = \mathcal{N}(\mathbf{x}|\boldsymbol{\mu}, \Sigma) = \frac{1}{(2\pi)^{\frac{N}{2}} |\Sigma|^{\frac{1}{2}}} \exp[\frac{-(\mathbf{x}-\boldsymbol{\mu})^\top \Sigma^{-1} (\mathbf{x}-\boldsymbol{\mu})}{2}]. \tag{2.20}$$

Another commonly used continuous distribution for PGM is the **beta** distribution. It is a continuous probability distribution for an RV defined between 0 and 1. The beta distribution is parameterized by two positive parameters α and β. Its pdf can be written as

$$f_X(x) = Beta(x|\alpha, \beta) = \frac{x^{\alpha-1}(1-x)^{\beta-1}}{B(\alpha, \beta)}, \tag{2.21}$$

where $B(\alpha, \beta) = \frac{\Gamma(\alpha)\Gamma(\beta)}{\Gamma(\alpha+\beta)}$, and $\Gamma()$ is the gamma function. The beta distribution is generic and represents a family of distributions. By varying the values of α and β, the beta distribution can become certain standard distributions such as Gaussian or uniform distribution ($\alpha = \beta = 1$). The beta distribution is often used as the conjugate for the binomial distribution. As a generalization of the beta distribution, the **Dirichlet** distribution is a continuous multivariate probability distribution for a random vector $\mathbf{X}=(X_1, X_2, \ldots, X_K)^\top$, where X_k is an RV whose value ranges between 0 and 1, that is, $0 \le x_k \le 1$. The Dirichlet distribution is denoted as $\mathbf{X} \sim Dir(\mathbf{x}|\boldsymbol{\alpha})$, where $\boldsymbol{\alpha} = (\alpha_1, \alpha_2, \ldots, \alpha_k)$ is a vector of K positive real numbers, which are often referred to as the hyperparameters. Given $\boldsymbol{\alpha}$, the Dirichlet distribution can be defined by the multivariate pdf

$$f_{\mathbf{X}}(\mathbf{x}) = Dir(\mathbf{x}|\boldsymbol{\alpha}) = \frac{1}{B(\boldsymbol{\alpha})} \prod_{k=1}^{K} x_k^{\alpha_k - 1}, \tag{2.22}$$

where $B(\boldsymbol{\alpha})$ is the normalization constant expressed in terms of the gamma function $\Gamma()$ as

$$B(\boldsymbol{\alpha}) = \frac{\prod_{k=1}^{K} \Gamma(\alpha_k)}{\Gamma(\sum_{k=1}^{K} \alpha_k)}. \tag{2.23}$$

The Dirichlet distribution is often used as the prior distribution, in particular, as the conjugate prior of the multinomial distribution since the multiplication of the Dirichlet prior with a multinomial likelihood yields a posterior that also follows the Dirichlet distribution. By varying its parameters the Dirichlet distribution can form different distributions, including the uniform distribution, where $\alpha_k = 1$.

Like the categorical distribution, a **uniform** distribution also exists for a continuous RV X. Often denoted as $X \sim U(x|a,b)$, where a, b represent the lower and upper bound values for X, $U(a,b)$ is specified by the pdf

$$f_X(x) = \begin{cases} \frac{1}{b-a} & \text{if } a \leq x \leq b, \\ 0 & \text{else.} \end{cases}$$

For many applications, a and b are typically chosen to be 0 and 1, respectively, leading to an RV that follows the uniform distribution between 0 and 1. Finally, we want to introduce the **exponential** distribution for a positive continuous RV X, where X measures the time between two consecutive events in a Poisson process, which involves events that happen continuously and independently. Let λ represent the event arrival rate within a time interval. Then the exponential distribution, denoted as $X \sim \exp(x|\lambda)$, has the pdf

$$f_X(x) = \exp(x|\lambda) = \lambda e^{-\lambda x}, \quad \text{x} > 0.$$

The exponential distribution is related to the Poisson distribution as both characterize the properties of a Poisson process. Different from the Poisson distribution, which captures the distribution of the number of arrival events in an interval, the exponential distribution captures the distribution of the interarrival time between two consecutive events. The exponential distribution belongs to the family of exponential distributions, which also includes the Gaussian distribution, binomial distribution, Dirichlet distribution, and Poisson distribution. Distributions of exponential family are widely used for PGM modeling.

2.3 Basic estimation methods

For learning any probabilistic models such as PGMs, there are two well-established point estimation methods: the maximum likelihood method and the Bayesian method. The maximum likelihood method can be further divided into the maximum joint likelihood method, maximum conditional likelihood method, and maximum marginal likelihood method. All methods formulate the PGM model learning as an optimization problem; they differ only in the objective function for the optimization.

2.3.1 Maximum likelihood

The maximum likelihood method can be divided into the maximum joint likelihood, maximum conditional likelihood, and maximum marginal likelihood methods discussed further.

2.3.1.1 Maximum joint likelihood estimation

For the maximum joint likelihood method (a.k.a. maximum likelihood estimation (MLE)), the goal is to learn the model parameters $\mathbf{\Theta}$ by maximizing the joint likelihood of $\mathbf{\Theta}$ given the training data $\mathbf{D}$. MLE represents an approximation to minimizing the KL divergence between the empirical distribution represented by the training data and the distribution represented by the model (e.g., the PGM model). Let $\mathbf{D}=\{D_m\}$, where m is the index to the mth training sample and $m = 1, 2, \ldots, M$. It is often assumed that D_m are independent identically distributed (i.i.d.). The joint likelihood for $\mathbf{\Theta}$ given $\mathbf{D}$ can then be expressed as

$$L(\mathbf{\Theta} : \mathbf{D}) = p(\mathbf{D}|\mathbf{\Theta}) = \prod_{m=1}^{M} p(D_m|\mathbf{\Theta}). \tag{2.24}$$

Due to the monotonicity of the logarithm function, maximizing the likelihood function is equivalent to maximizing the logarithm of the likelihood function. Furthermore, since the likelihood function is typically represented in the form of a log-linear distribution (i.e., a distribution that takes the natural exponential form), the log-likelihood function can simplify the optimization as the likelihood typically belongs to the exponential family. The joint log likelihood can be expressed as

$$LL(\mathbf{\Theta} : \mathbf{D}) = \log p(\mathbf{D}|\mathbf{\Theta}) = \sum_{m=1}^{M} \log p(D_m|\mathbf{\Theta}). \tag{2.25}$$

According to the MLE method, the parameters $\mathbf{\Theta}$ can be solved by maximizing the joint log likelihood, that is,

$$\theta^* = \arg\max_{\theta} LL(\theta : \mathbf{D}). \tag{2.26}$$

MLE is also referred to as generative learning in the literature since it maximizes the joint probability distribution. MLE is optimal in the sense that its estimate is unbiased and asymptotically approaches the true parameter values as the amount of training data tends to infinity.

2.3.1.2 Maximum conditional likelihood estimation

For classification and regression problems, to make the learning criterion consistent with the testing criterion and to yield better performance, learning is often done by maximizing the joint log conditional likelihood $LCL(\mathbf{\Theta} : \mathbf{D})$, that is,

$$\theta^* = \arg\max_{\theta} LCL(\theta : \mathbf{D}), \tag{2.27}$$

where

$$LCL(\mathbf{\Theta}:\mathbf{D}) = \sum_{m=1}^{M} \log p(\mathbf{y}_m|\mathbf{x}_m, \mathbf{\Theta}),$$

and $D_m = \{\mathbf{x}_m, \mathbf{y}_m\}$ represents the mth training sample with $\mathbf{x}_m$ as its inputs and $\mathbf{y}_m$ as its outputs. Note that unlike the MLE, where the parameters $\mathbf{\Theta}$ capture the joint probability distribution $p(\mathbf{x}, \mathbf{y}, \mathbf{\Theta})$, here $\mathbf{\Theta}$ capture the joint conditional distribution $p(\mathbf{y}|\mathbf{x}, \mathbf{\Theta})$. By maximizing the conditional likelihood $p(\mathbf{y}|\mathbf{x}, \mathbf{\Theta})$ the learning criterion is the same as the inference criterion, that is, given $\mathbf{x}$, finding $\mathbf{y}$ that maximizes $p(\mathbf{y}|\mathbf{x}, \mathbf{\Theta})$. Such learning is also referred to as discriminative learning. Compared to the generative learning by maximum likelihood, discriminative learning requires less data and typically performs better for classification tasks, in particular, when the amount of training data is not large. Discriminative learning, however, cannot handle incomplete inputs and cannot obtain the marginal distribution of the inputs.

2.3.1.3 Maximum marginal likelihood estimation

For learning with missing data or with latent variables $D_m = \{\mathbf{x}_m, \mathbf{z}_m\}$, where $\mathbf{z}_m$ is either missing or latent, learning can be formulated as finding $\mathbf{\Theta}$ that maximizes the marginal log-likelihood $p(\mathbf{x}|\mathbf{\Theta})$, that is,

$$\theta^* = \arg\max_{\theta} \sum_{m=1}^{M} \log p(\mathbf{x}_m|\theta), \tag{2.28}$$

where the marginal likelihood for discrete $\mathbf{z}_m$ can be written as

$$p(\mathbf{x}_m|\mathbf{\Theta}) = \sum_{\mathbf{z}_m} p(\mathbf{x}_m, \mathbf{z}_m|\mathbf{\Theta}). \tag{2.29}$$

As a result, we have

$$\theta^* = \arg\max_{\theta} \sum_{m=1}^{M} \log \sum_{\mathbf{z}_m} p(\mathbf{x}_m, \mathbf{z}_m|\theta), \tag{2.30}$$

where $\mathbf{\Theta}$ captures the joint probability distribution $p(\mathbf{x}, \mathbf{z})$. Equation (2.30) is hard to maximize because of the presence of the log sum term; hence, one commonly used method is maximizing its lower bound by applying Jensen's inequality. We address this topic in Section 3.5 of Chapter 3 when discussing learning under incomplete data.

2.3.2 Maximum A Posterior (MAP) estimation

The likelihood estimation method estimates $\mathbf{\Theta}$ purely from data, assuming no prior knowledge of the parameters $\mathbf{\Theta}$. For some applications, we may know the prior probability of $\mathbf{\Theta}$, that is, the prior knowledge of $\mathbf{\Theta}$. In this case, we can use the MAP estimation method. The

MAP method estimates $\mathbf{\Theta}$ by maximizing the posterior probability of $\mathbf{\Theta}$ given the training data $\mathbf{D}$. Let $\mathbf{D}=\{D_m\}_{m=1}^M$ be the training data. Then the posterior probability of $\mathbf{\Theta}$ given $\mathbf{D}$ can be written as

$$p(\mathbf{\Theta}|\mathbf{D}) = \alpha p(\mathbf{\Theta})p(\mathbf{D}|\mathbf{\Theta}), \tag{2.31}$$

where α is a normalization constant and can be dropped since it is independent of the parameters $\mathbf{\Theta}$. The first term is the prior probability of $\mathbf{\Theta}$, and the second term is the joint likelihood of $\mathbf{\Theta}$. The conjugate prior probability is often used to parameterize $p(\mathbf{\Theta})$. According to Eq. (2.24), $p(\mathbf{D}|\mathbf{\Theta}) = \prod_{m=1}^M p(D_m|\mathbf{\Theta})$. Similarly, the log posterior probability can be written as

$$\log p(\mathbf{\Theta}|\mathbf{D}) = \log p(\mathbf{\Theta}) + \sum_{m=1}^{M} \log p(D_m|\mathbf{\Theta}). \tag{2.32}$$

Given the log posterior probability, the MAP parameter estimation can be stated as follows:

$$\boldsymbol{\theta}^* = \arg\max_{\boldsymbol{\theta}} \log p(\boldsymbol{\theta}|\mathbf{D}). \tag{2.33}$$

MAP estimation degenerates to MLE when the prior probability $p(\mathbf{\Theta})$ follows a uniform distribution.

2.4 Optimization methods

As per the discussion above, we often need to maximize or minimize an objective function. The optimization can be done with respect to continuous or discrete target variables. For PGM parameter learning and continuous PGM inference, the problem is typically formulated as a continuous optimization problem. The PGM structure learning and discrete PGM inference (in particular, MAP inference) are typically formulated as a discrete optimization problem. We further briefly discuss the conventional techniques for continuous and discrete optimizations. We refer the readers to [2] for a detailed and systematic treatment of convex optimization topics.

2.4.1 Continuous optimization

For some objective functions, the continuous optimization can be done analytically with a closed-form solution. However, the objective functions are often nonlinear, and there are no closed-form solutions. Consequently, an iterative numerical solution is needed. This is increasingly becoming the preferred solution for optimization-particularly for large models and large amounts of data. Here we briefly introduce a few commonly used iterative solutions to perform continuous optimization. The non-linear iterative optimization methods typically start with an initial estimate of the parameters $\mathbf{\Theta}_0$, and then iteratively improve the parameter estimate until convergence. Specifically, given the previous parameter estimate $\mathbf{\Theta}^{t-1}$, the non-linear iterative optimization methods improve the parameter's current estimate as follows

$$\Theta^t = \Theta^{t-1} + \eta \Delta\Theta$$

where $\Delta\Theta$ are the parameter update and η is the learning rate, which typically decreases as the iteration continues. The non-linear iterative optimization methods differ in their approach to computing $\Delta\Theta$. They can be classified into first order and second order methods.

The first order methods assume that the objective function is first order differentiable. They typically belong to variants of the gradient-based approach. The first order methods can be classified into the gradient ascent (descent) method and the Gauss-Newton method. Given an objective function such as the log likelihood $LL(\Theta)$ for maximization, the gradient ascent method computes the parameter update $\Delta\Theta$ as the gradient of the objective function with respect to the parameters Θ at each iteration t, that is, $\Delta\Theta = \nabla_\Theta LL(\Theta^{t-1})$, which can be computed as

$$\nabla_\Theta LL(\Theta) = \frac{\partial LL(\Theta^{t-1})}{\partial \Theta} \tag{2.34}$$

and the gradient is then added to current parameter estimates,

$$\Theta^t = \Theta^{t-1} + \eta \nabla_\Theta LL(\Theta) \tag{2.35}$$

where $\frac{\partial LL(\Theta^{t-1})}{\partial \Theta}$ is often called the Jacobian matrix. This update continues until convergence. For an objective function that involves minimization of the sum of squared function values (i.e., the non-linear least-squares minimization problems), the Gauss-Newton method or its improved variant, the Levenberg Marquardt algorithm [10] may be employed. Assume the objective function can be written as $(f(\Theta)^\intercal(f(\Theta))$, Gauss-Newton method computes its parameter update by

$$\Delta\Theta = -[(\frac{\partial f(\Theta^{t-1})}{\partial \Theta})(\frac{\partial f(\Theta^{t-1})}{\partial \Theta})^\intercal]^{-1}(\frac{\partial f(\Theta^{t-1})}{\partial \Theta})f(\Theta) \tag{2.36}$$

The Levenberg Marquardt method improves the Gauss-Newton method by introducing a damping factor into the parameter update, that is,

$$\Delta\Theta = -[(\frac{\partial f(\Theta^{t-1})}{\partial \Theta})(\frac{\partial f(\Theta^{t-1})}{\partial \Theta})^\intercal + \alpha \mathbf{I}]^{-1}(\frac{\partial f(\Theta^{t-1})}{\partial \Theta})f(\Theta) \tag{2.37}$$

where $\mathbf{I}$ is an identity matrix and α is a damping factor that varies with each iteration. The iteration starts with a large α and gradually reduces the α value as the iteration progresses. With a small α, the Levenberg Marquardt algorithm becomes Gauss-Newton method and with a large α, it becomes the gradient descent method. Details on the first order gradient based methods can be found in Section 19.2 and Appendix A 5.2 of [11].

The second-order methods include the Newton method (and its variants). It require a twice differentiable objective function. The Newton method computes the parameter update $\Delta\Theta$ by first taking a second-order Taylor expansion of the objective function around current value of the parameters Θ^{t-1}, that is, $LL(\Theta) \approx LL(\Theta^{t-1}) + \Delta\Theta\frac{\partial LL(\Theta)}{\partial \Theta} + \frac{1}{2}\frac{\partial^2 LL(\Theta)}{\partial \Theta^2}(\Delta\Theta)^\top(\Delta\Theta)$. It then takes the derivative of the Taylor expansion with respect to $\Delta\Theta$ and sets it to zero, yielding

$$\Delta\Theta = -(\frac{\partial^2 LL(\Theta)}{\partial \Theta^2})^{-1}\frac{\partial LL(\Theta)}{\partial \Theta},$$

where $\frac{\partial^2 LL(\Theta)}{\partial \Theta^2}$ is often called the Hessian matrix. Finally, Θ^t is updated as follows:

$$\Theta^t = \Theta^{t-1} + \Delta\Theta.$$

One variant of the Newton method is the Broyden–Fletcher–Goldfarb–Shanno (BFGS) algorithm, which performs second-order descent without explicit computation of the second-order Hessian matrix. Compared to the first-order methods, the second-order methods are more accurate. However, they require the objective functions to be twice differentiable and are less robust to noise because of the second-order derivative. As a result, gradient ascent (or descent) is nowadays the most widely used method for solving the nonlinear optimization problems in machine learning.

Both first- and second-order methods require summation over all training data. They become computationally expensive when the amount of training data is large. The stochastic gradient (SG) method [3] was introduced to appropriately perform gradient-based methods. Instead of performing summation over all training data, it randomly selects a subset of training data at each iteration to compute the gradient. The size of the subset (called batch size) is fixed, ranging from 20 to 100 samples, regardless of the amount of training data. It has been proven theoretically that the SG method converges to the same point as the full gradient method. Empirical experiments have demonstrated the effectiveness of the SG method. In fact, it is becoming the preferred method for learning in big data. Details on SD and its properties can be found in [3]. For the nondifferentiable objective functions, we can use the subgradient method [4]. A more detailed discussion on various convex optimization techniques can be found in [4].

2.4.2 Discrete optimization

For discrete optimization, the optimization becomes combinatorial. Solutions to discrete optimization can be exact methods and approximate methods, depending on the objective function (e.g., sub-modular), the size of the model, and its structure (e.g., chain or tree structures). With respect to discrete PGM MAP inference, the exact methods include the variable elimination, message passing, graph cuts, and dynamic programming. For models with a large number of nodes and complex topology, exact methods become intractable as the solution space increases exponentially with respect to the number of variables. In this case, approximate methods are often employed. The commonly used approximate methods include the coordinate ascent and loopy belief propagation. In addition, discrete optimization is often converted into continuous optimization through target variable relaxation. For example, linear programming relaxation is often employed in CV to solve discrete minimization problems. We refer the readers to [5] for a thorough discussion on discrete optimization techniques for discrete graphical models.

2.5 Sampling and sample estimation

Sampling and sample estimation techniques are widely employed for PGM learning and inference. Sampling (also called Monte Carlo simulation) takes samples from an underlying distribution and uses the acquired samples as surrogates to represent the underlying distribution. Sample estimation aims at estimating the probabilities of the variables from the collected samples. The estimated probabilities represent an approximation to the underlying true probabilities. In this section, we first introduce the basic techniques to perform sampling and then discuss sample estimations and their confidence intervals.

2.5.1 Sampling techniques

Sampling is becoming a preferred method for performing probabilistic learning and inference. In this section, we briefly review the standard approach to performing univariate sampling from standard probability distributions. In Chapter 3, we will discuss more sophisticated multivariate sampling techniques, including Gibbs sampling. First, for univariate sampling, we must have a uniform sampler that can generate a sample from a uniform distribution $U \sim \mathbf{U}(0, 1)$, which can be done with a pseudorandom number generator, which is available with many existing software packages. Second, for an RV X with a standard distribution $p(X)$, let $F(X)$ be the corresponding cumulative distribution function (CDF) of X. We can obtain a sample of X by first generating a sample u from $U \sim \mathbf{U}(0, 1)$, and then a sample x of X can be computed as $x = F^{-1}(u)$, where $F^{-1}(u)$ represents the inverse of $F(x)$. This method can be applied to different standard families of distributions. For example, when applied to sampling Gaussian distribution, it produces the Box–Muller method [6] for sampling multivariate Gaussian distributions. For a categorical RV X with K values, $x \in \{c_1, c_2, \ldots, c_K\}$, and K parameters $\alpha_1, \alpha_2, \ldots, \alpha_K$, we can divide the region between 0 and 1 into K regions: $(0, \alpha_1), (\alpha_1, \alpha_1 + \alpha_2), \ldots, (\alpha_1 + \alpha_2 + \cdots + \alpha_{K-1}, 1)$. Then we can draw a sample u from $U(0, 1)$ and let $X = c_k$ if u is located in the kth region.

If the inverse of CDF cannot be estimated, the rejection sampling method may be employed. For the rejection sampling method, we first design a proposal distribution $q(X)$. We then sample x from $q(X)$ and u from $U(0, 1)$. If $u \geq \frac{p(x)}{q(x)}$, then we accept x and discard it otherwise. One choice for $q(x)$ is the prior probability distribution of X if it is known. If not, then we can choose the uniform distribution as $q(X)$. See Chapter 23.2 of [7] for details and additional sampling methods such as the importance sampling method.

For standard multivariate distribution such as the multivariate Gaussian distribution, we can perform direct sampling through the re-parameterization trick. Readers may refer to Appendix 2.6.1 for details.

2.5.2 Sample estimation

Given the collected samples, we can use them to estimate the probabilities of random variables. Such estimated probabilities are referred to as empirical probabilities. They can be used to approximate the population (true) probabilities. For discrete RVs, the empirical probabilities are calculated by counting the proportion $\frac{n}{N}$, where N is the total number of samples, and n is the number of samples that the random variable takes a particular value. This can be done individually for each probability. For example, to estimate the probability of a head appearing in a coin toss, we just count the number of times n of appearance of head a and then divide it by the total number of tosses N. For more accurate estimation or for a continuous RV, we can obtain the empirical probabilities through the kernel density estimation (KDE) method [9].

One major issue with empirical probabilities is their accuracies, that is, their closeness to the true probabilities. Two factors contribute to sampling accuracy: the sample number and sample representation. The sample number determines the number of samples to generate, whereas sample representation decides if the generated samples can accurately represent the population distribution. Here we address the effect of the sample number on sampling accuracy. The confidence interval is often used to measure the estimation accuracy. Let p be the true probability to estimate, and let $\hat{p}$ be the probability estimated from the samples. The confidence interval is defined as $p \in (\hat{p} - \epsilon, \hat{p} + \epsilon)$, where ϵ is called the interval bound or the margin of errors.

Various methods have been proposed to compute the confidence interval for binomial distributions of binary variables. The most widely used methods are the normal and exact methods.[2] The interval bound according to the normal method is

$$\epsilon = z_{\frac{1+\alpha}{2}} \sqrt{\frac{1}{N} \hat{p}(1 - \hat{p})}, \tag{2.38}$$

where α is the confidence level at which the estimated value is located within the interval, $z_{\frac{1+\alpha}{2}}$ is the $\frac{1+\alpha}{2}$ quantile of a standard normal distribution, and N is the total number of samples. For a confidence interval probability of 0.95, $z_{\frac{1+\alpha}{2}} = 1.96$. The normal method is simple, but it does not work well for small or large probabilities. In general, if $Np < 5$ or $N(1 - p) < 5$, the normal method cannot produce accurate bounds.

To overcome this problem, the exact method was introduced. It consists of two separate bounds: the lower bound ϵ_{lower} and the upper bound ϵ_{upper}, that is, $\hat{p} \in (\epsilon_{\text{lower}}, \epsilon_{\text{upper}})$. The upper and lower bounds are hard to compute analytically, and they are typically computed numerically with a computer program. They can be approximated by the inverse beta distribution as follows:

$$\begin{aligned} \epsilon_{\text{lower}} &= 1 - BetaInv(\frac{1-\alpha}{2}, N - n, n + 1), \\ \epsilon_{\text{upper}} &= 1 - BetaInv(\frac{1+\alpha}{2}, N - n + 1, n), \end{aligned} \tag{2.39}$$

[2] See https://en.wikipedia.org/wiki/Binomial_proportion_confidence_interval.

where α is the significance level (e.g., 0.95), N is the total number of samples, and n is the number of samples for which the binary variable has a value of 1.

For a multinomial distribution of multivariate discrete RVs, their intervals are often independently approximated by binary confidence intervals for each probability. To simultaneously estimate the intervals for all probabilities, sophisticated methods are often used [8]. Standard statistical programming packages such as the R package often include functions/routines for simultaneous computation of confidence intervals for multinomial probabilities.

2.6 Appendix

2.6.1 Multivariate Gaussian Sampling

Let $\mathbf{X} = \{X_1, X_2, ..., X_N\}$ be a random vector that follows multivariate Gaussian distribution, i.e., $\mathbf{X} \sim \mathcal{N}(\boldsymbol{\mu}, \boldsymbol{\Sigma})$. Directly sampling from the multivariate Gaussian distribution may be challenging. This challenge can be mitigated with the re-parameterization trick. Let $\mathbf{Z}$ be a random vector that follows the standard multivariate Gaussian distribution, i.e., $\mathbf{Z} \sim \mathcal{N}(0, \mathbf{I})$, where $\mathbf{I}$ is the identity covariance matrix. We can easily prove that $\mathbf{x} = \boldsymbol{L}\mathbf{z} + \boldsymbol{\mu}$, where $\boldsymbol{L}$ is the lower triangle matrix resulted from Cholesky decomposition of $\boldsymbol{\Sigma}$, i.e., $\boldsymbol{\Sigma} = \boldsymbol{L}\boldsymbol{L}^T$. As $\mathbf{Z}$ follows standard multivariate Gaussian distribution, we can sample each element of $\mathbf{Z}$ independently, yielding a sample $\mathbf{z}_s$, based on which we obtain the corresponding sample for $\mathbf{X}$ as $\mathbf{x}_s = \boldsymbol{L}\mathbf{z}_s + \boldsymbol{\mu}$.

References

[1] A. Papoulis, S.U. Pillai, Probability, Random Variables, and Stochastic Processes, Tata McGraw-Hill Education, 2002.
[2] S. Boyd, L. Vandenberghe, Convex Optimization, Cambridge University Press, 2004.
[3] L. Bottou, Large-scale machine learning with stochastic gradient descent, in: Proceedings of COMPSTAT, Springer, 2010, pp. 177–186.
[4] D.P. Bertsekas, A. Scientific, Convex Optimization Algorithms, Athena Scientific, Belmont, 2015.
[5] B. Savchynskyy, Discrete graphical models – an optimization perspective, https://hci.iwr.uni-heidelberg.de/vislearn/HTML/people/bogdan/publications/papers/book-graphical-models-submitted-17-08-79.pdf.
[6] G.E. Box, M.E. Muller, et al., A note on the generation of random normal deviates, The Annals of Mathematical Statistics 29 (2) (1958) 610–611.
[7] K.P. Murphy, Machine Learning: A Probabilistic Perspective, MIT Press, 2012.
[8] C.P. Sison, J. Glaz, Simultaneous confidence intervals and sample size determination for multinomial proportions, Journal of the American Statistical Association 90 (429) (1995) 366–369.
[9] V.A. Epanechnikov, Non-parametric estimation of a multivariate probability density, Theory of Probability & Its Applications 14 (1) (1969) 153–158.
[10] D.W. Marquardt, An algorithm for least-squares estimation of nonlinear parameters, Journal of the Society for Industrial and Applied Mathematics 11 (2) (1963) 431–441.
[11] D. Koller, N. Friedman, Probabilistic Graphical Models: Principles and Techniques, MIT Press, 2009.

3 Directed probabilistic graphical models

3.1 Introduction

In Chapter 2, for the joint probabilities $p(X_1, X_2, \ldots, X_N)$ of a set of random variables X_1, $X_2, \ldots, X_n$, we can assume that they follow parametric distributions such as a multivariate Gaussian distribution for continuous RVs or a multinomial distribution for integer RVs. Although these parametric distributions are concise and can be represented by a small number of parameters, they may not realistically capture the complex joint probability distribution $p(X_1, X_2, \ldots, X_N)$. Alternatively, we can make a strong assumption regarding the joint probability distribution, for example, that X_n are independent of each other. In this case, $p(X_1, X_2, \ldots, X_N) = \prod_{n=1}^{N} p(X_n)$. However, this assumption is too strong. Without any assumption, we can also apply the chain rule to compute the joint probability as the conditional probability of each variable:

$$p(X_1, X_2, \ldots, X_N) = p(X_1|X_2, X_3, \ldots, X_N)p(X_2|X_3, X_4, \ldots, X_N) \ldots p(X_{N-1}|X_N)p(X_N).$$

The chain rule can factorize the joint probability into a product of local conditional probabilities for each random variable. However, an exponential number of local conditional probabilities are needed to fully characterize the joint probability distribution. For example, for N binary random variables, to fully specify their joint probability distribution, we need $2^N - 1$ parameters (probabilities).

3.2 Bayesian Networks

Fortunately, we can alleviate the above problem using a graphical model such as Bayesian networks (BNs). By capturing the inherent independencies in variables, a BN allows us to efficiently encode the joint probability distribution of a set of random variables.

3.2.1 BN representation

Let $\mathbf{X}=\{X_1, X_2, \ldots, X_N\}$ be a set of N random variables. A BN is a graphical representation of the joint probability distribution $p(X_1, X_2, \ldots, X_N)$ over a set of random variables $\mathbf{X}$. A BN over $\mathbf{X}$ can be defined as a two-tuple graph $\mathcal{B} = (\mathcal{G}, \mathbf{\Theta})$, where $\mathcal{G}$ defines the qualitative part of the BN, whereas $\mathbf{\Theta}$ defines its quantitative part. The graph $\mathcal{G}$ is a directed acyclic graph (DAG). As a DAG, a BN does not allow directed cycles, but undirected cycles are fine. The graph $\mathcal{G}$ can be further defined as $\mathcal{G} = (\mathcal{V}, \mathcal{E})$, where $\mathcal{V}$ represents the nodes of $\mathcal{G}$ corre-

Probabilistic Graphical Models for Computer Vision. https://doi.org/10.1016/B978-0-12-803467-5.00008-3

sponding to the variables in $\mathbf{X}$, and $\mathcal{E} = \{e_{nm}\}$ the directed edges (links) between nodes m and n. They capture the probabilistic dependencies between variables. Such dependencies are often causal, though this is not necessarily the case. The causal semantics are an important property of BNs and partially contribute to their widespread use.

Specifically, X_n represents the nth node, and e_{nm} represents the directed edge (link) between nodes n and m. If a directed edge exists from node X_n to node X_m, then X_n is a parent of X_m, and X_m is a child of X_n. This definition can be extended to ancestor nodes and descendant nodes. A node X_n is an ancestor node of X_m, and X_m is a descendant of X_n if there is a directed path from X_n to X_m (see the definition of a directed path in Section 3.2.2.3). A node without a child is called a leaf node, whereas a node without a parent is called a root node. Two nodes are adjacent to each other if they are directly connected by a link. If two nodes X_n and X_m have a common child and are not adjacent, they are spouses of each other, and their common child is referred to as a collider node. The structure formed by two nonadjacent X_n and X_m and their common child is referred to as a V-structure (see Fig. 3.4).

Fig. 3.1 shows a BN with five nodes, where nodes X_1 and X_2 are the root nodes, and X_4 and X_5 are the leaf nodes. Node X_1 is the parent of node X_3, whereas node X_3 is a child of node X_1. Similarly, X_1 is an ancestor of node X_4, and X_4 is a descendant of X_1. Node X_2 is a spouse of node X_1. The structure formed by nodes X_1, X_3, and X_2 is a V-structure, and X_3 is the collider node.

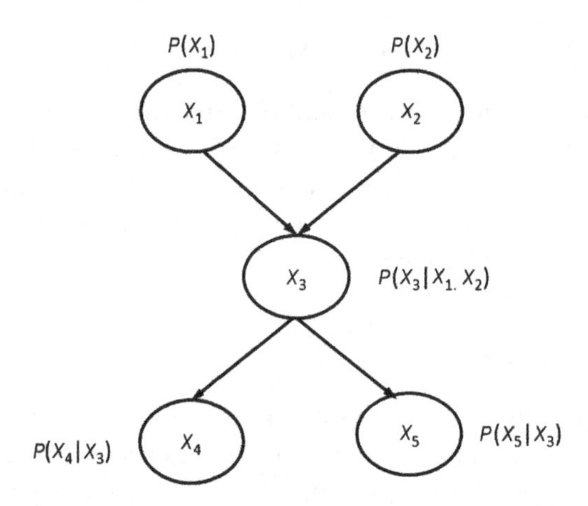

FIGURE 3.1 A simple Bayesian Network.

The quantitative part $\mathbf{\Theta} = \{\boldsymbol{\theta}_n\}_{n=1}^N$ consists of the BN parameters $\boldsymbol{\theta}_n$ for each node, where $\boldsymbol{\theta}_n$ is the conditional probability distribution (CPD) for node n given its parents, that is, $\boldsymbol{\theta}_n = p(X_n|\pi(X_n))$, where $\pi(X_n)$ represents a configuration set of nodes that are parents of X_n. The CPD measures the strength of the links between a node and its parents. For a root node X_n, the conditional probability degenerates to its prior probability $p(X_n)$. For the BN in Fig. 3.1, the probability next to each node gives the CPD for that node, including the prior probabilities for root nodes X_1 and X_2. Given a BN, because of the built-in conditional independencies as a result of the Markov condition (to be defined later), the joint probability distribution of N nodes can be written as

$$p(X_1, X_2, \ldots, X_N) = \prod_{n=1}^{N} p(X_n|\pi(X_n)). \tag{3.1}$$

The joint probability factorization in Eq. (3.1) is also called the *chain rule* of the BN. Following the BN's chain rule, the joint probability for the BN in Fig. 3.1 can be written as

$$p(X_1, X_2, \ldots, X_5) = p(X_1)p(X_2)p(X_3|X_1, X_2)p(X_4|X_3)p(X_5|X_3).$$

With this compact representation, the number of parameters needed to fully specify the joint probability distribution of a binary BN is upper-bounded by $2^K \times N$, where K is the maximum number of parents for all nodes. This represents a tremendous saving in the number of parameters to fully specify the joint probability distribution. For example, if $N = 10$ and $K = 2$, then the number of parameters for a BN is up to 40. In contrast, without a BN, to represent $p(X_1, X_2, \ldots, X_{10})$, we would need $2^{10} - 1 = 1023$ parameters. Therefore BNs can concisely represent the joint probability distribution. The compact representation and the causal semantics are two major advantages of BN.

Using CPD specification, the number of parameters for each node increases exponentially as the number of its parents increases. To alleviate this problem, other alternative specifications have been proposed. For example, CPD can be specified as a function of certain regression parameters, as a tree, through the noisy-or principle or as a function of the parental values. The most common such a CPD is the linear regression, where the CPD is specified as a weighted linear combination of parental values. These approximations can significantly reduce the number of BN parameters and make it feasible for BN learning with limited training data. The readers may refer to Chapter 5 of [7] for different methods to efficiently represent CPDs.

3.2.2 Properties of BNs

As previously stated, the joint probability distribution of a BN can be factorized into the product of the local conditional probabilities precisely because of its built-in conditional independencies. In this section, we explore the independence and other properties of a BN.

3.2.2.1 Markov condition

An important property of a BN is its explicit built-in conditional independencies. They are encoded through the Markov condition (MC). Also called the causal Markov condition or Markov assumption, the MC states that a node X_n is independent of its nondescendants given its parents. A node X_m is a descendant of another node X_n if there is a directed path from X_n to X_m. Mathematically, the MC can be represented as $X_n \perp \text{ND}(X_n)|\pi(X_n)$, where ND stands for the nondescendants. For the BN example in Fig. 3.1, we can derive the following conditional independencies using the MC: $X_4 \perp X_1|X_3$, $X_5 \perp X_2|X_3$, and $X_4 \perp X_5|X_3$. For a root node, MC still applies. In this case, a root node is marginally independent of its nondescendants. For the example in Fig. 3.1, node X_1 is marginally independent of node X_2, since with the switched link, X_5 is no longer a descendant of X_1. The MC directly leads to the chain rule of the BN in Eq. (3.1). We can easily prove the chain rule using the MC. In fact, a DAG becomes a BN if the DAG satisfies the MC. Note that the MC provides the structural independencies in a BN. The BN parameters, however, can produce additional independencies outside the BN structural independencies.

3.2.2.2 Markov blanket and moral graph

Using the MC, we can easily derive another useful local conditional independence property. The Markov blanket of a target variable T, MB_T, is a set of nodes conditioned on which all other nodes are independent of T, denoted as $X \perp T|MB_T, X \in \mathbf{X} \setminus \{T, MB_T\}$. For example, in Fig. 3.2, the MB of node T, MB_T, consists of the shaded nodes: namely, its parent node P, its child node C, and its spouse S. A moral graph of a DAG is an equivalent undirected graph, where each node of the DAG is connected to its Markov blanket.

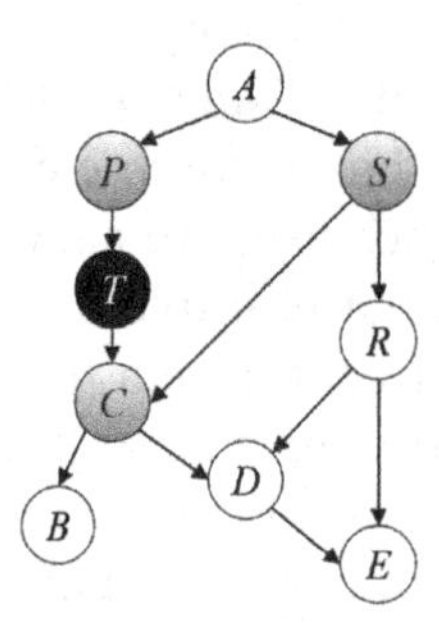

FIGURE 3.2 An example of a Markov blanket, where T is the target node, P is its parent, C is its child, and S is its spouse. Nodes P, C, and S collectively form the Markov blanket of T.

3.2.2.3 D-separation

Through the MC and Markov blanket, we can easily ascertain the local conditional independencies among variables. The MC can be extended to ascertain the independence relationships between variables far away from each other. This is accomplished through the so-called D-separation principle. Before we formally introduce this concept, we first define some notations.

An undirected path between two nodes X_m and X_n in $\mathcal{G}$ is a sequence of nodes between them such that any successive nodes are connected by a directed edge and no node appears in the sequence twice. A directed path of a DAG is a path with nodes $(X_1, X_2, \ldots, X_{N-1})$ such that, for $1 \le n < N$, X_n is a parent of X_{n+1}. A node X_k is defined as a collider node if it has two incoming edges from X_m and X_n in a path, regardless of whether

X_m and X_n are adjacent. Node X_k with nonadjacent parents X_m and X_n is an unshielded collider node for the path between X_m to X_n, as shown in Fig. 3.4.

Given a set of nodes $\mathbf{X}_E$, an undirected path J between node X_m and X_n is blocked by $\mathbf{X}_E$ if either of the following holds: 1) there is a noncollider node in J belonging to $\mathbf{X}_E$; or 2) there is a collider node X_c on J such that neither X_c nor any of its descendants belongs to $\mathbf{X}_E$. Otherwise, path J from X_m and X_n is unblocked or active. Nodes X_m and X_n are D-separated by $\mathbf{X}_E$ if every undirected path between X_m and X_n is blocked by $\mathbf{X}_E$, that is, $X_m \perp X_n | \mathbf{X}_E$. Fig. 3.3 provides examples of D-separation, where node X_6 is D-separated from node X_2 given X_3 and X_4 since both paths between X_6 and X_2 are blocked. Similarly, X_1 and X_2 are D-separated (marginally independent) since both paths between them are blocked. However, X_1 and X_2 are not D-separated given X_3, X_5, or X_6 since one of the two paths from X_1 to X_2 ($X_1 \rightarrow X_3 \leftarrow X_2$ and $X_1 \rightarrow X_3 \rightarrow X_5 \leftarrow X_4 \leftarrow X_2$) is unblocked by one of the two V-structures. An algorithm is introduced in Section 3.3.3 of [7] to perform automatic D-separation test.

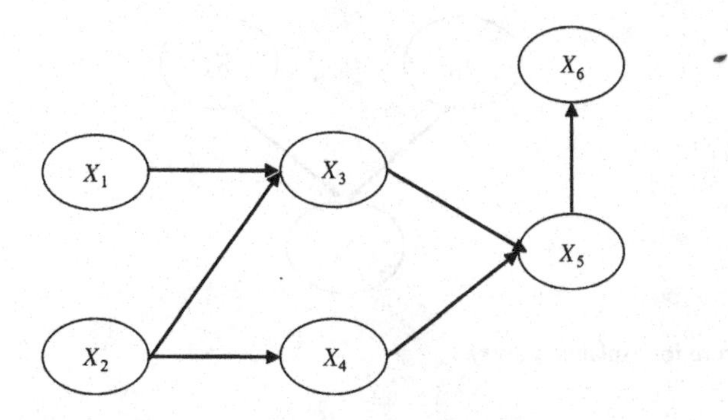

FIGURE 3.3 An example of D-separation, where node X_6 is D-separated from node X_2 given X_3 and X_4.

3.2.2.4 Faithfulness

The joint probability distribution of a BN is used to approximately capture the underlying data distribution p. A BN is completely faithful to p if its structural independencies (as a result of the MC) cover all and only independencies in p. Such a BN is called the perfect I-map of p. If, on the other hand, the BN only captures a subset of independencies in p, then it is only partially faithful and is called an I-map of p.

When learning a BN from data, faithfulness is a standard assumption. However, as previously stated, BN parameterizations may destroy some structural independencies or introduce additional independencies. It is well known that the Lebesgue measure of the unfaithfulness distributions is zero [8], which means that the probability of unfaithful parameterization is very low. Even when unfaithfulness distributions occur, the unfaithfulness relationships only consist of a small percent of the total independence and dependence relationships in the graph. Further information on BN faithfulness can be found in [9].

3.2.2.5 Explaining-away

An important and unique property of BNs is their explaining-away property. If three nodes form a V-structure, that is, two nonadjacent nodes share the same child as shown in Fig. 3.4, where node X_k is a child of both nodes X_m and X_n, then nodes X_m and X_n are spouses of each other. In this case, X_m and X_n are independent of each other not given X_k, but they become dependent on each other given X_k. This dependence between X_m and X_n given X_k forms the basis of the explaining-away principle, which states that knowing the state of X_m can alter the probability of X_n given X_k since they can both cause X_k. For example, let $X_m = Rain$, $X_n = Sprinkler$, and $X_k = Groundwet$. Given that the ground is wet, knowing it did not rain, the probability that the sprinkler is on increases. Explaining-away is a unique and powerful property for the BNs, and it cannot be represented by an undirected graphical model. Although it allows capturing additional dependencies for better representation, it can also cause computational difficulty during learning and inference.

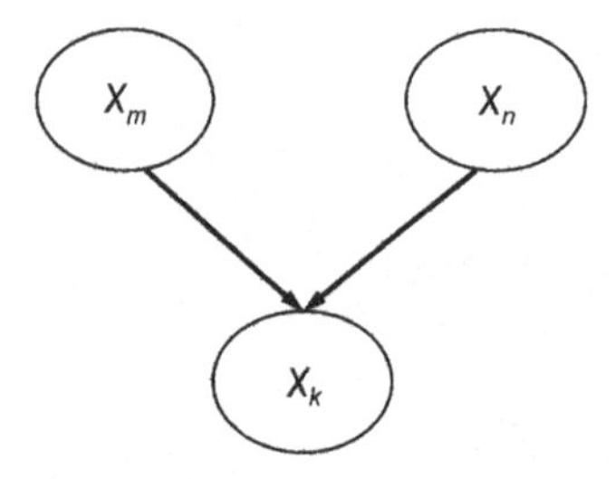

FIGURE 3.4 A V-structure for explaining-away.

3.2.2.6 Equivalent Bayesian networks

Two BNs of different topologies but over the same set of variables are probabilistically equivalent if they capture the same structural conditional independencies. To be equivalent, they must have the same skeleton, that is, links between the same pair of nodes but not necessarily in the same direction, and they must share the same V-structures [10]. For example, among the BNs in Fig. 3.5, the four BNs in the top row are equivalent, whereas the four BNs in the bottom row are not.

3.2.3 Types of Bayesian networks

There are three types of BNs: discrete BNs, continuous BNs, and hybrid BNs.

3.2.3.1 Discrete BNs

For a discrete BNs, all nodes represent discrete random variables, where each node may assume different but mutually exclusive values. Each node may represent an integer variable, a categorical variable, a Boolean variable, or an ordinal variable. The most commonly used discrete BN is the binary BN, where each node represents a binary variable. For a discrete

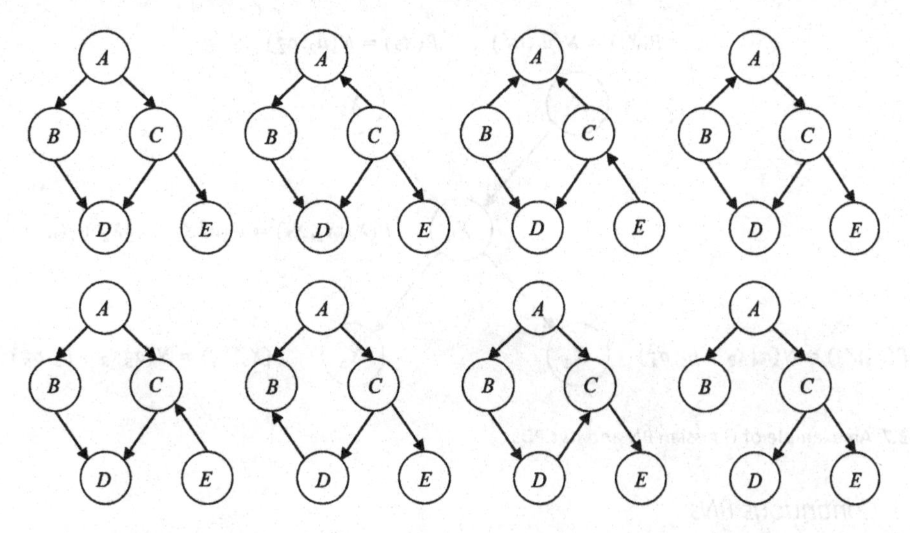

FIGURE 3.5 Equivalent BNs. The four top BNs are equivalent, but the four bottom BNs are not as they violate one of the two conditions.

BN, the CPD is specified in terms of the conditional probability table (CPT). This table lists the probabilities that a child node takes on each of its feasible values, given each possible configuration of its parents such as $\theta_{njk} = p(x_n = k|\pi(x_n) = j)$, which is the probability of node X_n assuming the value of k, given the jth configuration of its parents. Fig. 3.6 presents an example of a binary BN with CPTs for each node $X_n \in \{0, 1\}$.

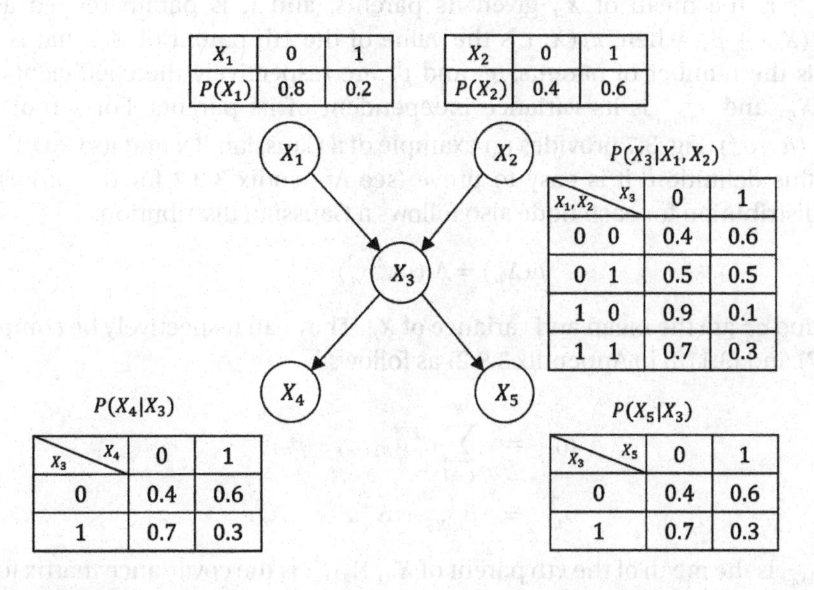

X_1 0 1
$P(X_1)$ 0.8 0.2
X_2 0 1
$P(X_2)$ 0.4 0.6
X_1
X_2
X_3
X_4
X_5
$P(X_3|X_1, X_2)$
X_1, X_2 \ X_3 0 1
0 0 0.4 0.6
0 1 0.5 0.5
1 0 0.9 0.1
1 1 0.7 0.3
$P(X_4|X_3)$
X_3 \ X_4 0 1
0 0.4 0.6
1 0.7 0.3
$P(X_5|X_3)$
X_3 \ X_5 0 1
0 0.4 0.6
1 0.7 0.3

FIGURE 3.6 A binary BN and its CPTs.

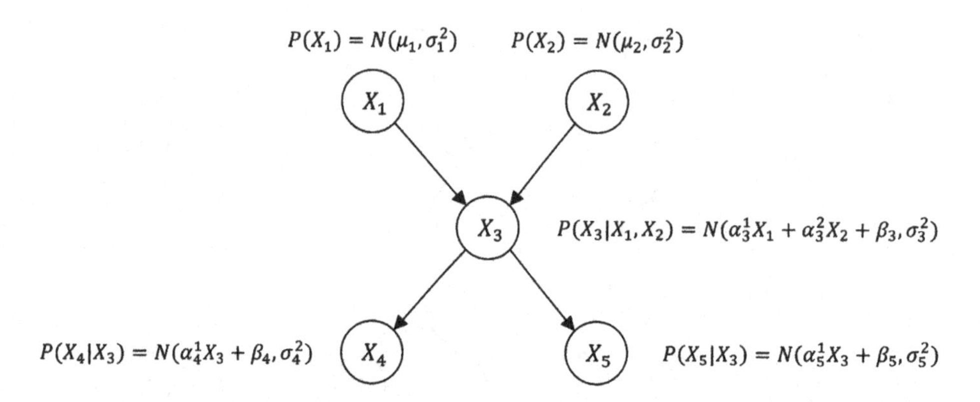

FIGURE 3.7 An example of Gaussian BN and its CPDs.

3.2.3.2 Continuous BNs

Despite the popularity with discrete BNs, continuous BNs are also useful. For a continuous BN, all nodes represent continuous random variables. The most common continuous BN is the linear Gaussian BN due to its simplicity in representation and the existence of analytic solutions for both learning and inference. For a linear Gaussian BN, the conditional probability of each node follows a Gaussian distribution, and the joint probability distribution of all nodes follows a multivariate Gaussian distribution. The CPD for each node is typically specified as a linear Gaussian, that is,

$$p(X_n|\pi(X_n)) = \mathcal{N}(\mu_{n|\pi_n}, \sigma^2_{n|\pi_n}), \tag{3.2}$$

where $\mu_{n|\pi_n}$ is the mean of X_n given its parents, and it is parameterized as $\mu_{n|\pi_n} = \sum_{k=1}^{K} \alpha_n^k \pi_k(X_n) + \beta_n$, where $\pi_k(X_n)$ is the value of the kth parent of X_n, that is, $\pi_k(X_n) \in \pi(X_n)$, K is the number of parents, α_n^k and β_n are respectively the coefficients and bias for node X_n, and $\sigma^2_{n|\pi_n}$ is its variance independent of its parents. For a root node X_n, $p(X_n) \sim \mathcal{N}(\mu_n, \sigma_n^2)$. Fig. 3.7 provides an example of a Gaussian BN and its CPDs.

Given this definition, it is easy to prove (see Appendix 3.9.2 for the proof) that the marginal distribution for each node also follows a Gaussian distribution:

$$p(X_n) = \mathcal{N}(\mu_n, \sigma_n^2), \tag{3.3}$$

where μ_n and σ_n^2 are the mean and variance of X_n. They can respectively be computed (see Eqs. (3.177) and (3.178) in Appendix 3.9.2) as follows:

$$\mu_n = \sum_{k=1}^{K} \alpha_n^k \mu_{\pi_k(x_n)} + \beta_n, \tag{3.4}$$

$$\sigma_n^2 = \sigma^2_{n|\pi_n} + \boldsymbol{\Lambda}_n^\top \Sigma_{\pi(x_n)} \boldsymbol{\Lambda}_n, \tag{3.5}$$

where $\mu_{\pi_k(x_n)}$ is the mean of the kth parent of X_n, $\Sigma_{\pi(x_n)}$ is the covariance matrix for parents of X_n, and $\boldsymbol{\Lambda}_n = (\alpha_n^1, \alpha_n^2, \ldots, \alpha_n^K)^\top$. Similarly, the joint probability distribution of a Gaussian

BN also follows a multivariate Gaussian distribution, that is,

$$p(X_1, X_2, \ldots, X_N) = \mathcal{N}(\boldsymbol{\mu}, \boldsymbol{\Sigma}), \tag{3.6}$$

where the joint mean vector is $\boldsymbol{\mu} = (\mu_1, \mu_2, \ldots, \mu_N)$. Assuming that $\mathbf{X}$ is arranged in the topological ordering, the joint covariance matrix can be computed from $\boldsymbol{\Sigma} = \mathbf{USU}^\top$, where $\mathbf{S}$ is an $N \times N$ diagonal matrix, $\mathbf{S} = \{\sigma_n^2\}$ contains the variances of nodes X_n, and $\mathbf{U} = (\mathbf{I} - \mathbf{W})^{-1}$, where $\mathbf{I}$ is the $N \times N$ identity matrix, and $\mathbf{W}$ is the lower triangle matrix with $w_{ii} = 0$ and $w_{ij} = \alpha_j^i$, the coefficient between parent node X_i and child X_j. Similarly, a multivariate Gaussian distribution can be equivalently represented by a linear Gaussian BN. Detailed derivations for the Gaussian BN and the conversion between multivariate Gaussian and a Gaussian BN can be found in Section 10.2.5 of [11] and Chapter 7.2 of [7].

3.2.3.3 Hybrid BNs

A BN is a hybrid if it has both discrete and continuous nodes. The most common configurations of hybrid BNs include discrete parents and continuous children nodes or hybrid parents (a combination of discrete and continuous parents) and continuous children. For discrete parents and continuous children, the CPD for each node can be specified with multiple Gaussians, one for each parental configuration, as follows:

$$p(X_n|\pi(X_n) = k) \sim \mathcal{N}(\mu_{nk}, \sigma_{nk}^2), \tag{3.7}$$

where μ_{nk} and σ_{nk}^2 are respectively the mean and variance of X_n given the kth configuration of its parents. For hybrid parents and continuous children, the conditional probability can be specified as a conditional linear Gaussian. Assume that X_n has K discrete parents and J continuous parents. Let $\pi^d(X_n)$ be the set of discrete parents of X_n, let $\pi^c(X_n)$ be the continuous parents, and let $l \in \pi^d(X_n)$ be the lth configuration of the discrete parents. The CPD for X_n can be specified by

$$p(X_n|\pi^d(X_n) = l, \pi^c(X_n)) = \mathcal{N}(\mu_{nl}, \sigma_{nl}^2), \tag{3.8}$$

where $\mu_{nl} = \sum_j \alpha_{njl}\pi_j^c(x_n) + \beta_{nl}$ and σ_{nl}^2 are the mean and variance of X_n under the lth configuration of its discrete parents, and $\pi_j^c(x_n)$ is the jth continuous parent of node X_n.

The configuration of continuous parents and discrete children is often transformed to discrete parents with continuous children through edge reversals [12].

3.2.3.4 Naive BNs

A special kind of BNs is the naive BN (NB). As shown in Fig. 3.8, an NB consists of two layers, with the top layer comprising one node Y and the bottom layer comprising N nodes $X_1, X_2, \ldots, X_N$, where X_n can be either discrete or continuous. Node Y is the parent of each X_n, and it can be discrete or continuous. An NB is parameterized by $p(Y)$ and $p(X_n|Y)$, which can be obtained from training data. Because of the topology, following the MC, we can easily prove that the variables in the bottom layer are conditionally independent of each other given the top node, that is, $X_m \perp X_n | Y, m \neq n$. This means that the joint probability of all the variables in a naive BN can be written as $p(X_1, X_2, \ldots, X_n, Y) = p(Y) \prod_{n=1}^{N} p(X_n|Y)$. The number of parameters for such a model is only $Nk + 1$, where k is the number of distinctive values for Y.

Given a fully specified NB, depending on whether Y is discrete or continuous, it can be used for classification or regression to estimate the most likely value of Y, given their features x_i, as follows:

$$y^* = \underset{y}{\operatorname{argmax}}\, p(Y = y|x_1, x_2, \ldots, x_N),$$

where Y typically represents the class label (value), whereas x_i represent the features. During training, we learn the conditional probability $p(X_n|Y)$ for each link as well as the probability $p(Y)$. During inference, we can find

$$\begin{aligned} y^* &= \underset{y}{\operatorname{argmax}}\, p(y|x_1, x_2, \ldots, x_N) \\ &= \underset{y}{\operatorname{argmax}}\, p(x_1, x_2, \ldots, x_n, y) \\ &= \underset{y}{\operatorname{argmax}}\, p(y) \prod_{n=1}^{N} p(x_n|y) \end{aligned} \tag{3.9}$$

Despite their simplicity and strong assumptions, NBs have been widely used in many applications, often with surprisingly good results.

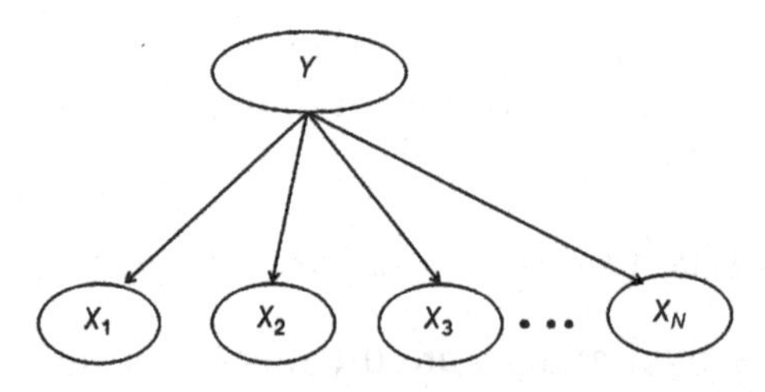

FIGURE 3.8 An example of a naive BN.

To relax the strong feature independence assumption of NBs, augmented naive Bayesian networks (ANBs) are introduced by allowing connections among the nodes in the bottom layer. To limit the complexity, in the bottom layer the internode connections from a feature node to another feature node are limited; for example, in Fig. 3.9, the number of connections from a feature node to another feature node is limited to one. Various studies have demonstrated that by allowing limited dependencies among the feature nodes, ANBs can often improve NBs' performance for classification. The readers are referred to [13,14] for a detailed discussion of ANBs and various variants and of their applications as classifiers.

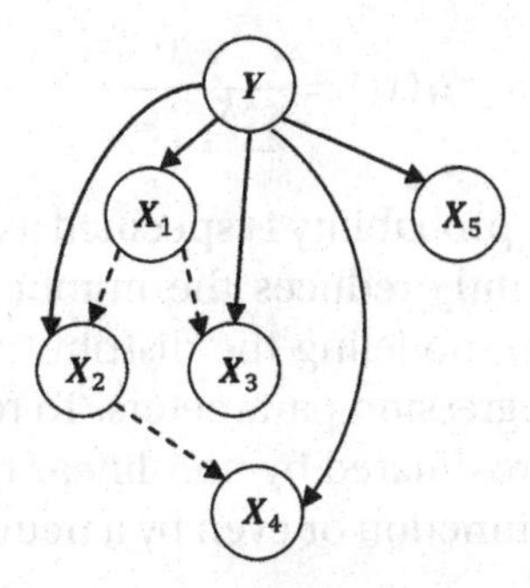

FIGURE 3.9 An example of an ANB, where the dotted links represent connections among features.

3.2.3.5 Regression BNs

A regression BN, as shown in Fig. 3.10, is a special kind of BN, where the CPD for each node is specified as the sigmoid or softmax function of the linear combination of its parent values. Specifically, for node X_n and its J parents $\pi(X_n)$, assuming that X_n is binary, its CPD can be expressed as

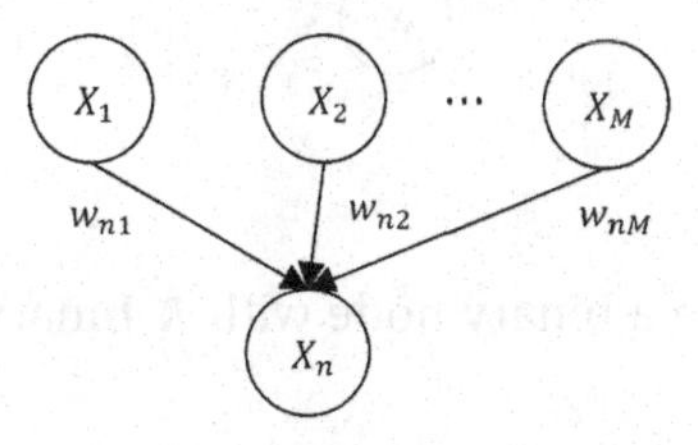

FIGURE 3.10 A regression BN, where the CPD for each node is specified as the sigmoid or softmax function of the linear combination of its parent values.

$$p(X_n = 1|\pi(X_n)) = \sigma(\sum_{m=1}^{J} w_{nm}\pi_m(X_n) + w_n), \tag{3.10}$$

where $\pi_m(X_n)$ is the value of the mth parent of X_n, w_{nm} is the weight for the mth parent, and $\sigma(x) = \frac{1}{1+e^{-x}}$ is the sigmoid function. For a root node, its prior probability is specified as $\sigma(\alpha_0)$. In this way, the CPD of X_n is specified as a function of regression parameters w_{nm}, and the number of parameters is equal to the number of links instead of its exponential. A binary regression BN with latent binary nodes is referred to as a sigmoid belief network [15] in the literature. For a categorical node with $K > 2$ values and J parents, Eq. (3.10) can be changed to

$$p(X_n = k|\pi(X_n)) = \sigma_M(\sum_{m=1}^{M} w_{nmk}\pi_m(X_n) + w_{nk}), \tag{3.11}$$

where $\sigma_M(x)$ is the multiclass sigmoid (a.k.a. the softmax function) defined as

$$\sigma_M(x_k) = \frac{e^{x_k}}{\sum_{k'=1}^{K} e^{x_{k'}}}.$$

If X_n is a root node, then its prior probability is specified as $\sigma_M(\alpha_k)$.

While regression BN significantly reduces the number of parameters to specify the CPTs, it also introduces errors in modeling the distributions as the exact CPTs are now approximated by a small set of regression parameters. To reduce the approximation error, the linear regression can be approximated by non-linear regression functions such as an nth order polynomial regression function or even by a neural network.

3.2.3.6 Noisy-OR BN

Another special kind of binary BN is the noisy-OR BN. It was introduced to reduce the number of CPD parameters for each node.

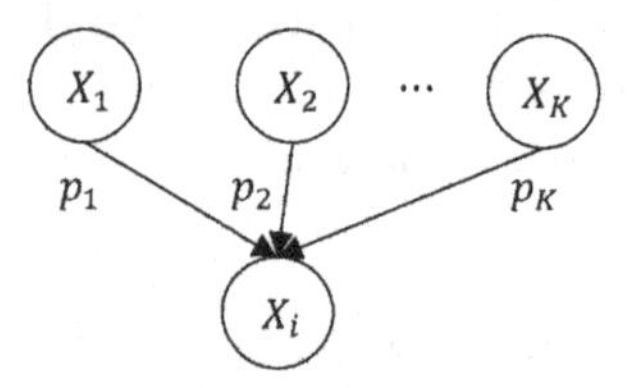

FIGURE 3.11 A noisy-OR BN.

In Fig. 3.11, let $x_n \in \{0, 1\}$ be a binary node with K binary parents $X_1, X_2, \ldots, X_K$, and $X_k \in \{0, 1\}$. Let

$$p_k = p(x_n = 0|x_1 = 0, x_2 = 0, \ldots, x_k = 1, \ldots, x_K = 0),$$

that is, the probability of $X_n = 0$ is given only if its kth parent is 1. Assuming that the mechanism for each $X_k = 1$ to cause $X_n = 0$ (or to inhibit X_n=1) is independent, we have

$$p(x_n = 0|x_1, x_2, \ldots, x_K) = \prod_{k \in \{x_k=1\}} p_k. \tag{3.12}$$

This way the number of parameters for each node is reduced to the number of parents instead of its exponential. Noisy-OR has since been extended to multivalue categorical nodes [16].

3.3 BN inference

Given a BN over the variable space $\boldsymbol{X}$, inference can be performed. BN inference aims at estimating the probability of a set of unknown variables or their most likely states, given the observations of other variables. Assume that the set of all variables is partitioned into $\boldsymbol{X} = \{\boldsymbol{X}_U, \boldsymbol{X}_E\}$, where $\boldsymbol{X}_U$ represents the set of unobserved variables, and $\boldsymbol{X}_E$ the set of observable variables (evidence). Four basic probabilistic inferences are often performed:

1. Posterior probability inference: compute the posterior probability of some unknown variables $X_Q \subseteq X_U$ given observed variables $\mathbf{X}_E = x_E$,

$$p(X_Q|X_E = x_E). \tag{3.13}$$

2. Maximum a posteriori (MAP) inference: identify the most likely configuration of all unobserved variables X_U,

$$x_U^* = \underset{x_U}{\operatorname{argmax}}\, p(X_U = x_U|X_E = x_E). \tag{3.14}$$

3. Marginal MAP inference: identify the most likely configuration of a subset of unobserved variables $X_Q \subseteq X_U$,

$$x_Q^* = \arg\max_{x_Q} p(X_Q = x_Q|X_E = x_E) = \arg\max_{x_Q} \sum_{x_U \backslash x_Q} p(X_U = x_U|X_E = x_E). \tag{3.15}$$

4. Model likelihood inference: estimate the likelihood of a BN, including its structure $\mathcal{G}$ and parameters Θ, given x_E, that is,

$$p(X_E = x_E|G, \Theta) = \sum_{x_U} p(\mathbf{x}_U, x_E|\mathcal{G}, \Theta). \tag{3.16}$$

Posterior probability inference is also referred to as sum-product inference in the literature since its operation involves summing over the irrelevant variables. MAP is referred to as max-product inference as it involves finding the best configuration for all unknown variables. It is also called MPE inference. Marginal MAP is also called max sum inference, as it involves finding the best configuration of only a subset of unknown variables. Summation is needed to marginalize out other unknown variables as shown in Eq. (3.15). In general, the MAP inference cannot be found by taking the most probable configuration of nodes individually. The best configuration of each node must be found collectively with other nodes. Furthermore, the marginal MAP cannot be found by taking the projection of the MAP onto the explanation set. For discrete BNs, both MAP and marginal MAP inferences are often formulated as a combinatorial optimization problem. The marginal MAP is usually computationally more expensive as it requires both marginalization and maximization. The MAP and marginal MAP inferences are built upon the posterior probability inference, since they both require computing the posterior probabilities. The MAP inference is widely used in CV. For example, for image labeling, the inference may be used to infer the best label configuration for all pixels in an image. For part-based object detection (e.g., facial or body landmark detection), the MAP inference may be used to simultaneously identify the optimal position for each object part (each landmark point). The model likelihood inference is used to evaluate different BNs and to identify the BN that is most likely to generate the observation. Complex models produce higher scores. Therefore, the likelihood should be normalized by the structure complexity before comparing different models. In CV, the likelihood inference may be used for model-based classification such as HMM-based classification. Theoretically, as long as the joint probability of any configuration of variables can be computed using the BN, all the inferences can be performed. However, the BN inference is shown to be NP-hard in the worst case [17], because computing the posterior probability and determining the best configuration often require summation and search over an exponential number of variable configurations. The worst case occurs when the BN is densely connected. In practice, we can always employ domain knowledge to simplify the model to avoid the worst-case scenario.

Despite these computational difficulties, many efficient inference algorithms have been developed. The key to improving inference efficiency is again exploiting the independencies embedded in the BN. These algorithms either take advantage of simple structures to reduce the complexity for exact inference or make some assumptions to obtain an approximate result.

Depending on the complexity and size of the BN, inference methods can be performed exactly or approximately. Inference can also be done analytically or numerically through sampling. The analytic inference often produces exact inference results, whereas the numerical sampling frequently leads to approximated inference results. In the sections to follow, we will introduce the exact and approximate inference methods, respectively.

3.3.1 Exact inference methods

Exact inference methods produce an exact estimation of the posterior probability, based on which MAP or marginal MAP inference can be performed. Due to their need to sum over all irrelevant variables, the exact inference methods are typically limited to simple structures or small BNs.

3.3.1.1 Variable elimination

One basic exact inference method for BNs is the variable elimination (VE) method [18]. VE is a simple inference method for performing the posterior probability inference, that is, computing $p(\mathbf{X}_Q|\mathbf{X}_E = \mathbf{x}_E)$. When computing the posterior probability, it is necessary to marginalize over irrelevant variables $\mathbf{X}_U \setminus \mathbf{X}_Q$. Naively marginalizing over all variables jointly requires K^M number of summations and $N-1$ products for each summation, where K is the number of values for each variable, M is the number of variables over which to marginalize, and N is the total number of variables in the BN. The computation soon becomes intractable for large K and M. Hence, the VE algorithm aims to significantly reduce the computations. Instead of marginalizing over all variables jointly or marginalizing each variable in a random order, the VE algorithm eliminates (marginalizes) variables one by one according to an *elimination order*. Following the order, the number of summations and the number of products for each summation can be reduced significantly by recursively using the previously computed results.

In the BN in Fig. 3.12, for example, we want to compute $p(A|E=e)$. According to the Bayes' rule, it can be written as

$$\begin{aligned} p(A|E=e) &= \alpha p(A, E=e) \\ &= \alpha \sum_{b,c,d} p(A,b,c,d,e), \end{aligned} \tag{3.17}$$

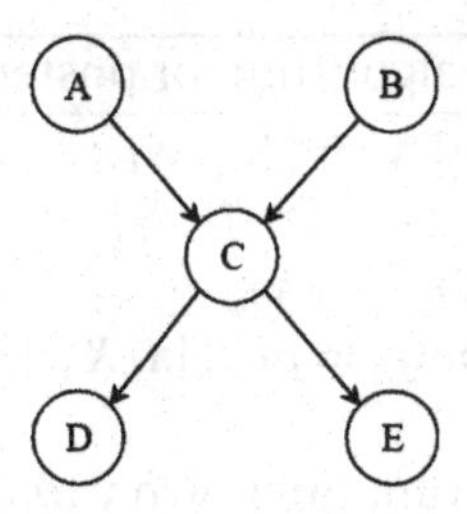

FIGURE 3.12 An example of a BN with five nodes.

where α is a normalization constant. With the BN chain rule, Eq. (3.17) can be rewritten as

$$p(A|E=e)=\alpha\sum_{b,c,d} p(A)p(b)p(c|A,b)p(d|c)p(e|c). \tag{3.18}$$

Computing Eq. (3.18) requires summation over irrelevant variables B, C, and D. Assuming that all variables are binary, naive summation will require eight summations, and each summation will require four products. With the VE algorithm, we can significantly reduce both the number of summations and the number of products for each summation. Taking, for example, Eq. (3.18), we can follow the order $C \to D \to B$ to perform the marginalization, that is, first eliminate C, then D, and finally B. Based on this order, Eq. (3.18) can be written as

$$\begin{aligned} &\sum_{c,b,d} p(A)p(b)p(c|A,b)p(d|c)p(e|c) \\ = \;& p(A)\sum_b p(b)\sum_d\sum_c p(c|A,b)p(d|c)p(e|c) \\ = \;& p(A)\sum_b p(b)\sum_d f_1(A,b,d,e) \\ = \;& p(A)\sum_b p(b)f_2(A,b,e) \\ = \;& p(A)f_3(A,e) \\ = \;& f_4(A), \end{aligned} \tag{3.19}$$

where $f_1(A,b,d,e)$ is a factor function of a, b, d, and e; $f_2(A,b,e)$ is a function of a, b, and e; and $f_3(A,e)$ is a function of a and e. Given $f_4(A)$, $p(A|e)$ can be obtained by normalizing over A. By following this order the number of summations is reduced from 8 (2^3) to 6 (2+2+2), and the total number of products is reduced to 7 by reusing the factors without having to recompute them. The specific saving for both summations and products depends on the complexity of the model and on the elimination order. Unfortunately, determining the optimal elimination order itself is NP-hard.

In general, for a BN over $\mathbf{X}$, let X_q be a query variable $X_q \in \mathbf{X}_U$, let $\mathbf{X}_E$ be the observed variables, and let $\mathbf{X}_I$ be the irrelevant variables, that is, $\mathbf{X}_I = \mathbf{X}_U \setminus X_q$. Our goal is to compute $p(X_q|\mathbf{x}_E) \propto p(X_q,\mathbf{x}_E) = \sum_{\mathbf{x}_I} p(X_q,\mathbf{x}_E,\mathbf{x}_I)$, which can be done using the VE algorithm as summarized in Algorithm 3.1.

Algorithm 3.1 Variable elimination algorithm for posterior inference.

▷ Let X_q be the query variable, and $\mathbf{X}_I = \{X_{I_1}, X_{I_2}, \ldots, X_{I_m}\}$ be irrelevant
▷ variables, i.e., $\mathbf{X}_I = \mathbf{X}_U \setminus X_q$
Order the variables in $\mathbf{X}_I$ as $X_{I_1}, X_{I_2}, \ldots, X_{I_m}$
Initialize the initial factors f_n to be $f_n = p(X_n|\pi(X_n))$, where $n = 1, 2, \ldots, N$, and N is the number of nodes
for j= I_1 to I_m **do** //for each irrelevant unknown variable in the order
  Search current factors to find factors $f_{j1}, f_{j2}, \ldots, f_{jk}$ that include X_{I_j}
  $F_j = \sum_{X_{I_j}} \prod_{i=1}^{k} f_{ji}$ // Generate a new factor F_j by eliminating X_{I_j}
  Replace factors $f_{j1}, f_{j2}, \ldots, f_{jk}$ by F_j
end for
$p(X_q, X_E) = \prod_{s \in X_q} f_s F_{I_m}$ // F_{I_m} is the factor for last irrelevant variable
Normalize $p(X_q, \mathbf{X}_E)$ to obtain $p(X_q|\mathbf{X}_E)$

Besides performing posterior probability inference, VE can also be applied to MAP inference, that is, identifying the most probable explanation for all unknown variables $\boldsymbol{X}_U$, $\boldsymbol{x}_U^* = \text{argmax}_{\boldsymbol{x}_U} p(\boldsymbol{X}_U = \boldsymbol{x}_U | \boldsymbol{X}_E = \boldsymbol{x}_E)$. Different from the VE for posterior probability inference, where variables are eliminated via marginalization, VE for MAP inference eliminates variables via maximization and it is hence called max-product. MAP can be written as $\boldsymbol{x}_U^* = \text{argmax}_{\boldsymbol{x}_U} p(\boldsymbol{X}_U = \boldsymbol{x}_U, \boldsymbol{X}_E = \boldsymbol{x}_E)$. Letting $\mathbf{X}_U = \{X_{U_j}\}$, $j = 1, 2, \ldots, n$, the variables of $\mathbf{X}_U$ be ordered in $X_{U_1}, X_{U_2}, \ldots, X_{U_n}$. The VE algorithm for MAP can be summarized by Algorithm 3.2. For comparison, Algorithm 13.1 in [7] also provides a pseudocode for max-product inference with the VE method.

Algorithm 3.2 Variable elimination algorithm for MAP inference.

Forward process:
Order the unknown variables in $\mathbf{X}_U$, that is, $X_{U_1}, X_{U_2}, \ldots, X_{U_m}$, where U_m is the number of unknown variables
Initialize the initial factors f_n to be $f_n = p(X_n|\pi(X_n))$, $n = 1, 2, \ldots, N$, and N is the number of nodes
for j=1 to U_m **do** //for each unknown variable
  Search current factors to find factors $f_{j1}, f_{j2}, \ldots, f_{jk}$ that include X_{U_j}
  $F_j = \max_{X_j} \prod_{k=1}^{jk} f_{jk}$ // Generate a new factor F_j by eliminating X_j
  Replace factors $f_{j1}, f_{j2}, \ldots, f_{jk}$ by F_j
end for
Trace back process:
$x_{U_m}^* = \text{argmax}_{x_{U_m}} F_{U_m}(x_{U_m})$
for j=U_{n-1} to 1 **do** //for each irrelevant unknown variable
  $x_{U_j}^* = \underset{x_{U_j}}{\text{argmax}}\, F_j(x_{U_m}^*, \ldots, x_{U_{j+1}}^*, x_{U_j})$
end for

Taking the BN in Fig. 3.12 as an example, we want to compute

$$
\begin{aligned}
a^*, b^*, c^*, d^* &= \underset{a,b,c,d}{\operatorname{argmax}}\, p(a,b,c,d|E=e) \\
&= \underset{a,b,c,d}{\operatorname{argmax}}\, p(a)p(b)p(c|a,b)p(d|c)p(e|c).
\end{aligned} \tag{3.20}
$$

Following the order $C \to D \to B \to A$, we can perform the forward maximization:

$$
\begin{aligned}
&\max_{a,b,c,d}\, p(a)p(b)p(c|a,b)p(d|c)p(e|c) \\
&= \max_a p(a) \max_b p(b) \max_d \max_c p(c|a,b)p(d|c)p(E=e|c) \\
&= \max_a p(a) \max_b p(b) \max_d \max_c f_1(a,b,c,d,e) \\
&= \max_a p(a) \max_b p(b) \max_d f_2(a,b,d,e) \\
&= \max_a p(a) \max_b p(b) f_3(a,b,e) \\
&= \max_a p(a) f_4(a,e).
\end{aligned} \tag{3.21}
$$

Given the factor functions f_1, f_2, f_3, and f_4, the trace-back process can then be performed to identify the MAP assignment for each node:

- $a^* = \operatorname{argmax}_a p(a) f_4(a,e)$,
- $b^* = \operatorname{argmax}_b p(b) f_3(a^*,b,e)$,
- $d^* = \operatorname{argmax}_d f_2(a^*,b^*,d,e)$,
- $c^* = \operatorname{argmax}_c f_1(a^*,b^*,d^*,c,e)$.

For both posterior and MAP inference, following the elimination order, we can exploit the independencies in the BN such that some summations (maximizations) can be performed independently for a subset of variables, and their results can subsequently be reused. This again demonstrates the importance of built-in independencies of a BN for reducing the complexity of the BN inference. The VE method has been shown to have a complexity exponential to the tree width[1] of the induced graph resulting from a particular elimination order. In the extreme case, when all variables are independent of each other, the number of summations is reduced to $M \times K$, that is, it is linear with respect to the number of variables, where M is the number of variables, and K is the number of summations for each variable.

3.3.1.2 Belief propagation in singly connected Bayesian networks

Another exact inference method for BNs is the belief propagation (BP) algorithm. It was originally proposed by Judea Pearl [19] to perform the sum-product inference in BNs. The BP was developed to exactly compute the posterior probability for singly connected BNs (i.e., BNs with no undirected loops). It can also be extended to perform the MAP or

[1] The tree width of a graph is the minimum width among all possible tree decompositions of the graph.

max-product inference. Given the observed values of certain nodes, the BP updates the probabilities of each node through message passing. For each node, the BP first collects the messages from all its children and all its parents, based on which it updates its probability (belief). Based on its updated belief, the node then broadcasts messages to its parents via upward passes and to its children through downward passes. The process is repeated until convergence.

Specifically, as shown in Fig. 3.13, let X be a node, $\mathbf{V} = (V_1, V_2, \ldots, V_K)$ be its K parental nodes, and $\mathbf{Y} = (Y_1, Y_2, \ldots, Y_J)$ be its J child nodes, where all variables are assumed to be discrete. Let $\mathbf{E}$ be the set of observed nodes. The goal of BP is to propagate the effect of

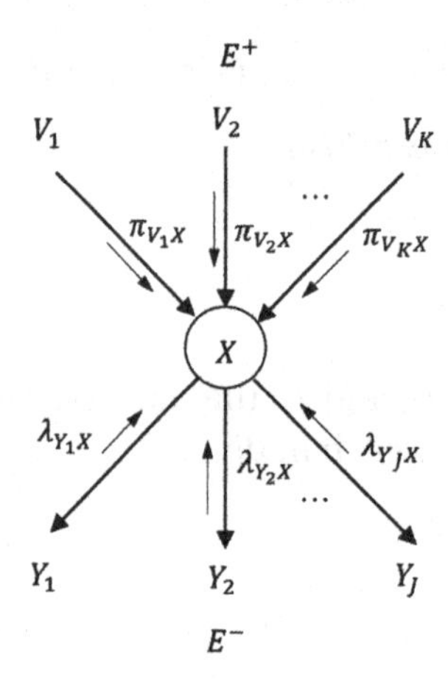

FIGURE 3.13 Incoming messages to node X from its neighbors.

$\mathbf{E}$ throughout the network to evaluate its impact on each node. As shown in Fig. 3.13, the effect of $\mathbf{E}$ on X is propagated into X through parent nodes $\mathbf{V}$ and child nodes $\mathbf{Y}$ of X. Let $\mathbf{E} = (E^+, E^-)$, where E^+ represents the portion of $\mathbf{E}$ that reaches X through its parents $\mathbf{V}$, and E^- is the portion of $\mathbf{E}$ that reaches X through its children $\mathbf{Y}$. We can then evaluate the impact of $\mathbf{E}$ on X, that is, $p(X|\mathbf{E})$ as follows:

$$\begin{aligned} p(X|\mathbf{E}) &= p(X|E^+, E^-) \\ &= \alpha p(X|E^+)p(E^-|X) \\ &= \alpha\pi(X)\lambda(X), \end{aligned} \tag{3.22}$$

where $\pi(X) = p(X|E^+)$ represents the total message X receives from all its parents $\mathbf{V}$ as a result of E^+, whereas $\lambda(X) = p(E^-|X)$ represents the total message X receives from all its children $\mathbf{Y}$ as a result of E^-. The belief of X, that is, $p(X|\mathbf{E})$, can be computed as a product of $\pi(X)$ and $\lambda(X)$ up to a normalization constant α, which can be recovered by using the fact that $\sum_x p(x|\mathbf{E}) = 1$.

We now investigate how we can compute $\pi(X)$ and $\lambda(X)$. Assuming that we are dealing with a singly connected BN, $\pi(X)$ can be proven to be

$$\pi(X) = \sum_{v_1,\ldots,v_K} p(X|v_1,\ldots,v_K) \prod_{k=1}^{K} \pi_{V_k}(X), \tag{3.23}$$

where $\pi_{V_k}(X)$ is the message that X receives from its parent V_k and can be defined as

$$\pi_{V_k}(X) = \pi(V_k) \prod_{C \in child(V_k) \backslash X} \lambda_C(V_k), \tag{3.24}$$

where $child(V_k) \setminus X$ represents the set of other children of V_k, and $\lambda_C(V_k)$ is the message that V_k receives from its other children as shown in Fig. 3.14.

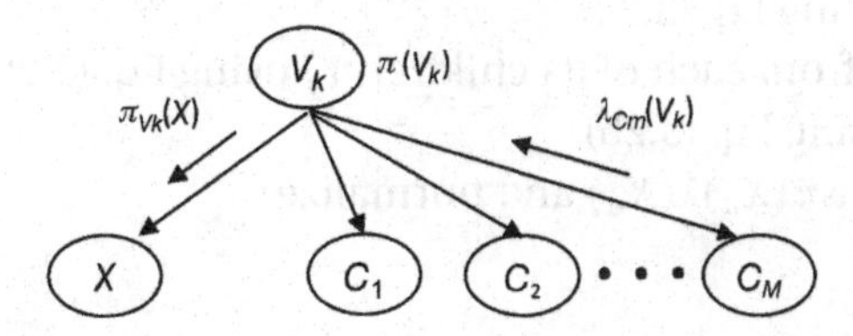

FIGURE 3.14 Other children of V_k.

Similarly, $\lambda(X)$ can be proven to be

$$\lambda(X) = \prod_{j=1}^{J} \lambda_{Y_j}(X), \tag{3.25}$$

where $\lambda_{Y_j}(X)$ is the message X receives from its child Y_j and is defined as

$$\lambda_{Y_j}(X) = \sum_{y_j} \lambda(y_j) \sum_{u_1,u_2,\ldots,u_p} p(y_j|X,u_1,u_2,\ldots,u_p) \prod_{k=1}^{p} \pi_{u_k}(y_j), \tag{3.26}$$

where $U_1, U_2, \ldots, U_p$ are the other parents of Y_j, except for X, as shown in Fig. 3.15. Detailed

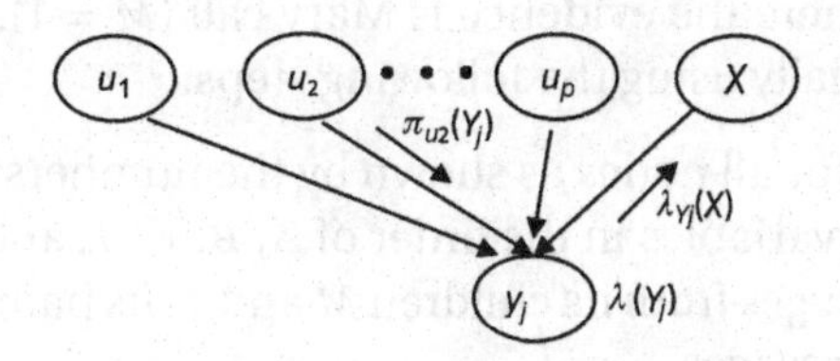

FIGURE 3.15 Other parents of Y_j.

proofs and derivations of Eq. (3.22) to Eq. (3.26) can be found in Section 3.2 of [2].

Before we can perform BP, we need to initialize the messages for each node. For a root node, $\pi(X_i) = p(X_i)$. For a leaf node, $\lambda(X_i) = 1$. For evidence nodes, $\lambda(X_i = e_i) = 1$, $\lambda(X_i \neq e_i) = 0$, $\pi(X_i = e_i) = 1$, and $\pi(X_i \neq e_i) = 0$. The entries of CPDs involving the evidence node are set to zero if the value for the evidence node is different from the observed value and unchanged otherwise. For other nodes, we can initialize their π and λ messages to 1. The BP algorithm can be summarized as the pseudocode in Algorithm 3.3.

Algorithm 3.3 Pearl's belief propagation algorithm.

Arrange the nonevidence nodes $\mathbf{X}_U$ in certain order $X_{U_1}, X_{U_2}, \ldots, X_{U_N}$
Initialize the messages for all nodes
while not converging **do**
  for n=U_1 to U_N **do** //for each node in $\mathbf{X}_U$
    Calculate $\pi_{V_k}(X_n)$ using Eq. (3.24) from each of its parents V_k
    Calculate $\pi(X_n)$ using Eq. (3.23)
    Calculate $\lambda_{Y_j}(X_n)$ from each of its children Y_j using Eq. (3.26)
    Calculate $\lambda(X_n)$ using Eq. (3.25)
    Compute $p(X_n) = \alpha\pi(X_n)\lambda(X_n)$ and normalize
  end for
end while

To produce valid and consistent probabilities for each node, message passing must follow the message passing protocol: a node can send a message to a neighbor only after it has received the messages from all its other neighbors. The belief updating can be conducted sequentially or in parallel. Sequential propagation involves updating one node at a time following an order, typically starting with the nodes closest to the evidence nodes. Following the order, each node collects all messages from its neighbors, updates its belief, and then sends messages to its neighbors. In contrast, parallel propagation updates all nodes simultaneously, that is, all nodes collect messages from their neighbors, update their beliefs, and send messages to their neighbors. This means belief propagation for each node can be performed in parallel and asynchronously. Choosing the best updating schedule requires trial and error. Moreover, additional care must be exercised to assess its convergence. For parallel updating, the messages for the nonboundary and nonevidence nodes are all initialized as 1. Fig. 3.16 gives an example of belief propagation for the burglary–alarm BN, where the evidence is Mary call ($M = 1$). The BP for this example can be performed sequentially using the following steps:

1. Initialize the messages for all nodes as shown by the numbers in Fig. 3.16B
2. Order the nonevidence variables in the order of A, B, E, J, and Ph
3. Node A collects all messages from its children M and J, its parents E and B, and updates its total π and total λ messages
4. Node B collects its messages from child A, and updates its total λ messages
5. Node E collects its messages from child A, and updates its total λ messages
6. Node J collects messages from its parents Ph and A, and updates its total π messages.

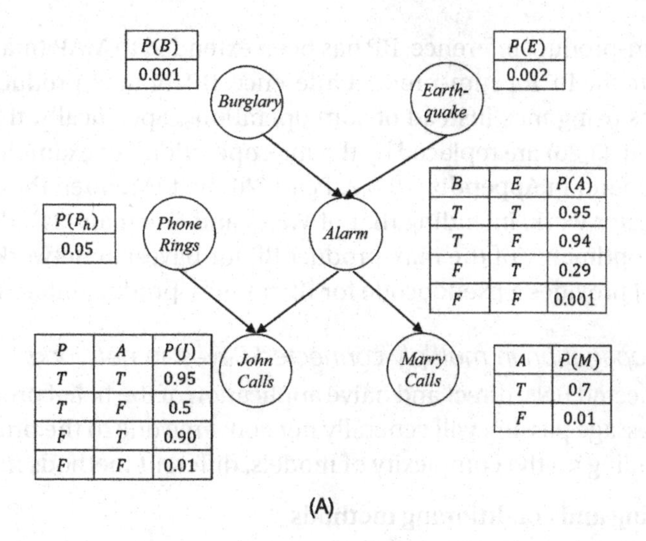

(A)

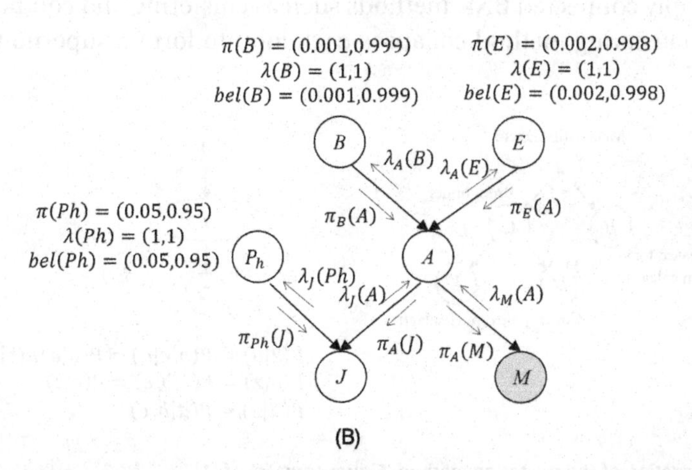

(B)

FIGURE 3.16 (A) An example BN and (B) belief propagation in example BN, where the numbers for each node represent the initial message values. Figure courtesy of [1].

7. Node Ph collects its message from its child J, and updates its total λ messages
8. Repeat steps 3–7 until convergence
9. Compute the final belief for each node and normalize.

BP can be viewed as the VE algorithm in all directions at once. Like the VE algorithm, the complexity of BP is exponential to the tree width of the graph. In addition, exact BP is limited to singly connected BNs. For multiply connected BNs (i.e., BNs with undirected loops), BP can still be applied to performing approximate inference. Refer to appendix 3.9.6.1 for additional examples of belief propagation, in particular for the case when the evidence nodes are non-boundary nodes.

Besides for sum-product inference, BP has been extended to MAP (max-product) inference. In contrast to the BP for sum-product inference, BP for max-product inference computes its messages using max instead of sum operations. Specifically, the sum operation in Eq. (3.23) and Eq. (3.26) are replaced by the max operation. An example of max-product inference can be found in Appendix 3.9.6.2. Pearl [20] first extended the method to singly connected BNs. Later work, including that of Weiss and Freeman [21], demonstrated the applicability and optimality of the max-product BP for Bayesian networks with loops. Algorithm 13.2 in [7] provides a pseudocode for BP for max-product inference.

3.3.1.3 Belief propagation in multiply connected Bayesian networks

For multiply connected BNs, direct and naive application of the belief propagation will not work since the message passing will generally not converge due to the presence of loops in the model. Depending on the complexity of models, different methods may be applied.

3.3.1.3.1 Clustering and conditioning methods

For simple multiply connected BNs, methods such as clustering and conditioning may be employed. The clustering method collapses each loop to form a supernode as shown in Fig. 3.17.

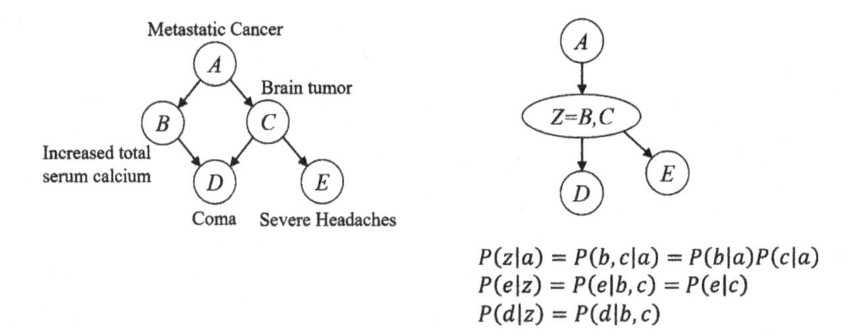

$$P(z|a) = P(b,c|a) = P(b|a)P(c|a)$$
$$P(e|z) = P(e|b,c) = P(e|c)$$
$$P(d|z) = P(d|b,c)$$

FIGURE 3.17 An illustration of the clustering method. Figure courtesy of [1].

Each super node consists of multiple variables. As a result of the clustering operation, the new BN comprises singleton nodes and supernodes but without loops. The CPTs in the original BN can be used to estimate the CPT in the new BN, based on which the traditional belief propagation method can be applied. Similarly, the conditioning method tries to eliminate the loops. It works by conditioning on a set of so-called loop-cutset nodes to break the loops as shown in Fig. 3.18.

Loop-cutset nodes such as node A in Fig. 3.18 are located on the loops such that by conditioning on them, the loops can be broken, producing multiple BNs (two BNs in Fig. 3.18), given each instantiation of the cutset nodes. BP can be performed individually in each network, and their results can then be combined to form the inference results for the original model. For example, in Fig. 3.18, by conditioning on node A we can break the original BN (A) into two BNs (B) and (C). We can perform BN inference in each BN separately and then

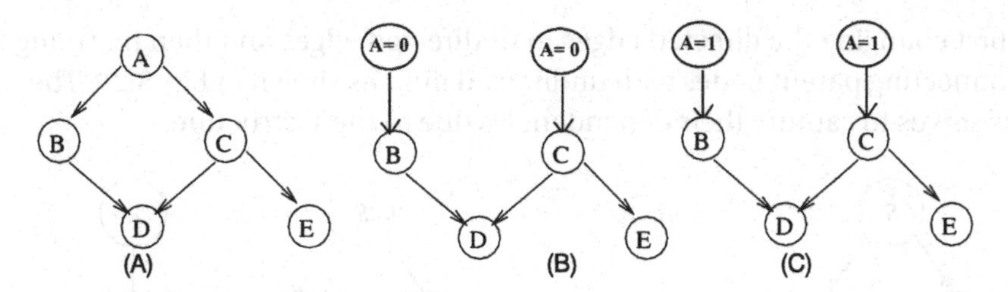

FIGURE 3.18 An illustration of the conditioning method, where node A is the loop-cutset node. Figure courtesy of Fig. 3.11 of [2].

combine their results. For example, $p(B|d)$ in the original BN can be written as

$$\begin{aligned} p(B|d) &= \sum_a p(B|d,a)p(a|d) \\ &= p(B|d,a=0)p(a=0|d) + p(B|d,a=1)p(a=1|d), \end{aligned} \tag{3.27}$$

where the first term can be computed using the first model B, whereas the second term can be computed using the second model C in Fig. 3.18. The worst-case complexity for this method remains NP-hard since the number of simplified networks is exponential to the number of loop-cutset nodes. Details on the conditioning algorithm can be found in Section 3.2.3 of [2] and Section 9.5 of [7].

3.3.1.3.2 Junction tree method

Clustering and conditioning algorithms are for special and simple BNs. For complex BNs with loops or for general BP algorithms in discrete multiply connected BNs, we employ the junction tree method [22–24]. Instead of directly performing the BP in the BN, this method first converts a BN model into a special undirected graph, called a junction tree, which eliminates loops in the original BN through node clustering. BN can then be performed on the junction tree. A junction tree is an undirected tree, and its nodes are clusters (ellipse nodes in Fig. 3.19), each of which consists of a set of nodes. Two cluster nodes are separated by a separator node (rectangular nodes in Fig. 3.19), which represents the intersection of two neighboring cluster nodes. To generate a valid junction tree, cluster nodes are arranged to satisfy the cluster intersection property.[2] Fig. 3.19 shows an example of a junction tree.

FIGURE 3.19 An example of a junction tree.

The junction tree method for BP consists of several steps. The first step is junction tree construction, which starts with constructing a moral graph from the original BN. This is

[2] For any two cluster nodes in the junction tree, their intersection variables must be contained in all cluster nodes on the unique path between the two nodes.

done by first changing the directed edges to undirected edges and then marrying the parents by connecting parent nodes with undirected links as shown in Fig. 3.20. The marriage of parents serves to capture their dependencies due to the V-structure.

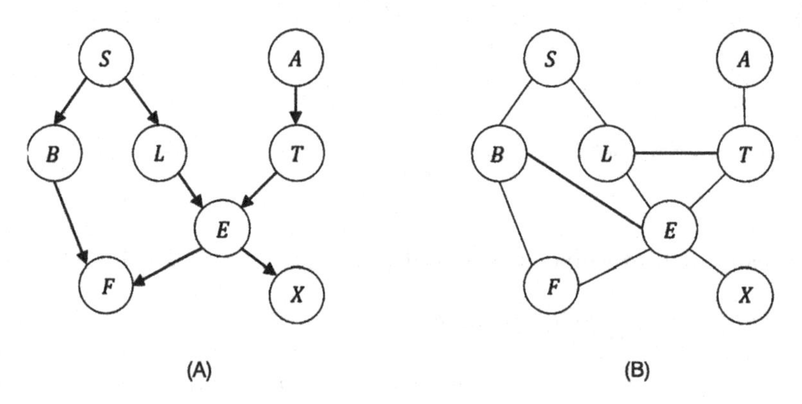

FIGURE 3.20 Constructing a moral graph (B) from a DAG (A). (A) DAG. (B) Marry parent and remove arrows.

Given the moral graph, the second step in junction tree construction is triangulation, which connects nodes in a loop greater than 3 with a chord (link) such that a loop greater than 3 can be broken down into loops of three nodes as shown in Fig. 3.21. In other words, links should be added so that a cycle of length > 3 consists of different triangles. Given

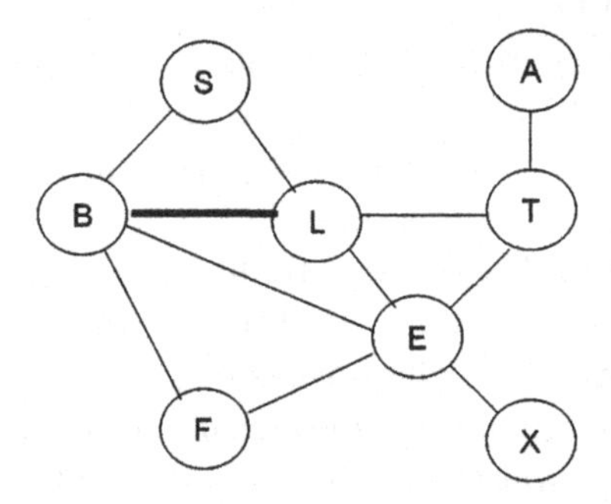

FIGURE 3.21 Triangulation, where the bold link breaks the loop S-B-F-E-L-S.

the triangulated graph, clustering is then performed to identify cliques of nodes (cluster nodes) that are fully connected. Cluster nodes are subsequently arranged following the running intersection property (RIP). RIP states that the cliques should be ordered $(C_1, C_2, \ldots, C_k)$ so that for all $1 < j \leq k$, there is an $i < j$ such that $C_j \cap (C_1 \ldots C_{j-1}) \subset C_i$ to form a junction tree as shown in Fig. 3.22. The RIP property is needed in order for the tree to satisfy the clustering intersection property.

Next, parameterizations of the junction tree can be performed based on the CPTs in the original BN. Each node and separator in the junction tree have an associated potential over its constituent variables. Specifically, for each cluster node in the junction tree,

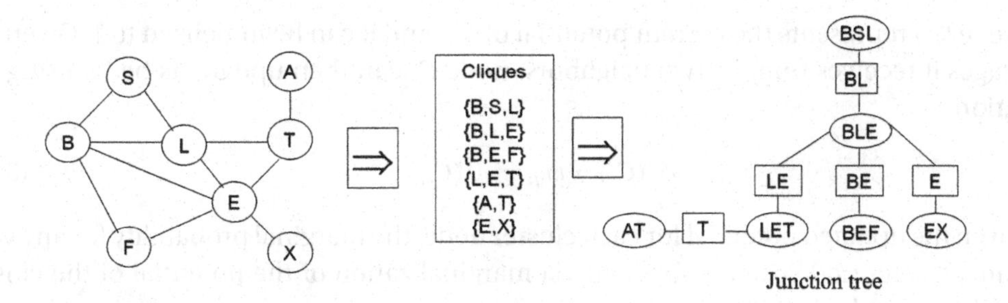

FIGURE 3.22 Clustering and junction tree. Figure courtesy of [1].

we identify all its variables and set its potential equal to the product of CPTs for each variable in the node minus the variables in the separator before it. Likewise, for each separator node, its potential can be computed as the product of CPTs for each variable in the separator node. For example, for nodes BSL and BEF, their respective potentials are $\phi_{BSL} = p(B|S)p(L|S)p(S)$ and $\phi_{BEF} = p(F|B, E)$. The joint probability of a junction tree is the product of potentials for all cluster nodes $\mathbf{X}$:

$$p(\mathbf{X}) = \prod_{c \in C} \phi(X_c), \tag{3.28}$$

where C represents the sets of cluster nodes.

Potential updating can then be performed on the junction tree, where each cluster node updates its potential based on the messages it receives from its neighbors. Specifically, each node first collects messages from its neighboring clustering nodes and then updates its potential using the received messages. During message collection, each cluster node collects messages from its two neighbors as can be seen in Fig. 3.23, where node C_i collects messages from its two neighboring nodes C_j and C_k. The message that node C_i receives from each of its neighboring nodes can be computed as follows.

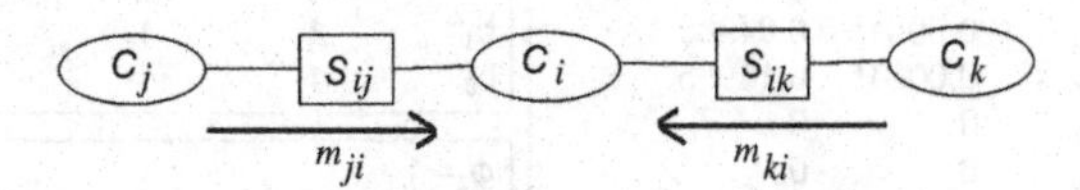

FIGURE 3.23 Message collection by node C_i.

Let m_{ji} represent the message cluster that node C_i receives from cluster node C_j, which can be computed as follows:

$$\begin{aligned} \phi^*(S_{ij}) &= \sum_{C_j \setminus S_{ij}} \phi(C_j), \\ m_{ji} &= \frac{\phi^*(S_{ij})}{\phi(S_{ij})}, \end{aligned} \tag{3.29}$$

where $\phi(S_{ij})$ represents the current potential of S_{ij}, and it can be initialized to 1. Given the messages it receives from its two neighbors, node C_i can then update its belief using the equation

$$\phi^*(C_i) = m_{ji} m_{ki} \phi(C_i). \tag{3.30}$$

Given the updated potential for each cluster node, the marginal probability for any variable in a cluster node can be obtained via marginalization of the potential of the cluster node. For example, $p(X)$ for $X \in C_i$ can be computed as

$$p(X) = \alpha \sum_{C_i \setminus X} \phi(C_i). \tag{3.31}$$

To illustrate the junction tree method, we use the following example. Given the BN in Fig. 3.24A, where each node is binary with value 0 or 1, we can produce the corresponding junction tree in Fig. 3.24B by following the steps for junction tree construction.

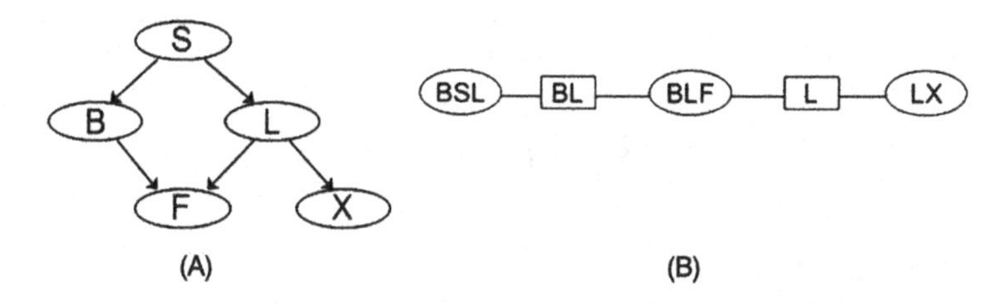

FIGURE 3.24 Illustration of the junction tree method. (A) a binary BN; (B) the corresponding junction tree.

Given $S = 1$ and $F = 1$, we want to compute $p(B = 1|S = 1, F = 1)$. Fig. 3.25 provides the values of the initial potential function for each node of the junction tree in Fig. 3.24B. Note that the potential functions for all separators are initialized to 1. We choose to update

ϕ_{BSL} = P(B | S=1)P(L | S=1)P(S=1)

	l_1	l_0
s_1,b_1	0.00015	0.04985
s_1,b_0	0.00045	0.14955
s_0,b_1	0	0
$s_0\ b_0$	0	0

ϕ_{BL} = 1

	l_1	l_0
b_1	1	1
b_0	1	1

ϕ_{L} = 1

l_1	l_0
1	1

ϕ_{BLF} = P(F=1 | B,L)

	l_1	l_0
f_1,b_1	0.75	0.1
f_1,b_0	0.5	0.05
f_0,b_1	0	0
f_0,b_0	0	0

ϕ_{LX} = p(X=1|L)

	l_1	l_0
x_1	0.6	0.02
x_0	0	0

FIGURE 3.25 Initial potential function values for each node in the junction tree in Fig. 3.24B.

the potential for node BLF. First, we compute the message it receives from node BSL using Eq. (3.29), producing

$m_{BSL}(BLF)$	L=1	L=0
B=1	0.00015	0.04985
B=0	0.00045	0.14955

Similarly, we can compute the message from node LX to node BLF as follows:

$m_{LX}(BLF)$	L=1	L=0
	0.6	0.02

Given the messages, node BLF can update its potential using Eq. (3.30), producing the updated potential:

$\phi^*(BLF)$	L=1	L=0
F=1, B=1	0.0000675	0.00009970
F=1, B=0	0.0001350	0.00014955
F=0, B=1	0	0
F=0, B=0	0	0

Using the updated $\phi^*(BLF)$, we can compute $p(B=1|S=1, F=1)$ as

$$p(B=1|S=1,F=1)=\frac{\sum_{L=0}^{1}\phi^*(B=1,L,F=1)}{\sum_{B=0}^{1}\sum_{L=0}^{1}\phi^*(B,L,F=1)}=0.37.$$

In general, the junction tree algorithm can be implemented with the Shafer–Shenoy algorithm [23]. In Fig. 3.26, let A and B be two neighboring cluster nodes. According to the

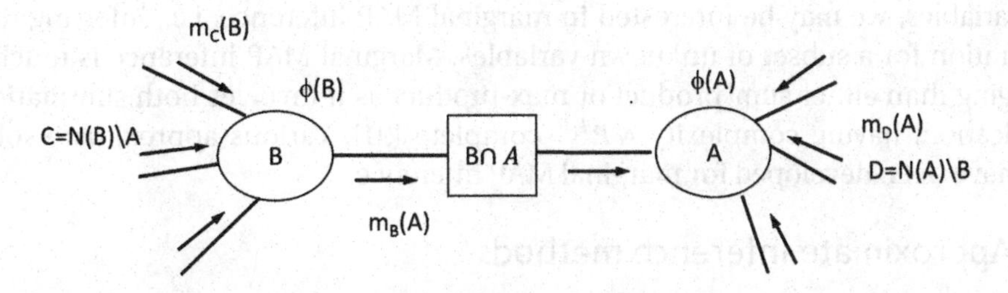

FIGURE 3.26 Illustration of the Shafer–Shenoy algorithm.

Shafer–Shenoy algorithm, the message that B sends to A can be computed as follows:

$$m_B(A)=\sum_{B\setminus A\cap B}\phi(B)\prod_{C\in N(B)\setminus A}m_C(B), \tag{3.32}$$

where $N(B)$ represents the neighbors of B. Procedurally, node B first computes its separator (the separator between A and B) potential by marginalizing its current potential $\phi(B)$

over variables that are not in the separator, and then it computes the product of its separator potential and the messages it receives from all other neighbors except for A. Given the message A receives from each of its neighbors B, node A can then update its belief:

$$\phi^*(A) = \phi(A) \prod_{B \in N(A)} m_B(A). \tag{3.33}$$

The message passing must follow the message passing protocol, which states that each node sends a message to its neighbor only after it has received all messages from all its other neighbors. Given a DAG, multiple valid junction trees may be constructed; they all yield the same inference results, but their computational complexity can vary. The optimal junction tree can lead to the most efficient inference.

Two variants of the junction tree algorithms are the Shafer–Shenoy algorithm [23] and the Hugin algorithm [25]. Lepar and Shenoy [26] provided a detailed comparison of different variants of the junction tree method. Junction tree methods for discrete BNs have been extended to Gaussian BNs and to max-product inference [27]. The junction tree algorithms generalize variable elimination to the efficient simultaneous execution of a large class of queries. Their computational complexity is determined by the triangulation procedure and by the message passing, both of which are NP-hard for nontree-structured models. Hence, the worst-case computational complexity remains NP-hard. In practice, approximations can be made to solve for the intractability by approximately computing the messages, yielding efficient yet approximate junction tree inference methods. Further discussion on the junction tree algorithm can be found in Section 10.4 of [7].

So far, we have discussed the exact methods for sum-product and max-product inferences. Sum-product (posterior probability) inference infers the posterior for one variable, while max-product inference (MAP) infers the most probable configuration for all unknown variable (max-product). In many real world applications such as presence of the latent variables, we may be interested in marginal MAP inference, i.e., inferring the best configuration for a subset of unknown variables. Marginal MAP inference is much more challenging than either sum-product or max-product as it involves both summation and multiplication, having complexity NP^{PP}-complete [91]. Various approximate solutions [92,93] have been developed for marginal MAP inference.

3.3.2 Approximate inference methods

For large complex BNs consisting of many loops, exact inference is computationally expensive. Approximate inference methods can be used instead. For posterior inference, approximate methods obtain an imprecise estimation of the posterior probability value $p(X|\mathbf{E})$. They trade accuracy for efficiency. For applications that do not need exact values of $p(X|\mathbf{E})$, approximate methods are applicable. Despite their gain in efficiency, it is proved in [28] that there is no polynomial-time algorithm for approximate inference with a tolerance value (bound) less than $1/2$, which means that accurate approximate inference remains NP-hard in the worst case. The most widely used approximate methods include loopy belief propagation, Monte Carlo sampling methods, and variational approaches discussed below. In addition, we can also perform approximate MAP inference using the Iterated conditional modes (ICM) method. Details about the ICM method may be found in Section 4.5.2.1.

3.3.2.1 Loopy belief propagation

For a multiply connected BN, instead of converting it to a junction tree and then performing BP, we can also directly apply either sum-product or max-product BP, leading to the so-called loopy belief propagation (LBP) algorithm. In this case however, there is no guarantee that the exact belief propagation will converge, since messages may circulate indefinitely in the loops. Even though there is no guarantee of convergence or correctness, LBP has achieved good empirical success. In practice, if the solution is not oscillatory but converges (though possibly to a wrong solution), LBP usually produces a good approximation. If not converging, LBP can be stopped after a fixed number of iterations or when there are no significant changes in beliefs. In either case, LBP often provides sufficiently good approximation. Murphy et al. [29] performed an empirical evaluation of the LBP under a wide range of conditions. Tatikonda and Jordan [30] evaluated the convergence of the LBP and determined sufficient conditions for its convergence.

3.3.2.2 Monte Carlo sampling

The analytic inference methods we have discussed so far are based on mathematical derivations. Although theoretically correct, they often require complex theoretical derivations and come with strong assumptions. Inference via Monte Carlo sampling represents a very different alternative. It avoids theoretical derivations needed to derive closed-form analytic inference methods. It obtains random samples via Monte Carlo simulation and uses the sample distribution to approximate the underlying distribution of the BN. For discrete BN, posterior inference with samples becomes a counting problem. The key for this method is generating sufficiently representative samples to reflect the underlying distribution. With growing computing power and better sampling strategies, inference by random sampling is becoming increasingly popular. The main challenge with the sampling method is efficiently generating sufficient and representative samples, in particular, for high-dimensional variable space. Various sampling strategies have been proposed to tackle this challenge.

3.3.2.2.1 Logic sampling

If the inference does not involve evidence, then we can employ the ancestral sampling technique, which samples each variable following the topological order from root nodes to their children and then to their descendants until the leaf nodes. By following the topological order of the BN, the sampler always first visits the parents of a node before visiting the node itself. At each node, we can employ the standard sampling methods introduced in Chapter 2. Ancestral sampling can be extended to inference with evidence, yielding the logic sampling method [31]. The latter works by performing ancestral sampling from the root nodes. Upon reaching the observed nodes, if their sampled values differ from the observed values, then we can reject the whole sample and start over. This sampling strategy is highly inefficient, however, especially when the probability of the evidence is low. To improve efficiency, the weighted logic sampling method [32] was introduced. Instead of rejecting all inconsistent samples, for nodes with observed values, the weighted logic sampling method uses their observed values as sampled values and associates each sample with a weight in terms of its likelihood, leading to weighted samples. Fig. 3.27 shows an example of weighted logic sampling with a BN of five binary nodes, where we want to obtain samples given Radio $= r$ and Alarm $= \bar{a}$.

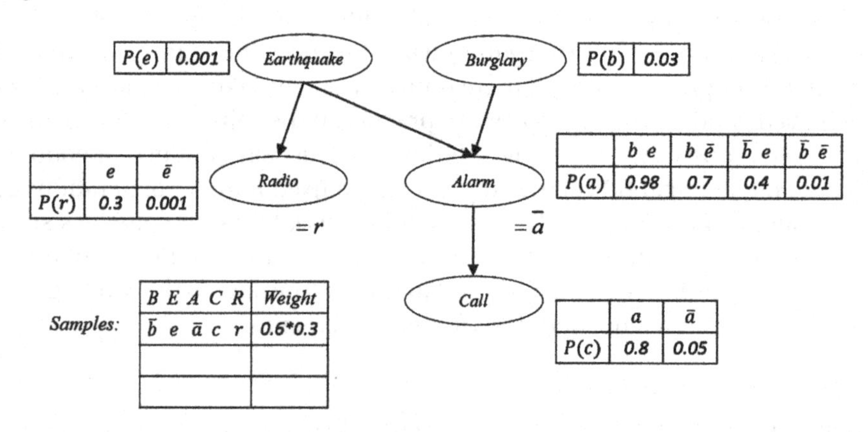

FIGURE 3.27 An example of weighted logic sampling. Figure courtesy of [1].

Following the order Burglary (B), Earthquake (E), Alarm (A), Radio (R), and Call (C), the sampling process starts with sampling node B to obtain $\bar{b}$, followed by sampling node E to obtain e, taking given values $\bar{a}$ and r for A and R, respectively, and finally sampling node C given Alarm $= \bar{a}$ to obtain c. This produces the first sample vector $\bar{b}, e, \bar{a}, r, c$, whose weight equals $p(\bar{a}|\bar{b}, e) \times p(r|e)(0.6 \times 0.3)$. This process continues to generate additional samples. Algorithm 3.4 provides the pseudocode for weighted logic sampling.

According to the central limit theorem, the estimation by sampling asymptotically approaches the true values as the number of samples increases. We can employ the normal method or the exact method introduced in Chapter 2 to determine the minimum number of samples needed to obtain the estimates within a certain confidence interval. See Chapter 15 of [33] for additional information on this topic. Logic sampling and weighted logic sampling have limitations: they are limited to discrete BNs and are inefficient for inference with evidence far away from the root node. We further introduce the Markov chain Monte Carlo sampling to overcome these limitations.

3.3.2.2.2 MCMC sampling

Traditional Monte Carlo sampling methods work well for low-dimensional spaces, but they do not scale up well to high-dimensional spaces. Markov chain Monte Carlo (MCMC) sampling can address this limitation and represents the most important sampling approach today. Its strengths lie in its ability to sample in the high-dimensional space, theoretical guarantees, parallelizability, and hardware implementability. The idea is that if the sam-

Algorithm 3.4 Weighted logic sampling algorithm.

E: evidence nodes
Order BN variables $X_1, X_2, \ldots, X_N$ according to their topological order from the root nodes until leaf nodes
Initialize weights $w_1, w_2, \ldots, w_T$ to 1
for t=1 to T **do** t: index to the number of samples
  for n=1 to N **do** n: index to the node number
    if $X_n^t \notin$ **E then**
      sample x_n^t from $p(X_n^t|\pi(X_n^t))$
    else
      $x_n^t = e_n$
      $w_t = w_t * p(X_n^t = e_n|\pi(X_n^t))$
    end if
  end for
  Form sample $\mathbf{x}^t = \{x_1^t, x_2^t, \ldots, x_N^t\}$ and compute its weight w_t
end for
Return $(\mathbf{x}^1, w_1), (\mathbf{x}^2, w_2), \ldots, (\mathbf{x}^T, w_T)$

pling follows a Markov chain and the Markov chain is ergodic,[3] then the samples of the chain after a certain burn-in period will closely follow the underlying true distribution $p(\mathbf{X})$, independent of where the chain starts. Furthermore, because of the Markov property, the next sample is determined by a transition probability, which depends only on the current sample, independent of all previous samples. Among various MCMC sampling methods, Gibbs sampling [34] is the most popular approach due to its simplicity, efficiency, and theoretical guarantees. The basic idea of Gibbs sampling is generating the next sample of a chain conditioned on the values of the previous sample, with each new sample differing from the previous sample by only one variable. Gibbs sampling can therefore scale up to models with a large number of variables.

For Gibbs sampling with BN, we can construct a Markov chain of samples where the next sample $\mathbf{x}^{t+1}$ is obtained from the transition probability $p(\mathbf{x}^{t+1}|\mathbf{x}^t)$ computed from the BN and differs from $\mathbf{x}^t$ by only one variable. Specifically, let $\mathbf{X} = \{X_1, X_2, \ldots, X_N\}$ represent all the variables in a BN, and let $\mathbf{X}_{-n} = \mathbf{X} \setminus X_n$, Gibbs sampling randomly initializes $\mathbf{X}$ to $\mathbf{x}^0$, then randomly selects a variable X_n, and finally begins to obtain a sample of X_n based on

$$x_n^{(t)} \sim p(X_n|\mathbf{x}_{-n}^{(t-1)}), \tag{3.34}$$

where t is the number of sampling iterations, starting with $t = 1$, and $p(X_n|\mathbf{x}_{-n}^{(t-1)})$ can be computed from the joint probability $p(X_n, \mathbf{x}_{-n}^{(t-1)})$ with normalization. Furthermore, since X_n is independent of all other variables, given its Markov blanket, Eq. (3.34) can be rewrit-

[3] An ergodic chain is defined as a chain of samples with positively recurrent and aperiodic states.

ten as

$$x_n^{(t)} \sim p(X_n|MB(X_n^{t-1})), \tag{3.35}$$

where $MB(X_n^{t-1})$ represents the set of variables belonging to the Markov blanket of X_n. In fact, as shown in Eq. 12.23 of [7], $p(X_n|MB(X_n^{t-1}))$ can be computed using only CPTs that involve X_n and its children:

$$p(X_n|MB(X_n)) = \frac{p(X_n|\pi(X_n)) \prod_{k=1}^{K} p(Y_k|\pi(Y_k)))}{\sum_{x_n} p(x_n|\pi(X_n)) \prod_{k=1}^{K} p(Y_k|\pi(Y_k))},$$

where Y_k is the kth child of X_n. Since the MB size is typically small, computation of Eq. (3.35) is simple and efficient. The main idea is to sample one variable at a time, with all other variables assuming their last values. As a result, the new sample $\mathbf{x}^t$ differs from the previous sample $\mathbf{x}^{t-1}$ by only x_n. This process repeats until mixing (after the burn-in period), after which we can collect samples. Algorithm 3.5 provides a pseudocode that summarizes the major steps of the Gibbs sampling algorithm.

There are a few issues about Gibbs sampling. First, one challenge is to determine the exact mixing timing, that is, the amount of time it takes for a Gibbs chain to converge to its stationary distribution. This varies with the distributions and the initializations. While various heuristics have been proposed, there is no way to ascertain when the sampling has completed the burn-in period; this is typically determined by trial and error in practice. One possible technique is comparison of the statistics in two consecutive windows for a single chain. If the statistics are close, this may mean convergence, or enough samples have been burned. Second, we can perform the sampling using either one long chain (i.e., one initialization) or multiple chains, each of which starts from a different initialization. In practice, we may choose a hybrid approach, that is, run a small number of chains of medium length and collect samples independently from each chain. This is where parallelization may be used. Third, when collecting samples from a chain, to avoid correlations among samples, a number of samples may be skipped before collecting another sample.

Various software has been developed to perform effective Gibbs sampling, including the Bayesian inference using Gibbs sampling (BUGs) software [35] at http://www.openbugs.net/w/FrontPage, Stan [36] at http://mc-stan.org/, and the Just Another Gibbs Sampler (JAGS) software at http://mcmc-jags.sourceforge.net/. In addition, there are also several Matlab toolboxes for MCMC sampling at http://helios.fmi.fi/~lainema/mcmc/.

Besides conventional Gibbs sampling, variants of Gibbs sampling techniques also exist, such as collapsed Gibbs sampling. While sampling for each variable, instead of conditioning on all the remaining variables, collapsed Gibbs sampling conditions on a subset of remaining variables by integrating out some of the remaining variables. Because of its conditioning on a subset of variables, this technique yields a more accurate sample of the variable, and the mixing can be faster. It has been applied to hierarchical Bayes models such as LDA by integrating out the hyperparameters. See Section 12.4.2 of [7] for further information on collapsed Gibbs sampling.

Algorithm 3.5 A single chain Gibbs sampling algorithm.

Initialize $\mathbf{X} = \{X_1, X_2, \ldots, X_N\}$ to $\mathbf{x}^0 = \{x_1, x_2, \ldots, x_N\}$
t=0
while not end of burn-in period **do**
  Randomly select an n
  Obtain a new sample $x_n^{t+1} \sim p(X_n|\mathbf{x}_{-n}^t)$
  Form a new sample $\mathbf{x}^{t+1} = \{x_1^t, x_2^t, \ldots, x_n^{t+1}, \ldots, x_N^t\}$
  t=t+1
end while//end of burn-in period
$\mathbf{x}^0 = \{x_1^t, x_2^t, \ldots, x_N^t\}$
for t=0 to T **do** //start collecting T samples
  Randomly select an n
  Sample $x_n^{t+1} \sim p(X_n|\mathbf{x}_{-n}^t)$
  Form a new sample, $\mathbf{x}^{t+1} = \{x_1^t, x_2^t, \ldots, x_n^{t+1}, \ldots, x_N^t\}$
end for
Return $\mathbf{x}^1, \mathbf{x}^{1+k}, \mathbf{x}^{1+2k} \ldots, \mathbf{x}^T$ //k is sample skip step

3.3.2.2.3 Metropolis Hastings sampling

Gibbs sampling assumes the availability of the joint probability $p(\mathbf{X})$ to construct a Markov chain. However, for some probability distributions, it may be difficult to obtain exactly the joint probability $p(\mathbf{X})$. Metropolis–Hastings sampling [37] was developed to overcome this limitation. It can produce a Markov chain for any distribution $p(\mathbf{X})$ if we can compute another density function $p'(\mathbf{X})$ proportional to $p(\mathbf{X})$, that is, $p'(\mathbf{X})$ is an unnormalized probability density function. As an MCMC sampling method, Metropolis–Hastings (MH) sampling represents a generalization of the Gibbs sampling method; it follows the basic idea of rejection sampling to construct the chain. It has a proposal distribution $q(\mathbf{X}^t|\mathbf{X}^{t-1})$, the probability of $\mathbf{X}^t$ given $\mathbf{X}^{t-1}$, which is typically assumed to be symmetric, that is, $q(\mathbf{X}^t|\mathbf{X}^{t-1}) = q(\mathbf{X}^{t-1}|\mathbf{X}^t)$. A common choice of q is the Gaussian distribution. At each iteration of the sampling, a sample $\mathbf{x}^t$ is first obtained from the proposal distribution conditioned on $\mathbf{x}^{t-1}$. Instead of just accepting $\mathbf{x}^t$ as the Gibbs sampling does, MH accepts $\mathbf{x}^t$ with probability p determined as follows:

$$p = \min(1, \frac{p'(\mathbf{x}^t)}{p'(\mathbf{x}^{t-1})}). \tag{3.36}$$

Eq. (3.36) shows that $\mathbf{x}^t$ is accepted if its probability is larger than that of $\mathbf{x}^{t-1}$; otherwise, $\mathbf{x}^t$ is accepted with probability $\frac{p'(\mathbf{x}^t)}{p'(\mathbf{x}^{t-1})}$. It can easily be proven that the Metropolis–Hastings sampling generalizes the Gibbs sampling, where the proposal distribution is $p(\mathbf{X}^t|\mathbf{X}^{t-1})$, and its sample is always accepted.

Original Metropolis–Hastings assumes symmetric proposal distribution. Later it was extended to non-symmetric proposal distribution by using the following equation to compute the acceptance probability.

$$p = \min(1, \frac{p'(x^t)q(x^{t-1}|x^t)}{p'(x^{t-1})q(x^t|x^{t-1})}) \tag{3.37}$$

While simple and easy to implement, Metropolis–Hastings method does not scale up well as it's proposal distribution typically follows a random walk and is hence slow. To speed up, Hamiltonian Monte Carlo was introduced. The key idea is to use Hamiltonian dynamics instead of a random walk to propose a new sample. Details about the Hamiltonian Monte Carlo method may be found in Appendix 3.9.5.

3.3.2.3 Variational inference

Variational inference [38] is another increasingly popular approach to approximate inference. The basic idea is finding a simple surrogate distribution $q(\mathbf{X}|\boldsymbol{\beta})$ to approximate the original complex distribution $p(\mathbf{X}|\mathbf{E})$ such that inference can be performed with q. The surrogate distribution often assumes independence among the target variables for tractable inference. To construct the approximate distribution, one strategy is finding the variational parameters $\boldsymbol{\beta}$ to minimize the Kullback–Leibler divergence[4] $KL(q||p)$

$$\boldsymbol{\beta}^* = \underset{\boldsymbol{\beta}}{\operatorname{argmin}}\, KL(q(\mathbf{X}|\boldsymbol{\beta})||p(\mathbf{X}|\mathbf{e})). \tag{3.38}$$

The KL divergence in Eq. (3.38) can be expanded as

$$\begin{aligned} KL(q(\mathbf{X}|\boldsymbol{\beta})||p(\mathbf{X}|\mathbf{e})) &= \sum_{\mathbf{x}} q(\mathbf{x}|\boldsymbol{\beta}) \log \frac{q(\mathbf{x}|\boldsymbol{\beta})}{p(\mathbf{x}|\mathbf{e})} \\ &= \sum_{\mathbf{x}} q(\mathbf{x}|\boldsymbol{\beta}) \log q(\mathbf{x}|\boldsymbol{\beta}) - \sum_{\mathbf{x}} q(\mathbf{x}|\boldsymbol{\beta}) \log p(\mathbf{x}, \mathbf{e}) + \log p(\mathbf{e}). \end{aligned} \tag{3.39}$$

Since $p(\mathbf{e})$ is a constant, minimizing $KL(q(\mathbf{X}|\boldsymbol{\beta})||p(\mathbf{X}|\mathbf{e}))$ is equivalent to minimizing

$$F(\boldsymbol{\beta}) = \sum_{\mathbf{x}} q(\mathbf{x}|\boldsymbol{\beta}) \log q(\mathbf{x}|\boldsymbol{\beta}) - \sum_{\mathbf{x}} q(\mathbf{x}|\boldsymbol{\beta}) \log p(\mathbf{x}, \mathbf{e}), \tag{3.40}$$

where $F(\boldsymbol{\beta})$ is referred to as the free energy function. As the KL divergence is always nonnegative, $-F(\boldsymbol{\beta})$ is called the variational evidence lower bound (ELBO) of $\log p(\mathbf{e})$. ELBO can be written as

$$ELBO(\boldsymbol{\beta}) = -\sum_{\mathbf{x}} q(\mathbf{x}|\boldsymbol{\beta}) \log q(\mathbf{x}|\boldsymbol{\beta}) + \sum_{\mathbf{x}} q(\mathbf{x}|\boldsymbol{\beta}) \log p(\mathbf{x}, \mathbf{e}),$$

where the first term is the entropy of $\mathbf{X}$ with respect to q, and the second term is the expected log-likelihood of $\mathbf{X}$ and $\mathbf{E}$ also with respect to q. Compared to the KL divergence, $F(\boldsymbol{\beta})$ is much easier to compute because of using the joint probability $p(\mathbf{x}, \mathbf{e})$ instead of the

[4] Also called the relative entropy. The KL divergence is nonsymmetric, and computing $KL(p(\mathbf{X}|\mathbf{e})||q(\mathbf{X}|\boldsymbol{\beta}))$ is intractable for high-dimensional $\mathbf{X}$ since it requires computing the mean over $p(\mathbf{X})$ between q and p. Other divergence measures can be used, but they may not be efficiently computed.

conditional probability $p(\mathbf{x}|\mathbf{e})$. Furthermore, it can be proven that the functional achieves its minimum when $q(\mathbf{x}|\boldsymbol{\beta}) = p(\mathbf{x}|\mathbf{e})$. The parameters $\boldsymbol{\beta}$ can be obtained as

$$\boldsymbol{\beta}^* = \underset{\boldsymbol{\beta}}{\operatorname{argmin}} F(\boldsymbol{\beta}).$$

Note that minimizing $F(\boldsymbol{\beta})$ is the same as maximizing the ELBO and that it is nonconcave. Given the distribution q, the posterior probability inference and the MAP inference can then be trivially performed with q. It should be noted that a different q needs to be computed for each evidence instantiation $\mathbf{e}$. Variational inference effectively converts inference into an optimization problem.

The choice of the function q is the key to the variational approach. If q is chosen to be fully factorized as shown in Fig. 3.28, this leads to the well-known mean field algorithm [39]. The mean field concept, originally derived from physics, approximates the effect of the interactions among individual components in complex stochastic models by their average effect, effectively simplifying the many-body problem to a one-body problem. Specifically, for the mean field method, we have $q(\mathbf{X}) = \prod_{n=1}^{N} q(X_n|\boldsymbol{\beta})$. Substituting

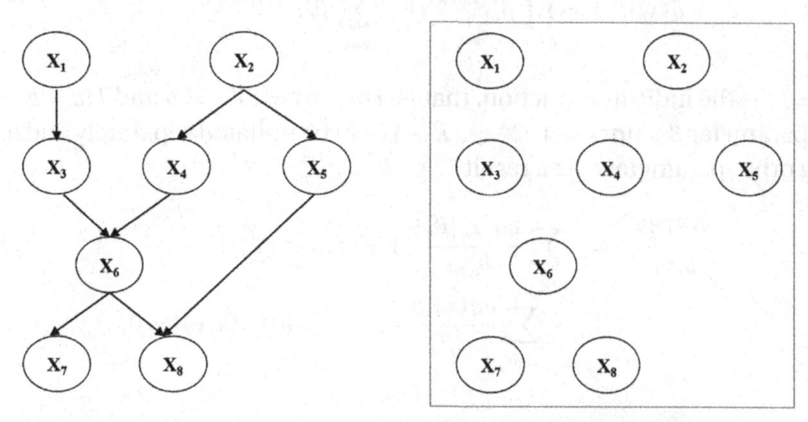

FIGURE 3.28 Fully factorized function q for the mean field method (figure from [3]).

this into Eq. (3.40) and some subsequent reorganizations yield

$$F(\boldsymbol{\beta}) = \sum_{x_n \in \mathbf{X}} \sum_{x_n} q(x_n|\beta_n) \log q(x_n|\beta_n) - \sum_{\mathbf{x}} \Big[\prod_{x_n \in \mathbf{X}} q(x_n|\beta_n) \Big] \log p(\mathbf{x}, \mathbf{e}), \tag{3.41}$$

where β_n is the parameter associated with each variable X_n. The second term of Eq. (3.41) can be rewritten as

$$\sum_{x_n} q(x_n|\beta_n) \sum_{\mathbf{x} \backslash x_n} \Big[\prod_{x_m \in \mathbf{x} \backslash x_n} q(x_m|\beta_m) \Big] \log p(\mathbf{x}, \mathbf{e}). \tag{3.42}$$

Substituting Eq. (3.42) into Eq. (3.41) yields

$$\begin{aligned} F(\boldsymbol{\beta}) &= \sum_{x_n \in \mathbf{X}} \sum_{x_n} q(x_n|\beta_n) \log q(x_n|\beta_n) \\ &\quad - \sum_{x_n} q(x_n|\beta_n) \sum_{\mathbf{x} \backslash x_n} \Big[\prod_{x_m \in \mathbf{x} \backslash x_n} q(x_m|\beta_m) \Big] \log p(\mathbf{x}, \mathbf{e}). \end{aligned} \tag{3.43}$$

Let $E_{q(\mathbf{x}\backslash x_n)}[\log p(\mathbf{x}, \mathbf{e})] = \sum_{\mathbf{x}\backslash x_n} \left[\prod_{x_m \in \mathbf{x}\backslash x_n} q(x_m|\beta_m) \right] \log p(\mathbf{x}, \mathbf{e})$. Eq. (3.43) can be rewritten as

$$F(\boldsymbol{\beta}) = \sum_{x_n \in \mathbf{X}} \sum_{x_n} q(x_n|\beta_n) \log q(x_n|\beta_n) - \sum_{x_n} q(x_n|\beta_n) E_{q(\mathbf{x}\backslash x_n)}[\log p(\mathbf{x}, \mathbf{e})]. \tag{3.44}$$

For a discrete BN, $x_n \in \{1, 2, \ldots, K\}$, $\beta_n = (\beta_{n1}, \beta_{n2}, \ldots, \beta_{nK})^\top, \beta_{nk} = p(x_n = k)$, and $\sum_{k=1}^{K} \beta_{nk} = 1$. As a result,

$$q(x_n|\beta_n) = \prod_{k=1}^{K-1} \beta_{nk}^{I(x_n=k)} (1 - \sum_{k=1}^{K-1} \beta_{nk})^{I(x_n=K)}, \tag{3.45}$$

where $I(a = b)$ is the indicator function, that is, $I(a = b) = 1$ if $a = b$ and $I(a = b) = 0$ otherwise. Each parameter β_{nk} (for $k = 1, 2, \ldots, K - 1$) can be updated separately and alternately while fixing other parameters. As a result,

$$\begin{aligned} \frac{\partial F(\boldsymbol{\beta})}{\partial \beta_{nk}} &= \sum_{x_n} \frac{\partial q(x_n|\beta_n)}{\partial \beta_{nk}} [1 + \log q(x_n|\beta_n)] \\ &\quad - \sum_{x_n} \frac{\partial q(x_n|\beta_n)}{\partial \beta_{nk}} E_{q(\mathbf{x}\backslash x_n)}[\log p(\mathbf{x}, \mathbf{e})] = 0, \end{aligned} \tag{3.46}$$

where

$$\frac{\partial q(x_n|\beta_n)}{\partial \beta_{nk}} = \begin{cases} 1, & x_n < K \,\&\, x_n = k, \\ 0, & x_n < K \,\&\, x_n \neq k, \\ -1, & x_n = K. \end{cases} \tag{3.47}$$

Note that we do not compute the gradient for β_{nK} as it can be computed using other parameters. Substituting Eq. (3.47) into Eq. (3.46) yields

$$\begin{aligned} \frac{\partial F(\boldsymbol{\beta})}{\partial \beta_{nk}} &= \log q(x_n = k|\beta_n) - \log q(x_n = K|\beta_n) \\ &\quad - E_{q(\mathbf{x}\backslash x_n)}[\log p(\mathbf{x} \setminus x_n, x_n = k, \mathbf{e})] \\ &\quad + E_{q(\mathbf{x}\backslash x_n)}[\log p(\mathbf{x} \setminus x_n, x_n = K, \mathbf{e})] = 0. \end{aligned} \tag{3.48}$$

Rearranging Eq. (3.48) yields

$$\begin{aligned}\log q(x_n = k|\beta_n) &= E_{q(\mathbf{x}\backslash x_n)}[\log p(\mathbf{x} \setminus x_n, x_n = k, \mathbf{e})] \\ &\quad + \log q(x_n = K|\beta_n) \\ &\quad - E_{q(\mathbf{x}\backslash x_n)}[\log p(\mathbf{x} \setminus x_n, x_n = K, \mathbf{e})]. \end{aligned} \tag{3.49}$$

Hence for $k = 1, 2, \ldots, K - 1$,

$$\begin{aligned}\beta_{nk} &= \exp(\log q(x_n = k|\beta_n)) \\ &= \exp(E_{q(\mathbf{x}\backslash x_n)}[\log p(\mathbf{x} \setminus x_n, x_n = k, \mathbf{e})]) \\ &\quad \beta_{iK} \exp(-E_{q(\mathbf{x}\backslash x_n)}[\log p(\mathbf{x} \setminus x_n, x_n = K, \mathbf{e})]) \\ &= \frac{\exp(E_{q(\mathbf{x}\backslash x_n)}[\log p(\mathbf{x} \setminus x_n, x_n = k, \mathbf{e})])}{\sum_{j=1}^{K} \exp(E_{q(\mathbf{x}\backslash x_n)}[\log p(\mathbf{x} \setminus x_n, x_n = j, \mathbf{e})])}. \end{aligned} \tag{3.50}$$

By applying the BN chain rule we can rewrite $E_{q(\mathbf{x}\backslash x_n)}[\log p(\mathbf{x} \setminus x_n, x_n = k, \mathbf{e})]$ as

$$\begin{aligned} & E_{q(\mathbf{x}\backslash x_n)}[\log p(\mathbf{x} \setminus x_n, x_n = k, \mathbf{e})] \\ = \; & E_{q(\mathbf{x}\backslash x_n)}[\log \prod_{l=1}^{N} p(x_l|\pi(x_l))] \\ = \; & E_{q(\mathbf{x}\backslash x_n)}[\sum_{l=1}^{N} \log p(x_l|\pi(x_l))] \\ = \; & \sum_{l=1}^{N} E_{q(\mathbf{x}\backslash x_n)}[\log p(x_l|\pi(x_l))]. \end{aligned} \tag{3.51}$$

Note that for $l = n$, $x_l = k$ and for $x_l \in \mathbf{e}$, $x_l = e_l$. Eq. (3.50) can be used to recursively update the parameters β_{nk} for each node n until convergence. As Eq. (3.41) is nonconcave, its minimization may vary with initialization and may lead to a local minimum. In Algorithm 3.6, we provide a pseudocode for the mean field inference.

Algorithm 3.6 The mean field algorithm.

▷ **Input**: a BN with evidence $\boldsymbol{e}$ and unobserved variables $\boldsymbol{X}$
▷ **Output**: mean field parameters $\boldsymbol{\beta} = \{\beta_{nk}\}$ for $k = 1, 2, \ldots, K_n$
Randomly initialize the parameters β_{nk} subject to $\sum_{k=1}^{K_n} \beta_{nk} = 1$
while $\boldsymbol{\beta} = \{\beta_{nk}\}$ not converging **do**
  for n=1 to N **do** //explore each node
    for k=1 to $K_n - 1$ **do** //K_n is the number of states for nth node
      Compute β_{nk} using Eq. (3.50)
    end for
  end for
end while

If q is chosen to be partially independent, then this leads to the structured variational methods, which vary in the complexity of the structured models such as those shown in Fig. 3.29 from simple tree models to fully feed-forward networks [40], in which a reversed network is used for efficient inference. Specifically, in [41], to perform inference in a deep belief network, a feed-forward network $Q(\mathbf{h}|\mathbf{x})$ is introduced to approximate the posterior probability inference of the latent variables $\mathbf{h}$, given input $\mathbf{x}$, that is, $p(\mathbf{h}|\mathbf{x})$. The network $Q(\mathbf{h}|\mathbf{x})$ assumes independencies among latent variables given input data.

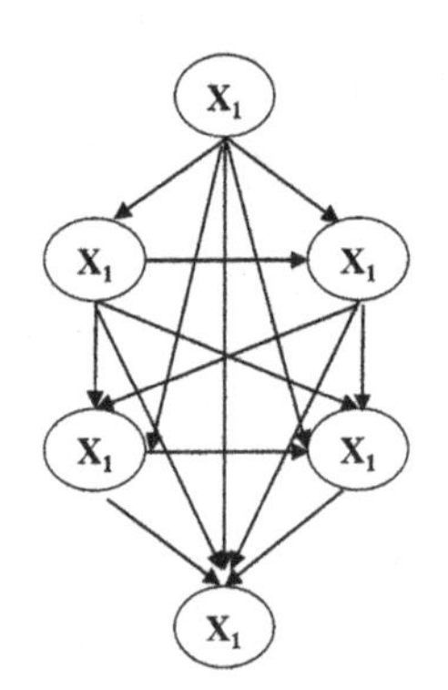

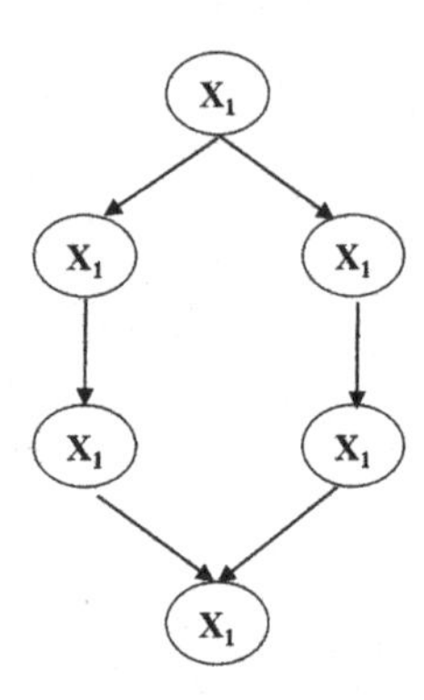

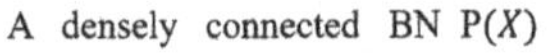

The corresponding sparsely connected BN Q(X)

FIGURE 3.29 Partially factorized q function for structured variational inference (figure from [3]).

Compared with MCMC methods, variational inference provides a deterministic solution. It is fast and provides a lower bound. It is suitable for the case, in which the true distribution has no explicit form or is intractable to handle. On the other hand, typical variational distributions assume certain independencies among the target variables, which destroys the dependencies among target variables and inevitably enlarges the gap to the true distribution. The accuracy of its solution depends on the difference between p and q. Besides the form of q, the initializations and convergence while estimating q also determine the quality of q. In contrast, MCMC methods are simple and can lead to a solution very close to those obtained by exact methods given enough samples. However, the uncertainty with mixing can produce variable results, and they could take a long time to converge.

Besides the aforementioned methods, the approximate inference can also be performed through structure simplification. Specifically, we learn a simplified structure (such as a tree) from the complex groundtruth structure and perform exact inference on the simplified structure such as the treewidth bounded method [42]. Although similar to variational inference in spirit, they can better preserve the dependencies among the variables.

3.3.3 Inference for Gaussian BNs

For Gaussian BNs, inference can be carried out directly from the joint covariance matrix. Specifically, for a GBN over $\mathbf{X} = (X_1, X_2, \ldots, X_N)^\top$, following the derivations in Appendix 3.9.2, we have $\mathbf{X} \sim \mathcal{N}(\boldsymbol{\mu}, \boldsymbol{\Sigma})$. For any subset of variables $\mathbf{X}_s \subset \mathbf{X}$, we have $p(\mathbf{X}_s) \sim \mathcal{N}(\boldsymbol{\mu}_s, \boldsymbol{\Sigma}_s)$, where $\boldsymbol{\mu}_s$ and $\boldsymbol{\Sigma}_s$ can be directly extracted from the corresponding elements in $\boldsymbol{\mu}$ and $\boldsymbol{\Sigma}$. Hence, for posterior probability inference $p(\mathbf{x}_Q|\mathbf{x}_E)$, where $\mathbf{x}_Q \subset \mathbf{x}$ are the query variables and $\mathbf{x}_E \subset \mathbf{x}$ are the evidence variables, applying the conditional probability rule, we have

$$p(\mathbf{x}_Q|\mathbf{x}_E) = \frac{p(\mathbf{x}_Q, \mathbf{x}_E)}{p(\mathbf{x}_E)}, \tag{3.52}$$

where $p(\mathbf{x}_Q, \mathbf{x}_E)$ and $p(\mathbf{x}_E)$ follow a Gaussian distribution, and their means and covariance matrix can be derived from the corresponding elements in $\boldsymbol{\mu}$ and $\boldsymbol{\Sigma}$. Furthermore, $p(\mathbf{x}_Q|\mathbf{x}_E)$ also follows a Gaussian distribution whose mean and covariance matrix can be derived using Eq. (3.184) in Appendix 3.9.2. Because computing the joint covariance matrix requires matrix inversion, such direct inference may be computationally costly for large GBNs. Given $p(\mathbf{x}_Q|\mathbf{x}_E)$, besides posterior inference, MAP inference of $\mathbf{x}_Q$ can also be obtained from the mode of $p(\mathbf{x}_Q|\mathbf{x}_E)$, which equals its mean for a Gaussian distribution. If $p(\mathbf{x}_Q|\mathbf{x}_E)$ does not follow a Gaussian distribution, its mode corresponds to the value of $\mathbf{x}_Q$, where $p(\mathbf{x}_Q|\mathbf{x}_E)$ achieves the maximum and the first-order derivative of the density function $p(\mathbf{x}_Q|\mathbf{x}_E)$ is zero.

3.3.4 Bayesian inference

The inference methods we have discussed so far are point-based in that they perform inference using a set of parameters (i.e., CPDs) either learned automatically from data or provided by humans. The point-based inference may suffer from overfitting as only one set of parameters is used during inference. Furthermore, they require an expensive and time-consuming training procedure to acquire the parameters. Bayesian inference, on the other hand, performs inference directly based on the training data and treats the model parameters as random variables. More importantly, its prediction is based on all parameters instead of only a set of parameters. We further discuss Bayesian inference for discrete BNs.

Let $\mathbf{X}$=$\{X_1, X_2, \ldots, X_N\}$ be the random variables defined over a discrete BN, where X_n is nodes of the BN, and let $\mathcal{D} = \{D_1, D_2, \ldots, D_M\}$ be the training data with $D_m = \{x_1^m, x_2^m, \ldots, x_N^m\}$. Furthermore, let $\boldsymbol{\alpha} = \{\boldsymbol{\alpha}_n\}, n = 1, 2, \ldots, N$, be the hyperparameters for each node, that is, $\boldsymbol{\alpha}_n$ specifies the prior probability distributions of $\boldsymbol{\theta}_n$, the parameters of the nth node. Given query data $\mathbf{x}'$, Bayesian posterior inference can be mathematically written as computing $p(\mathbf{x}'|\mathcal{D}, \boldsymbol{\alpha})$. To compute the latter, we need introduce parameters $\boldsymbol{\Theta}$:

$$\begin{aligned} p(\mathbf{x}'|\mathcal{D}, \boldsymbol{\alpha}) &= \int p(\mathbf{x}', \boldsymbol{\Theta}|\mathcal{D}, \boldsymbol{\alpha}) d\boldsymbol{\Theta} \\ &= \int p(\mathbf{x}'|\boldsymbol{\Theta}) p(\boldsymbol{\Theta}|\mathcal{D}, \boldsymbol{\alpha}) d\boldsymbol{\Theta} \\ &\propto \int p(\mathbf{x}'|\boldsymbol{\Theta}) p(\boldsymbol{\Theta}|\boldsymbol{\alpha}) p(\mathcal{D}|\boldsymbol{\Theta}) d\boldsymbol{\Theta}. \end{aligned} \tag{3.53}$$

From Eq. (3.53) it is clear that the prediction of the probability of query data is based on all model parameters $\boldsymbol{\Theta}$ instead of on only a set of point-estimated parameters. Bayesian inference, however, requires the integration over all parameters, which in general is computationally intractable. Nevertheless, by exploiting BN parameter independencies we can analytically solve Eq. (3.53) as follows. Specifically, exploiting parameter independencies among BN nodes and assuming that training data are i.i.d., Eq. (3.53) can be rewritten as

$$\begin{aligned} p(\mathbf{x}'|\mathcal{D},\boldsymbol{\alpha}) &\propto \int \prod_{n=1}^{N} p(x_n'|\pi(x_n'),\boldsymbol{\theta}_n) \prod_{n=1}^{N} \\ &\quad p(\boldsymbol{\theta}_n|\boldsymbol{\alpha}_n) \prod_{m=1}^{M}\prod_{n=1}^{N} p(x_n^m|\pi^m(x_n),\boldsymbol{\theta}_n) d\boldsymbol{\Theta} \\ &= \prod_{n=1}^{N} \int p(x_n'|\pi(x_n'),\boldsymbol{\theta}_n) p(\boldsymbol{\theta}_n|\boldsymbol{\alpha}_n) \prod_{m=1}^{M} p(x_n^m|\pi^m(x_n),\boldsymbol{\theta}_n) d\boldsymbol{\theta}_n. \end{aligned} \tag{3.54}$$

For each node, further exploiting the independencies between its parameters under different parental configurations, that is, independencies between $\boldsymbol{\theta}_{nj}$, yields

$$\begin{aligned} p(\mathbf{x}'|\mathcal{D},\boldsymbol{\alpha}) &\propto \prod_{n=1}^{N}\prod_{j=1}^{J_n} \int p(x_n'|\pi(x_n')=j,\boldsymbol{\theta}_{nj}) p(\boldsymbol{\theta}_{nj}|\boldsymbol{\alpha}_{nj}) \\ &\quad \prod_{m=1}^{M} p^{I(\pi^m(x_n)=j)}(x_n^m|\pi^m(x_n)=j,\boldsymbol{\theta}_{nj}) d\boldsymbol{\theta}_{nj}. \end{aligned} \tag{3.55}$$

Assuming the categorical distribution for x_n', Dirichlet distribution for $\boldsymbol{\theta}_{nj}$, and multinomial distribution for $\prod_{m=1}^{M} p(x_n^m|\pi^m(x_n)=j,\boldsymbol{\theta}_{nj})$, Eq. (3.55) can be rewritten as

$$\begin{aligned} p(\mathbf{x}'|\mathcal{D},\boldsymbol{\alpha}) &\propto \prod_{n=1}^{N}\prod_{j=1}^{J_n}\prod_{k=1}^{K_n} \int p^{I(x_n'=k\&\pi(x_n')=j)}(x_n'=k|\pi(x_n')=j|\theta_{njk})\theta_{njk}^{\alpha_{njk}-1} \\ &\quad \prod_{m=1}^{M} p^{I(x_n^m=k\&\pi^m(x_n)=j)}(x_n^m=k|\pi^m(x_n)=j,\theta_{njk}) d\theta_{njk} \\ &= \prod_{n=1}^{N}\prod_{j=1}^{J_n}\prod_{k=1}^{K_n} \int \{\theta_{njk}\theta_{njk}^{\alpha_{njk}+M_{njk}-1}\}^{I(x_n'=k\&\pi(x_n')=j\&x_n^m=k\&\pi^m(x_n)=j)} d\theta_{njk} \\ &= \prod_{n=1}^{N}\prod_{j=1}^{J_n}\prod_{k=1}^{K_n} \{\frac{\alpha_{njk}+M_{njk}}{\sum_{k'=1}^{K}(\alpha_{njk'}+M_{njk'})}\}^{I(x_n'=k\&\pi(x_n')=j\&x_n^m=k\&\pi^m(x_n)=j)}, \end{aligned} \tag{3.56}$$

where I is the indicator function, and M_{njk} is the count in $\mathcal{D}$ for node n with jth parental configuration and assuming value k. It is clear from Eq. (3.56) that for discrete BN, by replacing CPT $\frac{M_{njk}}{\sum_{k'=1}^{K} M_{njk'}}$ for each node with $\frac{\alpha_{njk}+M_{njk}}{\sum_{k'=1}^{K}(\alpha_{njk'}+M_{njk'})}$, the point-based posterior probability inference becomes Bayesian posterior inference.

In summary, point-based inference performs inference using a set of model parameters learned from the training data via a parameter maximization process. In contrast, Bayesian inference performs inference directly from the training data using all parameters via a parameter marginalization process. They are fundamentally different. Bayesian inference is robust to over-fitting, imbalanced, and insufficient data. But Bayesian inference is computationally expensive as it requires integration over the parameter space and hence does not scale up well.

3.3.5 Inference under uncertain evidences

In many real world problems, the evidences we observed are not certain, that is, the evidences are uncertain with a probability. Let $\mathbf{x}_E$ be the observed variables with a probability of $q(\mathbf{x}_E)$ and $\mathbf{x}_Q$ be the query variables. We want to infer $p(\mathbf{x}_Q|q(\mathbf{x}_E))$. This inference cannot be directly performed with a BN as it cannot handle uncertain evidence. Such uncertain evidence is called soft or virtual evidence. To overcome this problem, we can treat the uncertain evidence as soft evidence and then employ Jeffrey's update rule [94] to perform the expected inference, i.e., compute the expected value of $p(\mathbf{x}_Q|\mathbf{x}_E))$ over $q(\mathbf{x}_E)$. This can be accomplished as follows

$$p(\mathbf{x}_Q|p(\mathbf{x}_E)) = \sum_{\mathbf{x}_E} p(\mathbf{x}_Q|\mathbf{x}_E)q(\mathbf{x}_E) \tag{3.57}$$

where $p(\mathbf{x}_Q|\mathbf{x}_E)$ can be computed using the conventional BN inference methods. Jeffrey's rule cannot be used to combine multiple uncertain evidences as it's result depends on the combination order of the uncertain evidences. To address this issue, we may treat the uncertain evidence as virtual evidence. Further information about inference under uncertain evidence may be found in Appendix 3.9.4.

3.4 BN learning under complete data

An essential issue for BNs is learning. BN learning involves acquiring its parameters (e.g., CPDs) and its structure $\mathcal{G}$. BN learning can be specified manually for small BNs, in particular, for BNs that capture the causal relationships and for applications with easily accessible domain experts. Learning can also be performed automatically from data. We will first discuss automatic learning from data and then briefly discuss manual BN specifications in Section 3.6. BN learning from data in general is NP-hard. The key in improving learning efficiency is again exploiting the independencies embedded in the BN.

In general, automatic BN learning can be divided into four cases as shown in Table 3.1 [43]. Depending on whether $\mathcal{G}$ is given or not, BN learning can be divided into parameter learning when $\mathcal{G}$ is given and joint structure and parameter learning when $\mathcal{G}$ is not given. Depending on whether the training data are complete or not, learning in each category can be further divided into learning under complete training data and under incomplete data. If the structure $\mathcal{G}$ is given, then BN learning involves only learning its parameters for each node. Parameter learning can be divided into parameter learning under complete data and parameter learning under incomplete data.

Table 3.1 Four cases of BN Learning.

Case	Structure	Data	Learning Methods
1	Known	Fully observed	Parameter learning
2	Known	Partially observed	Incomplete parameter learning
3	Unknown	Fully observed	Structure and parameter learning
4	Unknown	Partially observed	Incomplete structure and parameter learning

In the following sections, we discuss BN learning under each of the four conditions. In this section, we focus on BN parameter learning. BN structure learning will be covered in Section 3.4.2.

3.4.1 Parameter learning

We first study the case where complete data are given, that is, there are no missing values for each training sample. In this case, BN parameter learning can be stated as follows. Let $\mathcal{D} = \{D_1, D_2, \ldots, D_M\}$ be a set of M i.i.d. distributed training samples, where D_m represents the mth training sample consisting of a vector of values for each node, that is, $D_m = \{x_1^m, x_2^m, \ldots, x_N^m\}$. We assume that the training samples are drawn from an underlying unknown distribution p^*. The goal of parameter learning is to estimate the BN parameter $\mathbf{\Theta} = \{\mathbf{\Theta}_n\}$ so that the joint distribution represented by the BN with the estimated θ^* best approximates the underlying distribution p^*, where $\mathbf{\Theta}_n$ is a vector of parameters for node X_n. Since p^* is unknown, this is typically done by replacing p^* with the empirical distribution $\hat{p}$ derived from the training samples. We can find $\mathbf{\Theta}$ by minimizing the KL-divergence between $\hat{p}$ and the distribution of the BN as parameterized by $\mathbf{\Theta}$. Minimizing the KL-divergence can be equivalently done by maximizing certain objective functions of the parameters and the data $\mathcal{D}$. Various objective functions have been used for parameter learning; the most commonly used objective functions are the maximum likelihood estimation (MLE) and Maximum Posterior Probability (MAP) as discussed in Section 2.3 of Chapter 2. In the following, we will cover BN parameter learning for each case. Besides generative learning, we will also briefly discuss the discriminative learning of BN when it is used for classification purposes.

3.4.1.1 Maximum likelihood estimation of BN parameters

As discussed in Chapter 2, MLE can be formulated as

$$\theta^* = \underset{\theta}{\operatorname{argmax}}\, LL(\theta : \mathcal{D}), \tag{3.58}$$

where $LL(\theta : \mathcal{D})$ represents the joint log-likelihood of θ, given the data $\mathcal{D}$. For a BN with N nodes and i.i.d. samples D_m, the joint log-likelihood can be written as

$$LL(\boldsymbol{\theta} : \mathcal{D}) = \log \prod_{m=1}^{M} p(x_1^m, x_2^m, \ldots, x_N^m|\boldsymbol{\theta}). \tag{3.59}$$

Following the BN chain rule, Eq. (3.59) can be rewritten as

$$\begin{aligned} LL(\boldsymbol{\theta} : \mathcal{D}) &= \log \prod_{m=1}^{M} p(x_1^m, x_2^m, \ldots, x_N^m|\boldsymbol{\theta}) \\ &= \log \prod_{n=1}^{N} \prod_{m=1}^{M} p(x_n^m|\pi(x_n^m), \boldsymbol{\theta}_n) \\ &= \sum_{n=1}^{N} \sum_{m=1}^{M} \log p(x_n^m|\pi(x_n^m), \boldsymbol{\theta}_n) \\ &= \sum_{n=1}^{N} LL(\boldsymbol{\theta}_n : \mathcal{D}), \end{aligned} \tag{3.60}$$

where $LL(\boldsymbol{\theta}_n : \mathcal{D}) = \log p(\mathcal{D}|\boldsymbol{\theta}_n) = \sum_{m=1}^{M} \log p(x_n^m|\pi(x_n^m), \boldsymbol{\theta}_n)$ is the marginal log-likelihood for the parameters of each node. It is clear from Eq. (3.60) that the joint log-likelihood for all parameters can be written as the sum of the marginal log-likelihood of the parameters of each node because of the BN chain rule. This greatly simplifies the learning since the parameters for each node can be learned separately, that is,

$$\boldsymbol{\theta}_n^* = \underset{\boldsymbol{\theta}_n}{\operatorname{argmax}}\, LL(\boldsymbol{\theta}_n : \mathcal{D}). \tag{3.61}$$

Furthermore, the log-likelihood function is concave, thus allowing obtaining the optimal solution either analytically or numerically. For discrete BNs, $\boldsymbol{\theta}_n$ can be further decomposed into $\boldsymbol{\theta}_n = \{\boldsymbol{\theta}_{nj}\}$, $j = 1, 2, \ldots, J$, where j is the index to the jth configuration of the parents for a total of J parental configurations. Assuming that the parameters $\boldsymbol{\theta}_{nj}$ are independent and each node has K values $x_n \in \{1, 2, .., K\}$, we can rewrite the likelihood for each node as

$$\begin{aligned} p(x_n^m|\pi(x_n^m)) &= \prod_{j=1}^{J} \boldsymbol{\theta}_{nj}^{I(\pi(x_n^m)=j)} \\ &= \prod_{j=1}^{J} \prod_{k=1}^{K} \theta_{njk}^{I(\pi(x_n^m)=j \& x_n^m=k)} \\ &= \prod_{j=1}^{J} \prod_{k=1}^{K-1} \theta_{njk}^{I(\pi(x_n^m)=j \& x_n^m=k)} \\ &\quad (1 - \sum_{l=1}^{K-1} \theta_{njl})^{I(\pi(x_n^m)=j \& x_n^m=K)}, \end{aligned} \tag{3.62}$$

where θ_{njk} is the conditional probability that X_n takes the value k given the jth configuration of its parents, and $\sum_{k=1}^{K} \theta_{njk} = 1$. Hence we have

$$
\begin{aligned}
LL(\boldsymbol{\theta}_n : \mathcal{D}) &= \sum_{m=1}^{M} \log p(x_n^m | \pi(x_n^m)) \\
&= \sum_{m=1}^{M} \sum_{j=1}^{J} \sum_{k=1}^{K-1} I(\pi(x_n^m) = j \& x_n^m = k) \log \theta_{njk} \\
&\quad + I(\pi(x_n^m) = j \& x_n^m = K) \log(1 - \sum_{l=1}^{K-1} \theta_{njl}) \\
&= \sum_{m=1}^{M} \sum_{j=1}^{J} \sum_{k=1}^{K-1} I(\pi(x_n^m) = j \& x_n^m = k) \log \theta_{njk} \\
&\quad + \sum_{m=1}^{M} \sum_{j=1}^{J} I(\pi(x_n^m) = j \& x_n^m = K) \log(1 - \sum_{l=1}^{K-1} \theta_{njl}) \\
&= \sum_{j=1}^{J} \sum_{k=1}^{K-1} M_{njk} \log \theta_{njk} + \sum_{j=1}^{J} M_{njK} \log(1 - \sum_{l=1}^{K-1} \theta_{njl}),
\end{aligned} \tag{3.63}
$$

where M_{njk} is the number of training samples with $X_n = k$ and with its parents assuming the jth configuration. As a result, we can compute θ_{njk} as

$$
\theta_{njk}^* = \underset{\theta_{njk}}{\operatorname{argmax}} \, LL(\boldsymbol{\theta}_n : \mathcal{D}),
$$

which can be solved by setting $\frac{\partial LL(\boldsymbol{\theta}_n : \mathcal{D})}{\partial \theta_{njk}} = 0$, yielding

$$
\theta_{njk} = \frac{M_{njk}}{\sum_{k'=1}^{K} M_{njk'}}. \tag{3.64}
$$

It is clear from Eq. (3.64) that θ_{njk} can be solved as a counting problem, that is, just counting the number of occurrences of the nth node with value k and the jth parental configuration and dividing the count by the total number of samples with the jth parent configuration.

To ensure a reliable estimate for the CPD for each node, we need a sufficient number of samples for each parameter. The specific number of samples needed to have a confident estimate of each parameter θ_{njk} can be computed using the confidence interval bounds as discussed in Section 2.5.2 of Chapter 2. In general, for a reliable estimation, we need at least five samples for each parameter θ_{njk}. For a BN, the amount of data necessary to learn the parameters reliably for a node is $(K-1)K^J \times 5$, where K is the number of states for the node, and J is the number of parents of the node. In addition, to deal with the cases where certain configurations have zero observation, that is, $M_{njk} = 0$, a small nonzero value is used to initialize the count for each configuration. Alternatively, this problem can be systematically solved through the Dirichlet prior as shown in Eq. (3.82) and further discussed in the MAP estimation.

For Gaussian BNs, following the CPD specification in Eq. (3.2) for a linear Gaussian, the marginal log-likelihood for the nth node can be written as

$$\begin{aligned} LL(\boldsymbol{\theta}_n : \mathcal{D}) &= \sum_{m=1}^{M} \log p(x_n^m | \pi(x_n^m), \boldsymbol{\theta}_n) \\ &= \sum_{m=1}^{M} \log \frac{1}{\sqrt{2\pi\sigma_n^2}} e^{-\frac{(x_n^m - \sum_{k=1}^{K} \alpha_n^k \pi_k^m(x_n) - \beta_n)^2}{2\sigma_n^2}} \\ &= -\frac{\sum_{m=1}^{M} (x_n^m - \sum_{k=1}^{K} \alpha_n^k \pi_k^m(x_n) - \beta_n)^2}{2\sigma_n^2} \\ &\quad + M \log \frac{1}{\sqrt{2\pi\sigma_n^2}}, \end{aligned} \tag{3.65}$$

where $\pi_k(X_n)$ is the kth parent of X_n. The numerator of the first term of Eq. (3.65) can be rewritten as

$$\sum_{m=1}^{M} (x_n^m - \sum_{k=1}^{K} \alpha_n^k \pi_k^m(x_n) - \beta_n)^2 = (\mathbf{x}_n - \mathbf{\Pi}_n \boldsymbol{\theta}_n)^\top (\mathbf{x}_n - \mathbf{\Pi}_n \boldsymbol{\theta}_n), \tag{3.66}$$

where $\mathbf{x}_n = (x_n^1, x_n^2, \ldots, x_n^M)^\top$, $\boldsymbol{\theta}_n = (\alpha_n^1, \alpha_n^2, \ldots, \alpha_n^K, \beta_n)^\top$, and $\mathbf{\Pi}_n$ is defined as follows:

$$\mathbf{\Pi}_n = \begin{bmatrix} \pi_1^1(x_n) & \pi_2^1(x_n) & \ldots & \pi_K^1(x_n) & 1 \\ \pi_1^2(x_n) & \pi_2^2(x_n) & \ldots & \pi_K^2(x_n) & 1 \\ \vdots & & & & \\ \pi_1^M(x_n) & \pi_2^M(x_n) & \ldots & \pi_K^M(x_n) & 1 \end{bmatrix}. \tag{3.67}$$

Substituting Eq. (3.66) into Eq. (3.65) yields

$$\begin{aligned} LL(\boldsymbol{\theta}_n : \mathcal{D}) &= -\frac{(\mathbf{x}_n - \mathbf{\Pi}_n \boldsymbol{\theta}_n)^\top (\mathbf{x}_n - \mathbf{\Pi}_n \boldsymbol{\theta}_n)}{2\sigma_n^2} \\ &\quad + M \log \frac{1}{\sqrt{2\pi\sigma_n^2}}. \end{aligned} \tag{3.68}$$

Eq. (3.68) is maximized by taking the partial derivatives of the log-likelihood with respect to $\boldsymbol{\theta}_n$ and setting them to zero, yielding

$$\boldsymbol{\theta}_n = (\mathbf{\Pi}_n^\top \mathbf{\Pi}_n)^{-1} \mathbf{\Pi}_n^\top \mathbf{x}_n. \tag{3.69}$$

Similarly, taking the derivative of the log-likelihood in Eq. (3.68) with respect to σ_n and setting it to zero leads to the solution to σ_n^2 as

$$\hat{\sigma}_n^2 = \frac{(\mathbf{x}_n - \mathbf{\Pi}_n \boldsymbol{\theta}_n)^\top (\mathbf{x}_n - \mathbf{\Pi}_n \boldsymbol{\theta}_n)}{M}. \tag{3.70}$$

Further details on the derivations for learning Gaussian BN parameters can be found in Theorem 7.4 and Section 17.2.4 of [7].

3.4.1.2 MAP estimation of BN parameters

MLE depends on data. When data is insufficient, MLE may not be reliable. In this case, we can use MAP estimation, which allows incorporating the prior knowledge on the parameters into the estimation process. For MAP parameter estimation, we assume that the parameters follow some prior distribution defined by hyperparameters $\boldsymbol{\gamma}$. Given the training data $\mathcal{D} = \{D_m\}_{m=1}^M$, the goal of MAP parameter estimation is to estimate $\boldsymbol{\theta}$ by maximizing the posterior probability of $\boldsymbol{\theta}$:

$$\boldsymbol{\theta}^* = \underset{\boldsymbol{\theta}}{\operatorname{argmax}}\, p(\boldsymbol{\theta}|\mathcal{D}), \tag{3.71}$$

where $p(\boldsymbol{\theta}|\mathcal{D})$ can be expressed as

$$\begin{aligned} p(\boldsymbol{\theta}|\mathcal{D}) &\propto p(\boldsymbol{\theta}, \mathcal{D}) \\ &= p(\boldsymbol{\theta}) \prod_{m=1}^{M} p(D_m|\boldsymbol{\theta}). \end{aligned} \tag{3.72}$$

Maximizing the posterior probability is the same as maximizing the log posterior probability:

$$\boldsymbol{\theta}^* = \underset{\boldsymbol{\theta}}{\operatorname{argmax}} \log p(\boldsymbol{\theta}|\mathcal{D}), \tag{3.73}$$

where the log-posterior probability can be written as

$$\log p(\boldsymbol{\theta}|\mathcal{D}) \propto \sum_{m=1}^{M} \log p(D_m|\boldsymbol{\theta}) + \log p(\boldsymbol{\theta}), \tag{3.74}$$

where the first term is the joint log-likelihood, and the second term is the prior. Assuming that the parameter prior for each node is independent of each other, that is, $p(\boldsymbol{\theta}) = \prod_{n=1}^{N} p(\boldsymbol{\theta}_n)$, and applying the BN chain rule, we can rewrite the log posterior as

$$\log p(\boldsymbol{\theta}|\mathcal{D}) = \sum_{n=1}^{N} \sum_{m=1}^{M} \log p(x_n^m | \pi(x_n^m), \boldsymbol{\theta}_n) + \sum_{n=1}^{N} \log p(\boldsymbol{\theta}_n). \tag{3.75}$$

It is clear from Eq. (3.75) that the joint posterior of the parameters is also decomposable, that is, the parameters for each node can be estimated separately. We can therefore perform MAP estimation of the parameters $\boldsymbol{\theta}_n$ for each node by maximizing the log posterior probability of $\boldsymbol{\theta}_n$:

$$\begin{aligned} \boldsymbol{\theta}_n^* &= \underset{\boldsymbol{\theta}_n}{\operatorname{argmax}} \log p(\boldsymbol{\theta}_n|\mathcal{D}) \\ &= \underset{\boldsymbol{\theta}_n}{\operatorname{argmax}} \left\{ \sum_{m=1}^{M} \log p(x_n^m|\pi(x_n^m), \boldsymbol{\theta}_n) + \log p(\boldsymbol{\theta}_n) \right\}. \end{aligned} \tag{3.76}$$

For discrete BNs, we can further decompose $\boldsymbol{\theta}_n$ into $\boldsymbol{\theta}_{nj}$, that is, $\boldsymbol{\theta}_n = \{\boldsymbol{\theta}_{nj}\}$, where j is the index to the jth configuration of the parents. Assuming that $\boldsymbol{\theta}_{nj}$ are independent of each other, $\boldsymbol{\theta}_{nj}$ can then be separately estimated as follows:

$$\begin{aligned} \boldsymbol{\theta}_{nj}^* &= \underset{\boldsymbol{\theta}_{nj}}{\operatorname{argmax}} \log p(\boldsymbol{\theta}_{nj}|\mathcal{D}) \\ &= \underset{\boldsymbol{\theta}_{nj}}{\operatorname{argmax}} \left\{ \sum_{m=1}^{M} \log p(x_n^m|\pi(x_n^m) = j, \boldsymbol{\theta}_{nj}) + \log p(\boldsymbol{\theta}_{nj}) \right\}. \end{aligned} \tag{3.77}$$

For a discrete BN, the joint likelihood function follows multinomial distribution, and according to Eq. (3.63), we can write the likelihood term in Eq. (3.77) as

$$\sum_{m=1}^{M} \log p(x_n^m|\pi(x_n^m) = j, \boldsymbol{\theta}_{nj}) = \sum_{k=1}^{K-1} M_{ijk} \log \theta_{njk} + M_{njK} \log(1 - \sum_{l=1}^{K-1} \theta_{njl}), \tag{3.78}$$

where θ_{njk} is the conditional probability of the nth node assuming a value of k, given the jth configuration of its parents, and M_{njk} is the number of training samples with $X_n = k$ and with its parents assuming the jth configuration. Given the multinomial distribution for the likelihood, the conjugate prior distribution for $\boldsymbol{\theta}_{nj}$ follows the Dirichlet distribution:

$$p(\theta_{nj}) = c_p \prod_{k=1}^{K} \theta_{njk}^{\alpha_{njk}-1}. \tag{3.79}$$

Introducing the constraint that $\sum_{k=1}^{K} \theta_{njk} = 1$, we can rewrite the Dirichlet prior as

$$p(\boldsymbol{\theta}_{nj}) = c_p \prod_{k=1}^{K-1} \theta_{njk}^{\alpha_{njk}-1} (1 - \sum_{l=1}^{K-1} \theta_{njl})^{\alpha_{njK}-1}. \tag{3.80}$$

Combining Eqs. (3.78) and (3.80) yields the log posterior for $\boldsymbol{\theta}_{nj}$:

$$\begin{aligned}\log p(\boldsymbol{\theta}_{nj}|\mathcal{D}) &= \log c_p + [\sum_{k=1}^{K-1}(\alpha_{njk}-1)+M_{njk}]\log\theta_{njk} \\ &+ (M_{njK}+\alpha_{njK}-1)\log(1-\sum_{l=1}^{K-1}\theta_{njl}). \end{aligned} \tag{3.81}$$

For $k = 1, 2, \ldots, K-1$, we maximize $\log p(\theta_{njk}|\mathcal{D})$ with respect to θ_{njk} by taking the partial derivative of $\log p(\boldsymbol{\theta}_{nj}|\mathcal{D})$ with respect to θ_{njk} and setting it to zero. This yields

$$\theta_{njk} = \frac{M_{njk}+\alpha_{njk}-1}{\sum_{k'=1}^{K} M_{njk'} + \sum_{k=1}^{K}\alpha_{njk'} - K}. \tag{3.82}$$

Eq. (3.82) shows that the parameters for each node depend on both its counts in the training data and its hyperparameters. When the counts are small, the hyperparameters are important in determining the value of a parameter. However, when the counts are large, the hyperparameters become negligible. In practice, $N_{ijk}+\alpha_{ijk} > 1$ and α_{ijk} are chosen as follows: $\alpha_{ijk} = M' \times p(x_i = k, \pi(x_i) = j)$, where M' is called equivalent sample size or prior strength and it can be tuned. $p(x_i = k, \pi(x_i) = j)$ is approximated by $\frac{1}{|x_i|\times|\pi(x_i)|}$, where $|\cdot|$ represents the cardinality of its argument. More details on selecting the Dirichlet prior can be found in [95].

For continuous Gaussian BN, the likelihood function follows the linear Gaussian as discussed in Section 3.2.3.2. The prior should also follow Gaussian distribution such that the posterior becomes a Gaussian distribution as well. The log posterior probability for the parameters for each node can be written as

$$\log p(\boldsymbol{\theta}_n|\mathcal{D}) = \sum_{m=1}^{M}\log p(x_n^m|\pi(x_n^m)) + \log p(\boldsymbol{\theta}_n), \tag{3.83}$$

where $p(x_n^m|\pi(x_n^m))$ can be written as a linear Gaussian,

$$\begin{aligned} p(x_n^m|\pi(x_n^m)) &= \mathcal{N}(\sum_k \alpha_n^k \pi_k^m(x_n) + \beta_n, \sigma_n^2) \\ &= \frac{1}{\sqrt{2\pi}\sigma_n} e^{-\frac{(x_n^m - \sum_{k=1}^{K}\alpha_n^k\pi_k^m(x_n)-\beta_n)^2}{2\sigma_n^2}} \end{aligned} \tag{3.84}$$

with parameters $\boldsymbol{\theta}_n = \{\alpha_n^k, \beta_n\}$ and $p(\boldsymbol{\theta}_n) \sim \mathcal{N}(\boldsymbol{\mu}_{\theta_n}, \boldsymbol{\Sigma}_{\theta_n})$. Substituting the Gaussian prior $p(\boldsymbol{\theta}_n)$ and the Gaussian likelihood in Eq. (3.84) into Eq. (3.83) yields

$$\begin{aligned} \log p(\boldsymbol{\theta}_n|\mathcal{D}) &= \sum_{m=1}^{M} -\frac{(x_n^m - \sum_{k=1}^{K}\alpha_n^k\pi_k^m(x_n) - \beta_n)^2}{2\sigma_n^2} \\ &- (\boldsymbol{\theta}_n - \boldsymbol{\mu}_{\theta_n})^\top \boldsymbol{\Sigma}_{\theta_n}^{-1}(\boldsymbol{\theta}_n - \boldsymbol{\mu}_{\theta_n}) + C, \end{aligned} \tag{3.85}$$

where C is a constant term. Replacing the log-likelihood term by Eq. (3.68) and ignoring the constant terms yield

$$\log p(\boldsymbol{\theta}_n|\mathcal{D}) = -\frac{(\mathbf{x}_n - \boldsymbol{\Pi}_n\boldsymbol{\theta}_n)^\top(\mathbf{x}_n - \boldsymbol{\Pi}_n\boldsymbol{\theta}_n)}{2\sigma_n^2} - (\boldsymbol{\theta}_n - \boldsymbol{\mu}_{\theta_n})^\top\boldsymbol{\Sigma}_{\theta_n}^{-1}(\boldsymbol{\theta}_n - \boldsymbol{\mu}_{\theta_n}). \tag{3.86}$$

Taking the derivative of $\log p(\boldsymbol{\theta}_n|\mathcal{D})$ with respect to $\boldsymbol{\theta}_n$ and setting it to zero produce

$$\frac{\boldsymbol{\Pi}_n^\top(\mathbf{x}_n - \boldsymbol{\Pi}_n\boldsymbol{\theta}_n)}{\sigma_n^2} + 2\boldsymbol{\Sigma}_{\theta_n}^{-1}(\boldsymbol{\theta}_n - \boldsymbol{\mu}_{\theta_n}) = 0. \tag{3.87}$$

The solution to $\boldsymbol{\theta}_n$ can be obtained by solving Eq. (3.87):

$$\boldsymbol{\theta}_n = \left(\frac{\boldsymbol{\Pi}_n^\top\boldsymbol{\Pi}_n}{\sigma_n^2} - 2\boldsymbol{\Sigma}_{\theta_n}^{-1}\right)^{-1}\left(\frac{\boldsymbol{\Pi}_n^\top\mathbf{x}_n}{\sigma_n^2} - 2\boldsymbol{\Sigma}_{\theta_n}^{-1}\boldsymbol{\mu}_{\theta_n}\right). \tag{3.88}$$

Since there is no prior on σ_n, its MAP estimate is the same as its maximum likelihood estimate in Eq. (3.70).

Besides conjugate priors, generic priors, such as sparseness, are often used as a generic prior. They are often implemented through the ℓ_1 or ℓ_2 norm regularization of $\boldsymbol{\theta}$, that is, replacing the prior $p(\boldsymbol{\theta})$ with either $\|\boldsymbol{\theta}\|_1$ or $\|\boldsymbol{\theta}\|_2$. The most commonly used such regularization is the least absolute shrinkage and selection operator (LASSO). By simultaneously performing variable selection and regularization with the ℓ_1 norm, LASSO can be used to improve both model prediction accuracy and its representation efficiency. Since the ℓ_1 norm is decomposable, MAP learning with the ℓ_1 norm as a prior remains decomposable, that is, each parameter can be estimated independently.

Finally, we briefly discuss Bayesian parameter learning. Different from MAP parameter learning, which uses the mode of the parameter posterior as the estimated parameters, Bayesian parameter learning uses the mean or expectation as the estimated parameters, that is, $\theta^* = E_{p(\theta|\mathcal{D})}(\theta)$. For discrete BN with Dirichlet prior, the estimated parameters are

$$\theta_{njk}^* = \frac{N_{njk} + \alpha_{njk}}{\sum_{k=1}^K N_{njk} + \sum_{k=1}^K \alpha_{njk}}. \tag{3.89}$$

The estimated parameters θ_{njk}^* are very similar to those obtained by the MAP estimate in Eq. (3.82), with the only difference being -1 and $-K$ are removed respectively from the numerator and denominator.

3.4.1.3 Discriminative BN parameter learning

When a BN is used for classification, it makes sense to learn the model using the same classification criterion. Let X_t be a node in the BN that represents the target node the class of which we want to infer, and let $\mathbf{X}_F = \mathbf{X} \setminus X_t$ be the remaining nodes of the BN that represent the features. The goal of a BN classifier learning is finding a BN that maps $\mathbf{X}_F$ to X_t.

This can be accomplished by learning the parameters $\boldsymbol{\theta}$ by maximizing the conditional log-likelihood, that is,

$$\boldsymbol{\theta}^* = \underset{\boldsymbol{\theta}}{\operatorname{argmax}} \sum_{m=1}^{M} \log p(x_t^m | \mathbf{x}_F^m, \boldsymbol{\theta}), \tag{3.90}$$

where $\log p(x_t^m | \mathbf{x}_F^m, \boldsymbol{\theta})$ can be rewritten as

$$\log p(x_t^m | \mathbf{x}_F^m, \boldsymbol{\theta}) = \log p(x_t^m, \mathbf{x}_F^m | \boldsymbol{\theta}) - \log \sum_{x_t} p(x_t, \mathbf{x}_F^m | \boldsymbol{\theta}), \tag{3.91}$$

where the first term is the joint log-likelihood of $\boldsymbol{\theta}$, given the data, and the second term is the marginal log-likelihood of $\boldsymbol{\theta}$, given $\mathbf{x}_F^m$ only. It is clear from Eq. (3.91) that because of the second term, the log conditional likelihood is no longer decomposable as for maximum likelihood estimation and MAP estimation. In other words, the parameters for all nodes must be jointly estimated. Therefore there is no closed-form solution to Eq. (3.90). When no closed-form solutions exist, we can employ the iterative solutions, as discussed in Section 2.4.1 of Chapter 2, to iteratively solve for the parameters.

3.4.2 Structure learning

In the previous sections, we discussed the BN parameter estimation with a known structure. In a more general case, neither the graph nor the parameters are known. We only have a set of examples generated from some underlying distribution. In such cases, structure learning is essential to constructing the BN from data. BN structure learning is to simultaneously learn the links among nodes and CPDs for the nodes. It is more challenging than BN parameter learning. Below, we first discuss the discrete BN structure learning methods. We then introduce continuous BN structure learning.

3.4.2.1 General BN structure learning

The problem of BN structure learning can be stated as follows. Given training data $\mathcal{D} = \{D_1, D_2, \ldots, D_M\}$, where $D_m = \{x_1^m, x_2^m, \ldots, x_N^m\}$, learn the structure $\mathcal{G}$ of the BN. Structure learning methods can be grouped into two major categories: the score-based approach and the independence-test-based approach. The former searches the BN structure space for the BN that yields the highest score, whereas the latter determines the BN structure by identifying the one that maximally satisfies the independencies among the variables in the training data. In the following, we first discuss the score-based approach and then the independence-test-based approach.

3.4.2.1.1 Score-based approach

For the score-based approach, we can formulate the learning as the maximum likelihood learning or maximum posterior probability (MAP) learning. For maximum likelihood learning, we can find the BN structure that maximizes the log structure likelihood:

$$\mathcal{G}^* = \underset{\mathcal{G}}{\operatorname{argmax}} \log p(\mathcal{D}|\mathcal{G}), \tag{3.92}$$

where $p(\mathcal{D}|\mathcal{G})$ is the marginal likelihood of $\mathcal{G}$ given the training data $\mathcal{D}$, which can be further expanded as

$$p(\mathcal{D}|\mathcal{G}) = \int_{\boldsymbol{\theta}} p(\boldsymbol{\theta}|\mathcal{G})p(\mathcal{D}|\mathcal{G},\boldsymbol{\theta})d\boldsymbol{\theta} = E_{\boldsymbol{\theta}\sim p(\boldsymbol{\theta}|\mathcal{G})}(p(\mathcal{G}|\mathcal{D},\boldsymbol{\theta})), \tag{3.93}$$

where $p(\mathcal{D}|\mathcal{G},\boldsymbol{\theta})$ represents the joint likelihood of the structure $\mathcal{G}$ and parameters $\boldsymbol{\theta}$, and $p(\boldsymbol{\theta}|\mathcal{G})$ is the prior probability of the parameter $\boldsymbol{\theta}$ given $\mathcal{G}$. Eq. (3.93) shows that the marginal likelihood can be expressed as the expected joint likelihood of $\mathcal{G}$ over the model parameters $\boldsymbol{\theta}$.

For discrete BNs, Heckerman et al. [96] shows that if we assume $p(\boldsymbol{\theta}|\mathcal{G})$ follows Dirichlet distribution and $P(\mathcal{D}|\mathcal{G},\boldsymbol{\theta})$ follows multinormal distribution, Eq. (3.93) can be solved analytically with a closed solution. In general, exactly computing Eq. (3.93) is intractable due to the integration over $\boldsymbol{\theta}$. Various methods have been proposed to approximate the integration, including the Laplace method, which approximates $p(\boldsymbol{\theta}|\mathcal{G},\mathcal{D})$ with a Gaussian distribution centered on one of its modes, and the variational Bayesian (VB) approach, which approximates the posterior distribution $p(\boldsymbol{\theta}|\mathcal{G},\mathcal{D})$ by a factorized distribution over $\boldsymbol{\theta}$. Additional methods for approximating the integral operation include Monte Carlo integration, which replaces the integral by the average of samples obtained via importance sampling or a maximization operation, assuming that the energy of $p(\boldsymbol{\theta}|\mathcal{G},\mathcal{D})$ is mostly centered on its mode. We adopt the Laplace approximation here. Additional details about the Laplace approximation can be found in Appendix 3.9.3, in Section 19.4.1 of [7], and in Eq. 41 of [44]. Following the Laplace approximation in Appendix 3.9.3, we can approximate the marginal likelihood of $\mathcal{G}$ as follows:

$$\begin{aligned}
p(\mathcal{D}|\mathcal{G}) &= \int_{\boldsymbol{\theta}} p(\mathcal{D},\boldsymbol{\theta}|\mathcal{G})d\boldsymbol{\theta} \\
&\approx \int_{\boldsymbol{\theta}} p(\mathcal{D},\boldsymbol{\theta}_0|\mathcal{G}) \exp -\frac{(\boldsymbol{\theta}-\boldsymbol{\theta}_0)^\top A(\boldsymbol{\theta}-\boldsymbol{\theta}_0)}{2} d\boldsymbol{\theta} \\
&= p(\mathcal{D},\boldsymbol{\theta}_0|\mathcal{G}) \int_{\boldsymbol{\theta}} \exp -\frac{(\boldsymbol{\theta}-\boldsymbol{\theta}_0)^\top A(\boldsymbol{\theta}-\boldsymbol{\theta}_0)}{2} d\boldsymbol{\theta} \\
&= p(\mathcal{D},\boldsymbol{\theta}_0|\mathcal{G})(2\pi)^{d/2}|A|^{-1/2} \\
&= p(\mathcal{D}|\boldsymbol{\theta}_0,\mathcal{G})p(\boldsymbol{\theta}_0|\mathcal{G})(2\pi)^{d/2}|A|^{-1/2},
\end{aligned} \tag{3.94}$$

where d is the degree of freedom of $\boldsymbol{\theta}$ (i.e., the total number of independent parameters), $\boldsymbol{\theta}_0$ is the maximum likelihood estimate of $\log p(\mathcal{D},\boldsymbol{\theta}|\mathcal{G})$, and A is the negative Hessian matrix of $\log p(\mathcal{D},\boldsymbol{\theta}|\mathcal{G})$. Hence the logarithm of the marginal likelihood can be written as

$$\begin{aligned}\log p(\mathcal{D}|\mathcal{G}) &\approx \log p(\mathcal{D}|\mathcal{G},\boldsymbol{\theta}_0)+\log p(\boldsymbol{\theta}_0|\mathcal{G})\\ &+ \frac{d}{2}\log 2\pi - \frac{1}{2}\log|A|,\end{aligned} \tag{3.95}$$

where

$$\begin{aligned}A &= -\frac{\partial^2 \log p(\mathcal{D},\boldsymbol{\theta}_0|\mathcal{G})}{\partial\boldsymbol{\theta}^2}\\ &= \frac{\partial^2 \log[p(\mathcal{D}|\boldsymbol{\theta}_0,\mathcal{G})p(\boldsymbol{\theta}_0|\mathcal{G})]}{\partial\boldsymbol{\theta}^2}\\ &= \frac{\partial^2 \log[\prod_{m=1}^{M} p(\mathcal{D}_m|\boldsymbol{\theta}_0,\mathcal{G})p(\boldsymbol{\theta}_0|\mathcal{G})]}{\partial\boldsymbol{\theta}^2}\\ &= \frac{\partial^2[\sum_{m=1}^{M}\log p(\mathcal{D}_m|\boldsymbol{\theta}_0,\mathcal{G})+\log p(\boldsymbol{\theta}_0|\mathcal{G})]}{\partial\boldsymbol{\theta}^2}\\ &= \frac{\sum_{m=1}^{M}\partial^2\log p(\mathcal{D}_m|\boldsymbol{\theta}_0,\mathcal{G})}{\partial\boldsymbol{\theta}^2}+\frac{\partial^2\log p(\boldsymbol{\theta}_0|\mathcal{G})}{\partial\boldsymbol{\theta}^2}.\end{aligned}$$

As $M\to\infty$, $\log|A|\approx d\log M$. By dropping all terms that are independent of M we arrive at the well-known Bayesian information score (BIC) [45] (for details, see [46])

$$S_{BIC}(\mathcal{G})=\log p(\mathcal{D}|\mathcal{G},\theta_0)-\frac{d}{2}\log M, \tag{3.96}$$

where the first term is the joint likelihood, which ensures $\mathcal{G}$ fits to the data, and the second term is a penalty term that favors simple structures. It has been proved that maximizing the joint likelihood (the first term) alone leads to overfitting since a complex BN structure always increases the joint likelihood [7]. Therefore the additional second term can prevent overfitting. Also note that because of the asymptotic requirement of BIC, that is, large M, caution should be exercised when training data is insufficient. In this case, we may use Eq. (3.95) instead of Eq. (3.96) to exactly compute the log marginal loglikelihood.

Applying the BN chain rule to the likelihood term, we can write the BIC score as

$$\begin{aligned}S_{BIC}(\mathcal{G}) &= \log p(\mathcal{D}|\mathcal{G},\hat{\boldsymbol{\theta}})-\frac{d(\mathcal{G})}{2}\log M\\ &= \sum_{n=1}^{N}\log p(\mathcal{D}|\mathcal{G}(x_n^m),\hat{\boldsymbol{\theta}}_n)-\frac{\log M}{2}\sum_{n=1}^{N}d(\mathcal{G}(X_n))\\ &= \sum_{n=1}^{N}\{\log p(\mathcal{D}|\mathcal{G}(x_n^m),\hat{\boldsymbol{\theta}}_n)-\frac{d(\mathcal{G}(X_n))}{2}\log M\}\\ &= \sum_{n=1}^{N}S_{BIC}(\mathcal{G}(X_n)),\end{aligned} \tag{3.97}$$

where $\mathcal{G}(X_n)$ is the structure for node n consisting of X_n and its parents $\pi(X_n)$. As can be seen, the learning of the global structure $\mathcal{G}$ using the BIC score (or, in fact, any score derived from the marginal likelihood) is again decomposable into learning the structure for each node subject to the acyclicity of the final structure $\mathcal{G}$. Using the BIC score, we can perform BN structure learning by searching the BN structure space to find $\mathcal{G}$ that maximizes the score function.

Variants of BIC scores have been proposed in [47–49]. The negative of the BIC score is referred to as the minimum description length (MDL) score. If we drop $\frac{\log M}{2}$ in the BIC score, it becomes the Akaike information criterion (AIC) score which renders the penalty term independent of the sample size. Depending on how $p(\boldsymbol{\theta}_0|\mathcal{G})$ in Eq. (3.95) is approximated, we may have variants of Bayesian (parameter) BIC scores. Specifically, if we assume that $p(\boldsymbol{\theta}_0|\mathcal{G})$ follows the full Dirichlet distribution without any assumption, the score becomes the Bayesian Dirichlet (BD) score. BD score is impractical as it is hard to fully specify the Dirichlet parameters. If we further assume that the hyperparameter for $p(\boldsymbol{\theta}_0|\mathcal{G})$ is $\alpha_{njk} = M' \times p(\theta_{njk}|\mathcal{G})$, the BD score becomes the Bayesian Dirichlet equivalence (BDe) score where M' is called equivalent sample size. In turn, if we assume the hyperparameters α_{njk} to be uniform, that is, $\alpha_{njk} = \frac{M'}{|x_n| \times |\pi(x_n)|}$, where M' is a tuning parameter for each node, the BDe score becomes the Bayesian Dirichlet equivalence uniform (BDeu) score because α_{njk} are the same for all j and k values of node n. Finally, if we assume that $\alpha_{njk} = 1$, then the BDeu score becomes the K2 score. Further information on different score functions can be found in [50]. Like the BIC score, these score functions are all decomposable, that is, the overall score can be written as the sum of local score functions,

$$S(\mathcal{G}) = \sum_{n \in N} S(\mathcal{G}(X_n)). \tag{3.98}$$

For MAP learning, the problem can be stated as finding the $\mathcal{G}$ that maximizes the log posterior probability of $\mathcal{G}$, given $\mathcal{D}$:

$$\mathcal{G}^* = \underset{\mathcal{G}}{\operatorname{argmax}} \log p(\mathcal{G}|\mathcal{D}), \tag{3.99}$$

where $\log p(\mathcal{G}|\mathcal{D})$ is the log posterior probability of $\mathcal{G}$, given the training data $\mathcal{D}$. It can be further decomposed into

$$\log p(\mathcal{G}|\mathcal{D}) \propto \log p(\mathcal{G}) + \log p(\mathcal{D}|\mathcal{G}), \tag{3.100}$$

where the first term is the prior probability of $\mathcal{G}$, and the second term is the log marginal likelihood of $\mathcal{G}$. The first term is the structure prior, which is different from the parameter prior in the BIC score. Assuming that the local structure for each node is independent of each other, the log posterior probability, often referred to as the Bayesian score, is also decomposable:

$$\begin{aligned} S_{Bay}(\mathcal{G}) &= \sum_{n=1}^{N} \log p(\mathcal{G}(X_n)) + \sum_{n=1}^{N} \log p(\mathcal{D}|\mathcal{G}(X_n)) \\ &= \sum_{n=1}^{N} S_{Bay}(\mathcal{G}(X_n)), \end{aligned} \tag{3.101}$$

where

$$S_{Bay}(\mathcal{G}(X_n)) = \log p(\mathcal{G}(X_n)) + \log p(\mathcal{D}|\mathcal{G}(X_n)).$$

The first term $p(\mathcal{G}(X_n))$ allows us to impose some prior knowledge on the parents for each node. The second term is the marginal log-likelihood of $\mathcal{G}(X_n)$, which can be further decomposed into a joint log-likelihood and a penalty term such as the BIC score, yielding the Bayesian BIC score

$$\begin{aligned} S_{BayBIC}(\mathcal{G}(X_n)) &= \log p(\mathcal{G}(X_n)) + \log p(\mathcal{D}|\mathcal{G}(X_n), \hat{\theta}_n) \\ &\quad - \frac{\log M}{2} d(\mathcal{G}(X_n)). \end{aligned} \tag{3.102}$$

Eq. (3.102) shows that the Bayesian BIC score is still decomposable. This allows structure learning to be done separately for each node subject to the DAG constraint:

$$\mathcal{G}^*(X_n) = \underset{\mathcal{G}(X_n)}{\operatorname{argmax}}\, S_{BayBIC}(\mathcal{G}(X_n)).$$

With the Bayesian BIC score, besides imposing a prior on the parameters by the BIC score, we can also add constraints on the local structure $\mathcal{G}(X_n)$ for each node. For example, we can restrict the presence or absence of certain links, the maximum number of parents for each node, or a generic sparseness prior. We can also use a manually constructed $\mathcal{G}_0$ and employ $p(\mathcal{G}|\mathcal{G}_0)$ as a structure prior; $p(\mathcal{G}|\mathcal{G}_0)$ can be quantified according to some measure of deviation between $\mathcal{G}$ and $\mathcal{G}_0$. Heckerman et al. [51] suggested one reasonable measure of deviation.

Given a score function, BN structure learning is then formulated as a combinatorial search problem, whereby we exhaustively search the BN structure space to identify a DAG that yields the highest score subject to the DAG constraint. However, this combinatorial is computationally intractable due to the exponentially large structure space. For a graph with N nodes, the number of possible directed graphs is $3^{\frac{N\times(N-1)}{2}}$. For example, the total number of directed graphs is 3^{45} for a graph with 10 nodes. As a result, the structure learning of a BN is NP hard due to the huge space of all DAGs. If we have some prior knowledge of the DAG, efficient exact solutions exist. For example, BN structure learning can be done efficiently if the DAG has a tree structure [52], or if the topological order of the variables is known [48], or if the in-degree (the number of parents) is limited. For general structures, exact methods typically handle relatively small networks, such as the branch and bound method [53], which eliminates many impossible structures by the properties of score functions, and the integer programming method [54], which takes advantage of the well-developed IP solvers. The latest effort [55] converts the combinatorial search problem to a continuous optimization problem by transforming the combinatorial DAG constraint into a continuous differentiable constraint.

Various approximation methods have been introduced, including the greedy approaches via heuristic search. One example is the hill-climbing algorithm [56], which explores the BN structure locally by adding, deleting, and reversing link directions for each

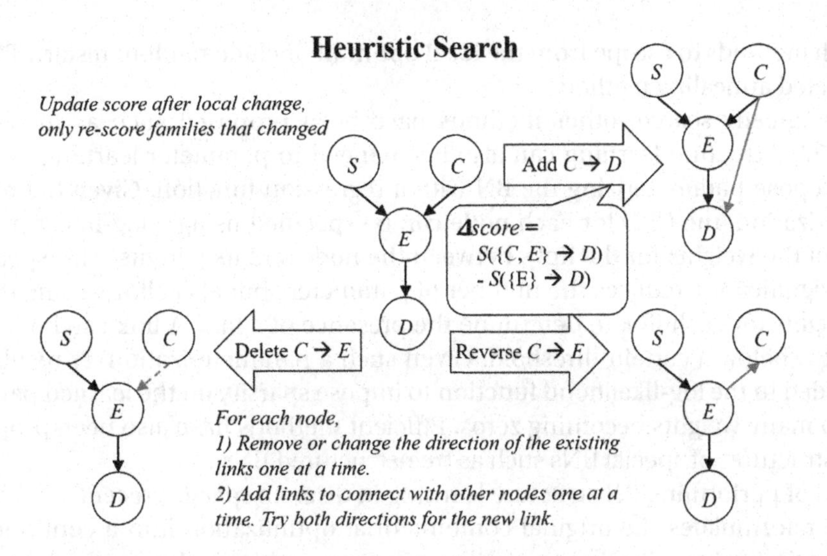

FIGURE 3.30 An example of the structural search for the hill-climbing method. Figure courtesy of [1].

node until convergence. As shown in Fig. 3.30, the hill-climbing approach starts with an initial network, which can be randomly or manually initialized or fully connected to a learned best tree. Following a certain topological order, for each node, the method performs a local search by systematically adding, removing, or changing directions of the existing links and then computes the change of score. For each node, it identifies the change that yields the maximum improvement of the score function subject to the acyclicity constraint.[5] This process is repeated until no further local change can improve the score function. Algorithm 3.7 provides the pseudocode for the hill-climbing method.

Algorithm 3.7 The hill-climbing algorithm.

Construct an initial BN $\mathcal{G}^0$ manually or randomly
Order nodes $X_1, X_2, \ldots, X_N$
repeat
  for n=1 to N **do** //explore each node, following the order
    Add, remove, or change direction of a link of X_n to maximize the BIC score
  end for
until no further changes on BIC score

This hill-climbing approach works well in practice. But as a greedy approach, it is nonoptimal and can easily become stuck in the local optima. It heavily depends on the structural initialization. One solution to this problem is network perturbation, whereby the network structure is randomly perturbed, and the process is then repeated. Other heuris-

[5] One way to impose the DAG constraint is to employ the Notear's constraint [97], which imposes the constraint on the weighted adjacent matrix that encodes the BN structure.

tics search methods to escape from the local optimum include random restart, TABU, and the simulated annealing method.

Besides greedy search, other methods have been proposed such as the A* search method [57]. Structure learning can also be changed to parameter learning. Schmidt et al. [58] propose parameterizing the BN with a regression function. Given the regression parameterization, the CPD for each node can be specified using a log-linear model as a function of the weights for the links between the node and its parents. This specification not only significantly reduces the number of parameters but also allows using the values of the weights for each link to determine the presence of a link. A link can be removed if its weight is below a certain threshold. Given such a parameterization, ℓ_1 regularization can be added to the log-likelihood function to impose sparsity on the learned parameters, leading to many weights becoming zeros. Efficient methods have also been proposed for learning structures of special BNs such as tree-structured BNs.

Instead of performing BN structure learning in discrete space, a recent work by Zheng et al. [97] reformulates the original combinatorial optimization into a continuous constrained optimization, allowing to employ non-linear optimization methods to perform BN structure learning. While avoiding the combinatorial search, the method encounters difficulties associated with non-linear non-convex optimization. In addition, the method is slow in convergence and requires a thresholding operation to produce the final structure.

3.4.2.1.2 Independence-test-based approach

Besides the score-based approach to BN structure learning, another option is the independence-test-based approach [59]. Also called constraint-based learning, independence-test-based algorithms discover the independencies among variables and use these independencies to build the network.

The necessary and sufficient condition for the presence of a link between variables X_i and X_j is

$$X_i \not\perp X_j | S, \forall S \subset \mathbf{X} \setminus X_i \setminus X_j \tag{3.103}$$

The necessary and sufficient condition for the absence of a link between variables X_i and X_j is

$$X_i \perp X_j | S, \exists S \subset \mathbf{X} \setminus X_i \setminus X_j \tag{3.104}$$

To reduce computational complexity, the independence-test-based approach may employ the local-to-global approach, which consists of three steps. First, it identifies the Markov blanket (MB) for each node through independence tests with mutual information. Second, it constructs the skeleton of the global BN by connecting the MBs for each node. Third, it determines the link directions. For the third step, the approach first identifies the V-structures and sets the directions of their links according to the V-structure. For the remaining links, certain heuristics are often employed to set their directions subject to the DAG constraint. See [60] for recent work on constraint-based BN structure learning.

Note that dependencies among variables are always assumed by default, that is, variables are assumed to be dependent unless proven to be independent. Reliably proving independence requires adequate data in terms of both quantity and quality. Inadequate data may fail to prove independence and hence lead to false dependencies and complex networks. Methods such as Bayesian estimation of the mutual information or Bayesian independence testing with the Bayes factor can lead to more robust and accurate independence tests, in particular, when training data are insufficient. Furthermore, the independence-test-based approach is also NP-hard since the number of independence tests that must be performed is exponential to the number of variables.

3.4.2.2 Gaussian BN structure learning

For a Gaussian BN, its structure learning can be done jointly with parameter learning. For a Gaussian BN defined over the variables $\mathbf{X}$, we can follow Eq. (3.182) in Appendix 3.9.2 to derive the joint covariance matrix $\mathbf{\Sigma}$. We can then define the information (precision) matrix $\mathbf{J} = \mathbf{\Sigma}^{-1}$. It can be shown that $p(\mathbf{x}) \propto \exp[\frac{1}{2}\mathbf{x}^\top \mathbf{J}\mathbf{x} + (\mathbf{J}\boldsymbol{\mu})^\top \boldsymbol{\mu}]$ (see Theorem (7.2) of [7]). The matrix $\mathbf{J}$ can be used to decide the presence or absence of a link in a Gaussian BN since $\mathbf{J}_{ij}$ measures the strength of the link between nodes X_i and X_j. It can be proven (see Theorem 7.2 of [7]) that $\mathbf{J}_{ij} = 0$ if and only if X_i and X_j are independent of each other, given $\mathbf{X} \setminus X_i \setminus X_j$. Hence $\mathbf{J}_{ij} = 0$ means that there is no link between nodes X_i and X_j, while a nonzero $\mathbf{J}_{ij}$ means the presence of a link between nodes X_i and X_j. To determine the directions of the links, the variables in $\mathbf{X}$ are often ordered in a list such that the parents of a variable can only be those variables that appear earlier in the list. Determining the optimal variable ordering can be computationally intractable, as the ordering space can be huge. Details on learning Gaussian BNs can be found in [61,49]. Various methods have been proposed to learn sparse Gaussian BNs. Huang et al. [62] proposed imposing an ℓ_1 constraint on entries of $\mathbf{J}$ during parameter learning to learn a sparse information matrix $\mathbf{J}$. In addition, they also introduced a recursive procedure to efficiently explore the variable ordering space subject to the DAG constraint to produce the optimal variable ordering.

3.5 BN learning under incomplete data

In the last sections, we dealt with the cases, in which all variables in a BN are observable during training. However, in many real-world applications the training data may be incomplete. Two sources contribute to the incompleteness of training data: data missing at random and data always (or deliberately) missing (e.g., latent variables). In the first case, some values for certain variables may be missing at random, leading to incomplete training samples. In the second case, because of the presence of latent variables, some variables in the BN are always missing. Both cases often happen in CV applications, where measurements for certain variables may be missing due to either poor or unavailable measurements as a result of occlusion or illumination change (e.g., at night). For the second case, many latent variable models have been introduced such as the latent SVM and variants of the deep models to capture the latent representation of the data. The latent variables in these models will never have any observations during both training and testing. In either case, it is necessary to learn the BN under incomplete data. In this section, we first discuss the methods for BN parameter learning under incomplete data. This is then followed by BN structure learning under incomplete data.

3.5.1 Parameter learning

Given the incomplete data, each training sample is decomposed into two parts, the visible part and the invisible part, that is, $D_m = \mathbf{X}^m = (\mathbf{Y}^m, \mathbf{Z}^m)$, $m = 1, 2, \ldots, M$, $\mathbf{Y}^m \subset \mathbf{X}$ is a subset of fully observed variables for the mth training sample, whereas $\mathbf{Z}^m \subset \mathbf{X}$ represents a subset of variables with missing values for the same training sample. For the case of missing at random, $\mathbf{Z}^m$ varies from sample to sample, whereas for the case of always missing, $\mathbf{Z}^m$ is missing for every sample. Like under complete data, BN parameter learning under incomplete data can also be accomplished with either the maximum likelihood or the MAP method.

3.5.1.1 Maximum likelihood estimation

For ML estimation, we need to find the BN parameters $\boldsymbol{\Theta} = \{\Theta_n\}_{n=1}^N$ by maximizing the marginal likelihood:

$$\begin{aligned} \boldsymbol{\theta}^* &= \underset{\boldsymbol{\theta}}{\operatorname{argmax}} \log p(\mathbf{y}|\boldsymbol{\theta}) \\ &= \underset{\boldsymbol{\theta}}{\operatorname{argmax}} \sum_{m=1}^{M} \log \sum_{\mathbf{z}^m} p(\mathbf{y}^m, \mathbf{z}^m|\boldsymbol{\theta}), \end{aligned} \tag{3.105}$$

where $\mathbf{y} = \{\mathbf{y}^m\}_{m=1}^M$ and $\mathbf{z}$ are assumed to be discrete. The main challenge with Eq. (3.105) is that the marginal log-likelihood is no longer decomposable because of the log sum term. Hence, we can no longer independently estimate the parameters for each node. Furthermore, unlike the joint likelihood function, which is typically concave, the marginal likelihood function is no longer concave. The number of local maximums depends on the number of missing variables. As a result, parameter learning under incomplete data becomes significantly more complex. There are typically two methods for solving Eq. (3.105): the direct method via the gradient ascent and the expectation maximization (EM) approach. In addition, general heuristic optimization methods such as the concave–convex procedure (CCCP) [63] can also be employed to solve nonconvex optimization problems.

3.5.1.1.1 Direct method

By directly maximizing the log marginal likelihood the direct method solves the maximization problem iteratively using one of the gradient ascent methods,

$$\boldsymbol{\theta}^t = \boldsymbol{\theta}^{t-1} + \eta \nabla \boldsymbol{\theta}, \tag{3.106}$$

where η is the learning rate, and the gradient of the parameters can be computed as follows:

$$\begin{aligned} \nabla\boldsymbol{\theta} &= \frac{\partial \sum_{m=1}^{M} \log \sum_{\mathbf{z}^m} p(\mathbf{y}^m, \mathbf{z}^m|\boldsymbol{\theta})}{\partial \boldsymbol{\theta}} \\ &= \sum_{m=1}^{M} \frac{\partial \log \sum_{\mathbf{z}^m} p(\mathbf{y}^m, \mathbf{z}^m|\boldsymbol{\theta})}{\partial \boldsymbol{\theta}} \end{aligned}$$

$$= \sum_{m=1}^{M} \sum_{\mathbf{z}^m} p(\mathbf{z}^m|\mathbf{y}^m, \boldsymbol{\theta}) \frac{\partial \log p(\mathbf{y}^m, \mathbf{z}^m|\boldsymbol{\theta})}{\partial \boldsymbol{\theta}}$$

$$= \sum_{m=1}^{M} E_{\mathbf{z}^m \sim p(\mathbf{z}^m|\mathbf{y}^m, \boldsymbol{\theta})}(\frac{\partial \log p(\mathbf{x}^m, \boldsymbol{\theta})}{\partial \boldsymbol{\theta}}). \quad (3.107)$$

Eq. (3.107) shows that the gradient of the parameters can be expressed as the sum of the expected log-likelihood gradient for all training samples. When the number of configurations of $\mathbf{z}^m$ is large due to a large number of missing or latent variables to marginalize over, it is not possible to enumerate all the configurations of $\mathbf{z}^m$ to exactly compute the expected gradient. Approximation can be done by sampling $p(\mathbf{z}^m|\mathbf{y}^m, \boldsymbol{\theta})$ to obtain samples $\mathbf{z}^s$, $s = 1, 2, \ldots, S$, and the sample average can then be used to approximate the expected gradient. Given $\mathbf{z}^s$, $\nabla\boldsymbol{\theta}$ can be approximately computed as follows:

$$\nabla\boldsymbol{\theta} = \sum_{m=1}^{M} \frac{1}{S} \sum_{s=1}^{S} \frac{\partial \log p(\mathbf{y}^m, \mathbf{z}^s, \boldsymbol{\theta})}{\partial \boldsymbol{\theta}}. \quad (3.108)$$

Furthermore, when the training data size M is very large, summing over all training samples becomes computationally expensive. In this case, the stochastic gradient method may be used.

The joint gradient in Eq. (3.107) can be performed individually for the parameters of each node $\boldsymbol{\theta}_n$ by rewriting Eq. (3.105) as a function of $\boldsymbol{\theta}_n$:

$$\begin{aligned} \log p(\mathbf{y}|\boldsymbol{\theta}) &= \sum_{m=1}^{M} \log \sum_{Z^m} p(\mathbf{y}^m, \mathbf{z}^m|\boldsymbol{\theta}) \\ &= \sum_{m=1}^{M} \log \sum_{Z^m} p(\mathbf{x}^m|\boldsymbol{\theta}) \\ &= \sum_{m=1}^{M} \log \sum_{\mathbf{z}^m} \prod_{n=1}^{N} p(x_n^m|\pi(x_n^m), \boldsymbol{\theta}_n), \end{aligned} \quad (3.109)$$

where $x_n^m \in \{\mathbf{y}^m, \mathbf{z}^m\}$. For discrete $x_n \in \{1, 2, \ldots, K_n\}$, we can further write Eq. (3.109) in terms of θ_{njk}:

$$\begin{aligned} \log p(\mathbf{y}|\boldsymbol{\theta}) &= \sum_{m=1}^{M} \log \sum_{\mathbf{z}^m} \prod_{n=1}^{N} p(x_n^m|\pi(x_n^m), \boldsymbol{\theta}_n) \\ &= \sum_{m=1}^{M} \log \sum_{\mathbf{z}^m} \prod_{n=1}^{N} \prod_{j=1}^{J_n} \prod_{k=1}^{K_n} [p(x_n^m = k|\pi((x_n^m) = j)]^{I(x_n^m = k \& \pi((x_n^m) = j))} \\ &= \sum_{m=1}^{M} \log \sum_{\mathbf{z}^m} \prod_{n=1}^{N} \prod_{j=1}^{J_n} \prod_{k=1}^{K_n} \theta_{njk}^{I(x_n^m = k \& \pi((x_n^m) = j))}. \end{aligned} \quad (3.110)$$

Given Eq. (3.110), the gradient of $\boldsymbol{\theta}_{njk}$ can be computed as

$$\begin{aligned} \nabla\theta_{njk} &= \frac{\partial \log p(\mathbf{y}|\boldsymbol{\theta})}{\partial\theta_{njk}} \\ &= \sum_{m=1}^{M} \frac{\sum_{\mathbf{z}^m} \prod_{n'=1,n'\neq n}^{N} \prod_{j=1}^{J_n} \prod_{k=1}^{K_n} \theta_{n'jk}^{I(x_n^m=k \& \pi((x_n^m)=j))}}{\sum_{\mathbf{z}^m} \prod_{n=1}^{N} \prod_{j=1}^{J_n} \prod_{k=1}^{K} \theta_{njk}^{I(x_n^m=k \& \pi((x_n^m)=j))}}, \end{aligned} \tag{3.111}$$

and θ_{njk} can be updated as

$$\theta_{njk}^{t} = \theta_{njk}^{t-1} + \eta\nabla\theta_{njk}. \tag{3.112}$$

It is clear that both the numerator and the denominator of Eq. (3.111) involve the values of the parameters for other nodes. This means that we cannot estimate θ_{njk} independently. In addition, we must ensure that θ_{njk} is a probability number between 0 and 1. This can be accomplished via reparameterization $\theta_{njk} = \sigma(\alpha_{njk})$. Finally, we need normalize the estimated θ_{njk} in each iteration to ensure that $\sum_{k=1}^{K} \theta_{njk} = 1$.

3.5.1.1.2 Expectation maximization method

As an alternative to the direct method, the expectation maximization (EM) procedure [64] is a widely used approach for parameter estimation under incomplete data. Instead of directly maximizing the marginal log-likelihood, the EM method maximizes the expected log-likelihood, as shown below. The expected log-likelihood is the lower bound of the marginal log-likelihood, which is a concave function. The marginal log-likelihood can be rewritten as

$$\begin{aligned} \log p(\mathcal{D}|\boldsymbol{\theta}) &= \sum_{m=1}^{M} \log p(\mathbf{y}^m|\boldsymbol{\theta}) \\ &= \sum_{m=1}^{M} \log \sum_{\mathbf{z}^m} p(\mathbf{y}^m, \mathbf{z}^m|\boldsymbol{\theta}) \\ &= \sum_{m=1}^{M} \log \sum_{\mathbf{z}^m} q(\mathbf{z}^m|\mathbf{y}^m, \boldsymbol{\theta}_q) \frac{p(\mathbf{y}^m, \mathbf{z}^m|\boldsymbol{\theta})}{q(\mathbf{z}^m|\mathbf{y}^m, \boldsymbol{\theta}_q)}, \end{aligned} \tag{3.113}$$

where $q(\mathbf{z}^m|\mathbf{y}^m, \boldsymbol{\theta}_q)$ is an arbitrary density function over $\mathbf{z}$ with parameters $\boldsymbol{\theta}_q$. Jensen's inequality states that for a concave function, the function of the mean is greater than or equal to the mean of the function, namely,

$$f(E(x)) \geq E(f(x)) \tag{3.114}$$

Applying Jensen's inequality to Eq. (3.113) and using the concavity of the log function yield

$$\log \sum_{\mathbf{z}^m} q(\mathbf{z}^m|\mathbf{y}^m, \boldsymbol{\theta}_q) \frac{p(\mathbf{y}^m, \mathbf{z}^m|\boldsymbol{\theta})}{q(\mathbf{z}^m|\mathbf{y}^m, \boldsymbol{\theta}_q)} \geq \sum_{\mathbf{z}^m} q(\mathbf{z}^m|\mathbf{y}^m, \boldsymbol{\theta}_q) \log \frac{p(\mathbf{y}^m, \mathbf{z}^m|\boldsymbol{\theta})}{q(\mathbf{z}^m|\mathbf{y}^m, \boldsymbol{\theta}_q)}. \tag{3.115}$$

Instead of directly maximizing the marginal log-likelihood on the left-hand side of Eq. (3.115), the EM method maximizes its lower bound, that is, the right-hand side of Eq. (3.115):

$$\boldsymbol{\theta}^* = \underset{\boldsymbol{\theta}}{\operatorname{argmax}} \sum_{m=1}^{M} \sum_{\mathbf{z}^m} q(\mathbf{z}^m|\mathbf{y}^m, \boldsymbol{\theta}_q) \log p(\mathbf{y}^m, \mathbf{z}^m|\boldsymbol{\theta}). \tag{3.116}$$

Note that the term $q(\mathbf{z}^m|\mathbf{y}^m, \boldsymbol{\theta}_q) \log q(\mathbf{z}^m|\mathbf{y}^m, \boldsymbol{\theta}_q)$ (the entropy of $\mathbf{z}^m$) is dropped from Eq. (3.116) since $q(\mathbf{z}^m|\mathbf{y}^m, \boldsymbol{\theta}_q)$ is often chosen to be independent of current $\boldsymbol{\theta}$. In fact, for the EM algorithm, $\boldsymbol{\theta}_q$ is chosen to be $\boldsymbol{\theta}^{t-1}$, that is, the parameters estimated in the last iteration. Hence $q(\mathbf{z}^m|\mathbf{y}^m, \boldsymbol{\theta}_q) = p(\mathbf{z}^m|\mathbf{y}^m, \boldsymbol{\theta}^{t-1})$. This selection of q has been shown to be a tight lower bound of p. Given the function q, maximizing Eq. (3.116) is often iteratively done in two steps, E-step and M-step. The E-step estimates the lower bound function Q based on $\boldsymbol{\theta}^{t-1}$.

E-step:

$$Q^t(\boldsymbol{\theta}^t|\boldsymbol{\theta}^{t-1}) = \sum_{m=1}^{M} \sum_{\mathbf{z}^m} p(\mathbf{z}^m|\mathbf{y}^m, \boldsymbol{\theta}^{t-1}) \log p(\mathbf{y}^m, \mathbf{z}^m|\boldsymbol{\theta}^t). \tag{3.117}$$

The main computation for the E-step is computing the weight $p(\mathbf{z}^m|\mathbf{y}^m, \boldsymbol{\theta}^{t-1})$ for each possible configuration of $\mathbf{z}^m$. Given $p(\mathbf{z}^m|\mathbf{y}^m, \boldsymbol{\theta}^{t-1})$, the M-step maximizes Q^t to obtain an estimate of $\boldsymbol{\theta}$.

M-step:

$$\boldsymbol{\theta}^* = \underset{\boldsymbol{\theta}^t}{\operatorname{argmax}}\, Q^t(\boldsymbol{\theta}^t|\boldsymbol{\theta}^{t-1}). \tag{3.118}$$

This function remains decomposable, and the parameters for each node can be separately estimated. The maximum likelihood methods we introduce in Section 3.4.1.1 for learning BN parameters under complete data can be applied here.

Specifically, given the weights computed in the E-step, the maximization in the M-step is decomposable, so that parameters for each node $\boldsymbol{\theta}_n$ can be estimated individually. If all variables are discrete, then the M-step produces

$$\theta^t_{njk} = \frac{M_{njk}}{M_{nj}}, \tag{3.119}$$

where M_{njk} is the total weighted count for node X_n having a value of k, given its jth parent configuration, that is,

$$M_{njk} = \sum_{m=1}^{M} w_{m,c} \times I((\mathbf{y}^m, \mathbf{z}^m) = njk),$$

where $w_{m,c} = p(\mathbf{z}^m = c|\mathbf{y}^m, \boldsymbol{\theta}^{t-1})$, where $\mathbf{z}^m = c$ means $\mathbf{z}^m$ takes it's cth configuration. $I((\mathbf{y}^m, \mathbf{z}^m) = njk)$ is the indicator function that is equals 1 when it's argument is true, i.e., node n takes njk configuration, with $x_n = k$ and its parents taking on jth configuration. See Section 19.2.2.3 of [7] for details.

The E-step and M-step iterate until convergence. The EM method is proven to improve the likelihood estimate at each iteration, but it may converge to a local maximum, depending on the initialization. Algorithm 3.8 provides a pseudocode for the EM algorithm. An alternative pseudocode for the EM algorithm for discrete BN can be found in Algorithm 19.2 of [7].

Algorithm 3.8 The EM algorithm.

▷ $X_1, X_2, \ldots, X_N$ are nodes for a BN, with each node assuming K values
▷ $\mathbf{z}^m$ are the missing variables for the mth sample for $m = 1, 2, \ldots, M$.
$w_{m,l_m} = 1, l_m{=}1, 2, \ldots, K^{|\mathbf{z}^m|}$ // initialize the weights for each sample
$\boldsymbol{\theta} = \boldsymbol{\theta}_0$, //initialize the parameters for each node
t=0
while not converging **do**
  E-step:
  for m=1 to M **do**
    if $\mathbf{x}^m$ contains missing variables $\mathbf{z}^m$ **then**
      for l_m=1 to $K^{|\mathbf{z}^m|}$ **do**
        $w_{m,l_m} = p(\mathbf{z}^m_{l_m}|\mathbf{y}^m, \boldsymbol{\theta}^t)$ //$\mathbf{z}^m_{l_m}$ is the l_mth configuration of $\mathbf{z}^m$
      end for
    end if
  end for
  M-step:
  $\boldsymbol{\theta}^t = \text{argmax}_{\boldsymbol{\theta}} \sum_{m=1}^{M} \sum_{l_m=1}^{K^{|z^m|}} w_{m,l_m} \log p(\mathbf{y}^m, \mathbf{z}^m_{l_m}|\boldsymbol{\theta})$ using Eq. (3.119)
  t=t+1
end while

Besides the traditional EM, there are other variants of EM, including the hard EM, the Monte Carlo EM, and the variational EM. The hard EM represents an approximation to the traditional EM. Instead of providing soft estimates (weights) for the missing values in the E-step, the hard EM obtains the MAP estimate of the these values. Given the completed data, the M-step can be carried out like the traditional maximum likelihood method. Specifically, the E-step and M-step of the hard EM can be stated as follows.

E-step:

$$\hat{\mathbf{z}}^m = \underset{\mathbf{z}^m}{\text{argmax}}\, p(\mathbf{z}^m|\mathbf{y}^m, \boldsymbol{\theta}^{t-1}). \tag{3.120}$$

M-step:

$$\boldsymbol{\theta}^* = \underset{\boldsymbol{\theta}}{\text{argmax}} \sum_{m=1}^{M} \log p(\mathbf{y}_m, \hat{\mathbf{z}}^m|\boldsymbol{\theta}). \tag{3.121}$$

It is easy to prove that $\sum_{m=1}^{M} \log p(\mathbf{y}_m, \hat{\mathbf{z}}^m|\boldsymbol{\theta})$ is a lower bound of the expected log-likelihood. Instead of maximizing the expected log-likelihood, the hard EM maximizes its lower bound. One of the most widely used hard EMs is the famous K-means algorithm for data clustering.

The Monte Carlo EM [98] was developed to approximate the summation over $\mathbf{z}^m$ in the E-step in Eq. (3.117). When the number of incomplete variables in $\mathbf{z}^m$ is large as in a latent deep model, a brute force computing of the expected gradient by marginalization over $\mathbf{z}^m$ becomes intractable. This can be approximated by obtaining samples of $\mathbf{z}^s$ from $p(\mathbf{z}|\mathbf{y}^m, \boldsymbol{\theta}^{t-1})$ and using the sample average to replace the mean. Like Eq. (3.108), which uses sample average to approximate the expected parameter gradient, the expected log-likelihood $\log p(\mathbf{y}^m, \mathbf{z}^s)$ over the S in the E-step can be approximated by its sample average. Algorithm 3.9 provides the pseudocode for the Monte Carlo EM.

Algorithm 3.9 The Monte Carlo EM algorithm.

Initialize $\boldsymbol{\theta}$ to $\boldsymbol{\theta}^0$ //get an initial value for $\boldsymbol{\theta}$
t=0
while not converging **do**
  for m=1 to M **do** //go through each sample
    Obtain S samples of $\mathbf{z}_s^{m,t}$ from $p(\mathbf{z}|\mathbf{y}^m, \boldsymbol{\theta}^t)$
  end for
  $\boldsymbol{\theta}^{t+1} = \text{argmax}_{\boldsymbol{\theta}^t} \sum_{m=1}^{M} \frac{1}{S} \sum_{s=1}^{S} \log p(\mathbf{y}^m, \mathbf{z}_s^{m,t}|\boldsymbol{\theta}^t)$ // update $\boldsymbol{\theta}^t$
  t=t+1
end while

The variational EM was developed to handle the challenge of computing $p(\mathbf{z}^m|\mathbf{y}^m, \boldsymbol{\theta}^{t-1})$ since for some variables, this probability may not factorize over $\mathbf{z}^m$ and its computation can be intractable. To overcome this challenge, the variational EM designs a function $q(\mathbf{z}|\boldsymbol{\alpha})$ that factorizes over Z^m and is closest to $p(\mathbf{z}^m|\mathbf{y}^m, \boldsymbol{\theta}^{t-1})$. The simplest variational method is the mean field method, which has a function $q(\mathbf{z}|\boldsymbol{\alpha})$ that is fully factorized over $\mathbf{z}$ and whose parameters $\boldsymbol{\alpha}$ can be found by minimizing the KL divergence between the function $q(\mathbf{z}|\boldsymbol{\alpha})$ and $p(\mathbf{z}^m|\mathbf{y}^m, \boldsymbol{\theta}^{t-1})$. Given $q(\mathbf{z}|\boldsymbol{\alpha})$, we can use it to compute $p(\mathbf{z}^m|\mathbf{y}^m, \boldsymbol{\theta}^{t-1})$. Section 3.3.2.3 further discusses the variational methods.

Compared with the direct approach, the EM approach need not compute the gradient of the likelihood. Its M-step optimization is convex and admits a closed-form solution. It generally produces better estimates than the direct approach, but like the direct approach, it heavily depends on the initialization and can only find a local maximum. Furthermore, it only maximizes the lower bound of the marginal likelihood instead of the marginal likelihood itself. Its solution can hence be worse than that of the direct approach, depending on the gap between the marginal log likelihood and its lower bound.

Both the direct and the EM approaches require initialization of the parameters. Initializations can greatly affect the results, in particular, when the number of variables with missing values is large. The parameter initialization can be done in several ways: random initialization, manual initialization, or initialization based on the complete portion of the data. For random initialization, the learned parameters may vary in values and performance. One possible solution is performing multiple random initializations and selecting the one that yields the best parameters in terms of their performance. Manual initialization can be done based on any prior knowledge about the parameters. Although such an initialization may lead to consistent parameters, it tends to inject bias into the parameter estimation. Finally, the parameter initialization based on the complete portion of the data can produce good and consistent results, but the method assumes that there are sufficient complete data.

Further discussion on the differences between the direct method and the EM approach can be found in Section 19.2.3 of [7]. Table 3.2 summarizes different methods for BN parameter learning under incomplete data.

Table 3.2 Methods for BN parameter learning under incomplete data.

Specific methods	Objective functions
Direct method	Marginal log-likelihood
EM	Expected log-likelihood
Variational EM	Approximated expected log-likelihood
Monte Carlo EM	Approximated expected log-likelihood

3.5.1.2 Maximum Posterior Probability parameter estimation

Like the case with complete data, MAP estimation can also be applied to BN parameter learning under incomplete data. Instead of learning the BN parameters from only the data, MAP parameter estimation exploits the prior distribution of the parameters. Given the training data $\mathcal{D} = \{\mathbf{y}^m, \mathbf{z}^m\}$, $m = 1, 2, \ldots, M$, where $\mathbf{y} = \{\mathbf{y}^m\}$ is visible and $\mathbf{z} = \{\mathbf{z}^m\}$ is not observable, the goal is finding $\boldsymbol{\theta}$ by maximizing the log posterior probability of $\boldsymbol{\theta}$:

$$\boldsymbol{\theta}^* = \underset{\boldsymbol{\theta}}{\operatorname{argmax}} \log p(\boldsymbol{\theta}|\mathcal{D}), \tag{3.122}$$

where $p(\boldsymbol{\theta}|\mathcal{D})$ can be written as

$$\log p(\boldsymbol{\theta}|\mathcal{D}) = \log p(\boldsymbol{\theta}) + \log p(\mathcal{D}|\boldsymbol{\theta}) - p(\mathcal{D}), \tag{3.123}$$

where the first term is the log prior probability of $\boldsymbol{\theta}$, the second term is the log marginal likelihood of $\boldsymbol{\theta}$, and the third term $p(\mathcal{D})$ is a constant. As in the case with complete data, the prior probability can be chosen to be conjugate to the likelihood function. Given the objective function in Eq. (3.122), we can employ a gradient-based approach as for the maximum likelihood estimation to iteratively obtain an estimate of $\boldsymbol{\theta}$. The EM approach can also be extended to this case. Details can be found in Section 19.3 of [7].

One approximate implementation of MAP learning under incomplete data is MAP learning via sampling. Given a prior of the $\boldsymbol{\theta}$, we first obtain parameter samples from $p(\boldsymbol{\theta})$.

We can then weigh the samples by their likelihood given the observed data **y**. We can repeat this to obtain many samples. The final parameter is estimated by the weighted average of all parameter samples. A direct sampling of the prior may be inefficient, in particular, when the parameter space is large. Gibbs and collapsed Gibbs sampling methods are introduced in Section 19.3.2 of [7] to perform efficient sampling of the prior. Algorithm 3.10 provides the pseudocode for BN parameter learning under incomplete data with sampling. One potential problem with the sampling-based approach is that the parameter samples are actually generated from the joint probability distribution $p(\boldsymbol{\theta}, \mathbf{y})$ of the parameters and the available variables as shown in Algorithm 3.10, whereas strict MAP estimation of the parameters requires samples from the marginal posterior probability $p(\boldsymbol{\theta}|\mathbf{y})$ of the parameters, which may be difficult to obtain from the samples generated by the joint distribution.

Algorithm 3.10 MAP parameter learning under incomplete data with Gibbs sampling.

```
▷ Given y = {y^1, y^2, ..., y^M}
t=0
while t<T do
  Sample θ^t from p(θ) //use a Gibbs sampling method
  if t > T_0 then //burn-in threshold
    Collect θ^t
    Compute w^t = p(y|θ^t) //Compute the weight for the sample
  end if
  t=t+1
end while
θ̂ = (Σ_{t=T_0+1}^{T} w^t θ^t) / (T − T_0) and normalize
```

In addition to sampling methods, variational Bayesian learning method is also introduced in Section 19.3.3 of [7] for MAP parameter learning under incomplete data.

3.5.2 Structure learning

For BN structure learning under incomplete data, the problem is much more challenging. Given $\mathcal{D} = \{D_1, D_2, .., D_M\}$, where $D_m = \{\mathbf{y}^m, \mathbf{z}^m\}$, instead of maximizing the marginal likelihood of $\mathcal{G}$, we maximize its marginal log-likelihood:

$$\begin{aligned} \mathcal{G}^* &= \underset{\mathcal{G}}{\text{argmax}} \log p(\mathbf{y}|\mathcal{G}) \\ &= \underset{\mathcal{G}}{\text{argmax}} \log \sum_{\mathbf{z}} p(\mathbf{y}, \mathbf{z}|\mathcal{G}) \\ &= \underset{\mathcal{G}}{\text{argmax}} \log \sum_{\mathbf{z}} p(\mathcal{D}|\mathcal{G}), \end{aligned} \tag{3.124}$$

where $\mathbf{y} = \{\mathbf{y}^m\}_{m=1}^M$ and $\mathbf{z} = \{\mathbf{z}^m\}_{m=1}^M$. To maximize Eq. (3.124), we can follow the same strategy as parameter learning under incomplete data, that is, maximizing it directly through gradient ascent or maximizing its lower bound through the EM approach. The latter leads to the well-known structural EM (SEM) approach as discussed below. By introducing the function q as in the parameter EM, the marginal log likelihood in Eq. (3.124) can be rewritten as

$$\begin{aligned}
\log p(\mathbf{y}|\mathcal{G}) &= \log \sum_{\mathbf{z}} q(\mathbf{z}|\mathbf{y},\boldsymbol{\theta}_q) \frac{p(\mathcal{D}|\mathcal{G})}{q(\mathbf{z}|\mathbf{y},\boldsymbol{\theta}_q)} \\
&\geq \sum_{\mathbf{z}} q(\mathbf{z}|\mathbf{y},\boldsymbol{\theta}_q) \log \frac{p(\mathcal{D}|\mathcal{G})}{q(\mathbf{z}|\mathbf{y},\boldsymbol{\theta}_q)} \quad \text{Jensen's inequality} \\
&= \sum_{\mathbf{z}} q(\mathbf{z}|\mathbf{y},\boldsymbol{\theta}_q) \log p(\mathcal{D}|\mathcal{G}) - q(\mathbf{z}|\mathbf{y},\boldsymbol{\theta}_q) \log q(\mathbf{z}|\mathbf{y},\boldsymbol{\theta}_q) \\
&= E_q(\log p(\mathcal{D}|\mathcal{G})) - \sum_{\mathbf{z}} q(\mathbf{z}|\mathbf{y},\boldsymbol{\theta}_q) \log q(\mathbf{z}|\mathbf{y},\boldsymbol{\theta}_q),
\end{aligned} \tag{3.125}$$

where the first term is the expected log marginal likelihood, and the second term represents the entropy of $\mathbf{z}$, independent of $\mathcal{G}$. It is clear from Eq. (3.125) that instead of maximizing the log marginal likelihood, we maximize the expected log marginal likelihood for the structure EM. If we choose to approximate the marginal log likelihood by the BIC score, then Eq. (3.125) computes the expected BIC score. Like the parameter EM, SEM requires an initial structure and parameters. Given the initial guesses, the score-based SEM then iteratively refines the model until convergence. Algorithm 3.11 details the SEM procedure. Given an initial structure, the structural EM then performs the E-step of the parameter EM to compute the weights $w_{m,j} = p(\mathbf{z}_j^m|\mathbf{y}^m, \boldsymbol{\theta}^{t-1})$ for each sample. Given the weights, BN structure learning can then be applied in the M-step by maximizing the expected BIC score. This can be implemented by an exact method or an approximated method such as the hill-climbing method. The E-step and M-step alternate, and they iterate until convergence. Details can be found in Algorithm 3.11 below and in Algorithm 19.3 of [7].

3.6 Manual Bayesian Network specification

Besides automatic BN learning from data, it is also possible to construct a BN manually by exploiting the causal semantics in the BN. Specifically, to construct a BN for a given set of variables, we simply draw directed links from cause variables to their immediate effects, producing a causal model that satisfies the definition of BN. For example, we can manually construct the BN for image segmentation, shown in Fig. 1.1A of Chapter 1, based on the causal relationships between image regions, edges, and vertices. Given the structure, domain knowledge can also be used to specify the conditional probabilities for each node. The manual BN specification requires domain knowledge through knowledge representation. It becomes infeasible for domains lacking domain knowledge, and cannot scale up well to large models.

Algorithm 3.11 The structural EM algorithm.

Initialize BN structure to $\mathcal{G}^0$ and parameters to $\boldsymbol{\theta}^0$
t=0
while not converging **do**
  E-step:
  for m=1 to M **do**
    if $\mathbf{x}^m$ contains missing variables $\mathbf{z}^m$ **then**
      for j=1 to $K^{|\mathbf{z}^m|}$ **do** //K is the cardinality for each variable in $\mathbf{z}^m$
        $w_{m,j} = p(\mathbf{z}_j^m|\mathbf{y}^m, \boldsymbol{\theta}^t)$ //$\mathbf{z}_j^m$ is the jth configuration of $\mathbf{z}^m$
      end for
    end if
  end for
  M-step:
  $E_q(BIC(\mathcal{G})) = \sum_{m=1}^{M}\sum_{j=1}^{K^{|\mathbf{z}^m|}} w_{m,j}[\log p(\mathbf{y}^m, \mathbf{z}_j^m|\boldsymbol{\theta}^t, \mathcal{G}) - \frac{\log M}{2}Dim(\mathcal{G})]$
  $\mathcal{G}^{t+1}, \boldsymbol{\theta}^{t+1} = \text{argmax}_{\mathcal{G}}\, E_q(BIC)(\mathcal{G})$ // find $\mathcal{G}^{t+1}$ to maximize the expected BIC score through a search algorithm such as hill-climbing method
  t=t+1
end while

3.7 Dynamic Bayesian Networks

3.7.1 Introduction

In real-world applications, we often need to model a dynamic process that involves a set of RVs evolving over time. BNs are static by nature; they cannot be used to capture the dynamic relationships among RVs. Hence, dynamic BNs were developed to extend BNs to model dynamic processes. Instead of modeling the continuous evolution of a dynamic process, the dynamic process is discretized over time by sampling at consecutive time points (also called time slices) to obtain $\mathbf{X}^0$, $\mathbf{X}^1$, $\mathbf{X}^2$,..., $\mathbf{X}^T$, where $\mathbf{X}^t$ is a random vector at time t. The goal of dynamic modeling is to capture the joint probability distribution $p(\mathbf{X}^0, \mathbf{X}^1, \mathbf{X}^2, \ldots, \mathbf{X}^T)$ of $\mathbf{X}^0$, $\mathbf{X}^1$, $\mathbf{X}^2$,..., $\mathbf{X}^T$. Following the chain rule, the joint probability distribution can be written as

$$\begin{aligned} & p(\mathbf{X}^0, \mathbf{X}^1, \mathbf{X}^2, \ldots, \mathbf{X}^T) \\ = \; & p(\mathbf{X}^0)p(\mathbf{X}^1|\mathbf{X}^0)p(\mathbf{X}^2|\mathbf{X}^0, \mathbf{X}^1)\ldots p(\mathbf{X}^T|\mathbf{X}^0, \mathbf{X}^1, \ldots, \mathbf{X}^{T-1}). \end{aligned} \tag{3.126}$$

Assuming the first-order temporal Markov property, we can rewrite Eq. (3.126) as

$$p(\mathbf{X}^0, \mathbf{X}^1, \mathbf{X}^2, \ldots, \mathbf{X}^T) = p(\mathbf{X}^0)\prod_{t=1}^{T} p(\mathbf{X}^t|\mathbf{X}^{t-1}), \tag{3.127}$$

where the first term captures the static joint distribution, and the second term the dynamic joint distribution. Naively, we can represent each $\mathbf{X}^t$ by a BN, and the dynamic process can be represented by an extended BN resulting from concatenating BNs for each $\mathbf{X}^t$ over T time slices. However, such a representation is inefficient since it requires the specification of a BN for each $\mathbf{X}^t$.

Assuming a stationary first-order transition, the two-slice dynamic Bayesian network (DBN) was introduced to address this issue. First proposed by Dagum et al. [65], a DBN can be represented by a two-tuple $\mathcal{B} = (\mathcal{G}, \mathbf{\Theta})$, where $\mathcal{G}$ represents the DBN structure, and $\mathbf{\Theta}$ represents its parameters. Topologically, $\mathcal{G}$ is an extended BN consisting of two parts, a prior network $\mathcal{G}^0$ and a transition network $\overrightarrow{\mathcal{G}}$ as shown in Figs. 3.31A and B, respectively. $\mathcal{G}^0$ captures $p(\mathbf{X}^0)$, i.e., the joint distribution of $\mathbf{X}^0$, while $\overrightarrow{\mathcal{G}}$ captures the conditional joint

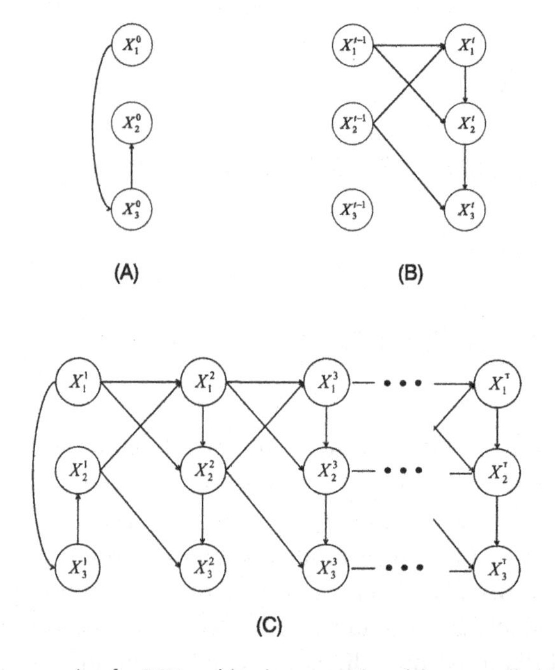

FIGURE 3.31 (A) and (B) An example of a DBN and its decomposition; (C) an unrolled DBN. (A) Prior network $\mathcal{G}^0$. (B) Transition network $\overrightarrow{\mathcal{G}}$. (C) An unrolled DBN with T time slices.

distributions of $\overrightarrow{\mathbf{X}}^t$, i.e., $p(\overrightarrow{\mathbf{X}}^t | \overrightarrow{\mathbf{X}}^{t-1})$. The prior network is simply a BN capturing the joint distribution of static random variables at the beginning of a dynamic process, whereas the transition network is also a BN that captures the temporal transition of dynamic random variables between two consecutive time slices $t-1$ and t. Specifically, the nodes in a DBN represent either static random variables in the prior network that do not vary over time or dynamic random variables in the transition network that do vary over time. We denote the static random variables in the prior network collectively by $\mathbf{X}^0$ and the dynamic random variables in the second layer of the transition network collectively by $\overrightarrow{\mathbf{X}}^t$. Depending on the variables to which they connect, the links capture either spatial dependencies or temporal dependencies. Specifically, as shown in Fig. 3.31A, the links in $\mathcal{G}^0$ among $\mathbf{X}^0$ capture the spatial dependencies among static variables. The links in a transition network can be intraslice or interslice links. Intraslice links connect nodes within the same time slice and capture the spatial relationships among temporal nodes. The interslice links, on the other hand, connect nodes at successive time slices and capture the temporal dependencies among temporal nodes at two different time steps. Note that whereas the intraslice links can appear in both the prior network and the transition network, the interlinks appear only in the transition network. Moreover, as shown in Fig. 3.31B, the intraslice links in the transition network only apply to nodes in the second time slice, and they can be different from those in the prior network. Furthermore, the interlinks in the transition network can only point from the first to the second slice. To construct a DBN for a dynamic process over T time steps, we can unroll the prior and transition network (i.e., by duplicating the transition network for each time step) to the desired number of time slices as shown in Fig. 3.31C.

Similarly, the DBN parameters $\mathbf{\Theta}$ can be decomposed into $\mathbf{\Theta^0}$ and $\overrightarrow{\mathbf{\Theta}}$, respectively representing the parameters for $\mathcal{G}^0$ and $\overrightarrow{\mathcal{G}}$; $\mathbf{\Theta^0}$ stands for the conditional probabilities for all nodes in the prior network, that is, $\mathbf{\Theta^0}$=$\{p(X_n^0|\pi(X_n^0))\}$, where $X_n^0 \in \mathbf{X}^0$, and $\overrightarrow{\mathbf{\Theta}}$ encode the conditional probabilities for the $\overrightarrow{\mathbf{X}^t}$ nodes in $\overrightarrow{\mathcal{G}}$, that is, $\overrightarrow{\mathbf{\Theta}} = \{p(X_n^t)|\pi(X_n^t))\}$, where $X_n^t \in \overrightarrow{\mathbf{X}^t}$. Besides the standard Markov condition assumptions, DBN has two additional assumptions. First, DBN follows the first-order Markov property

$$p(\mathbf{X}^t|\mathbf{X}^{t-1}, \mathbf{X}^{t-2}, \ldots, \mathbf{X}^0) = p(\mathbf{X}^t|\mathbf{X}^{t-1}). \tag{3.128}$$

This means that the nodes at time t only temporally depend on the nodes at the previous time slice, that is, the first-order Markov assumption. Second, DBN assumes that the dynamic transition is locally stationary across two consecutive time slices:

$$p(\mathbf{X}^{t_1}|\mathbf{X}^{t_1-1}) = p(\mathbf{X}^{t_2}|\mathbf{X}^{t_2-1}) \ \forall t_1 \neq t_2. \tag{3.129}$$

Such an assumption produces the stationary DBN. The assumption may not hold for time variant dynamic processes.

Given the aforementioned definitions and assumptions, the joint probability distribution $p(\mathbf{X}^0, \mathbf{X}^1, \mathbf{X}^2, \ldots, \mathbf{X}^T)$ of the RVs over time can be written as the product of CPDs for each node via the DBN chain rule:

$$\begin{aligned} p(\mathbf{X}^0, \mathbf{X}^1, \mathbf{X}^2, \ldots, \mathbf{X}^T) &= p(\mathbf{X}^0|\mathbf{\Theta}^0) \prod_{t=1}^{T} p(\overrightarrow{\mathbf{X}}^t, \overrightarrow{\mathbf{\Theta}}) \\ &= \prod_{n=1}^{N^0} p(X_n^0|\pi(X_n^0), \mathbf{\Theta}^0)) \prod_{t=1}^{T}\prod_{n=1}^{N^t} p(X_n^t|\pi(X_n^t), \overrightarrow{\mathbf{\Theta}}), \end{aligned} \tag{3.130}$$

where N^0 is the number of nodes in the prior network, and N^t is the number of nodes in the second layer of the transition network. The first product term captures the joint probability distribution for the static nodes in the prior network, whereas the second and third product terms capture the joint probability for the temporal nodes in the transition network. Hence, to fully parameterize a DBN, we need specify each individual CPD in both $\mathbf{\Theta}^0$ and $\vec{\mathbf{\Theta}}$.

3.7.2 Learning and inference

In this section, we briefly summarize learning and inference for DBN. Because of the significant overlap with BN learning and inference, we will not discuss DBN learning and inference in detail.

3.7.2.1 DBN learning

The methods we discussed for BN learning, including both the parameter learning and structure learning methods, can also be applied to DBN learning. Unlike BN learning, DBN learning involves learning the parameters and structures for both the prior network $\mathcal{G}^0$ and the transition network $\vec{\mathcal{G}}$. We first discuss methods for their learning under complete data and then under incomplete data.

Under complete data, we can perform DBN parameter learning using either the maximum likelihood method or the MAP estimation method. Let the training data be $\mathbf{D} = \{\mathbf{S}_1, \mathbf{S}_2, ..., \mathbf{S}_M\}$, where $\mathbf{S}_m$ is the m th training sequence of t_m time slices, that is, $\mathbf{S}_m = \{\mathbf{X}^{m,0}, \mathbf{X}^{m,1}, ..., \mathbf{X}^{m,t_m}\}$. The maximum likelihood estimation of the DBN parameters $\mathbf{\Theta}$ can be formulated as

$$\mathbf{\Theta}^* = \underset{\mathbf{\Theta}}{\operatorname{argmax}} \log p(\mathbf{D}|\mathbf{\Theta}), \tag{3.131}$$

where $\log p(\mathbf{D}|\mathbf{\Theta})$ can be further written as after applying DBN chain rule

$$\begin{aligned} \log p(\mathbf{D}|\mathbf{\Theta}) &= \sum_{m=1}^{M} \log p(\mathbf{S}_m|\mathbf{\Theta}) \\ &= \sum_{m=1}^{M}\sum_{n=1}^{N^0} \log p(X_n^{0,m}|\pi(X_n^{0,m}), \mathbf{\Theta}^0) \\ &\quad + \sum_{m=1}^{M}\sum_{t=1}^{t_m}\sum_{n=1}^{N^t} \log p(X_n^{t,m}|\pi(X_n^{t,m}), \vec{\mathbf{\Theta}}). \end{aligned} \tag{3.132}$$

Note that X_n^t represents the nodes in the second layer of the transition network. It is clear from Eq. (3.132) that $\mathbf{\Theta}^0$ and $\vec{\mathbf{\Theta}}$ can be learned separately. To learn them, $\mathbf{D}$ can be decomposed into $\mathbf{D} = \{\mathbf{D}^0, \vec{\mathbf{D}}\}$, $\mathbf{D}^0 = \{\mathbf{X}^{0,m}\}$ for $m = 1, 2, \ldots, M$, and $\vec{\mathbf{D}} = \{\mathbf{X}^{m,t-1}, \mathbf{X}^{m,t}\}$ for $m = 1, 2, \ldots, M$ and $t = 1, 2, \ldots, t_m$, where $S_{m,0}$ represents the data in the first time slice for all sequences, and $\{\mathbf{X}^{m,t-1}, \mathbf{X}^{m,t}\}$ the data from two consecutive time slices for all sequences. Given $\mathbf{D}^0$ and $\vec{\mathbf{D}}$, we can then apply the BN parameter learning techniques in Section 3.4.1 to learn the parameters for the prior network and transition network separately.

For DBN structure learning, following the score-based approaches, we can formulate the DBN structure learning using the BIC score as follows:

$$\mathcal{G}^* = \underset{\mathcal{G}}{\operatorname{argmax}}\, score_{BIC}(\mathcal{G}), \tag{3.133}$$

where

$$\begin{aligned} score_{BIC}(\mathcal{G}) &= \sum_{m=1}^{M} \log p(\mathbf{S}_m|\mathcal{G}, \hat{\boldsymbol{\Theta}}) - \frac{\log M}{2} d(\mathcal{G}) \\ &= \sum_{m=1}^{M}\sum_{n=1}^{N} \log p(X_n^0|\pi(X_n^0), \hat{\boldsymbol{\Theta}}^{\mathbf{0}}, \mathcal{G}^0) \\ &\quad + \sum_{m=1}^{M}\sum_{t=1}^{t_m}\sum_{n=1}^{N} p(X_n^t|\pi(X_n^t), \overrightarrow{\boldsymbol{\Theta}}, \overrightarrow{\mathcal{G}}) \\ &\quad - \frac{\log M}{2} d(\mathcal{G}^0) - \frac{\log M}{2} d(\overrightarrow{\mathcal{G}}) \\ &= score_{BIC}(\mathcal{G}^0) + score_{BIC}(\overrightarrow{\mathcal{G}}). \end{aligned} \tag{3.134}$$

It is clear that the structure for $\mathcal{G}^0$ and $\overrightarrow{\mathcal{G}}$ can be learned separately. Using data $\mathbf{D}^0$ and $\overrightarrow{\mathbf{D}}$, they can be learned using the BN structure learning techniques in Section 3.4.2. To learn the structure of $\overrightarrow{\mathcal{G}}$, we impose the constraints that there are no intraslice links in the first slice of $\overrightarrow{\mathcal{G}}$ and that interslice links between two slices always point from nodes at time $t-1$ to the nodes at time t.

For DBN learning under incomplete data, parameters and structures for prior and transition networks can no longer be learned separately, since they are coupled together as in BN learning under incomplete data; they need be learned together. Following the techniques discussed in Section 3.5.1.1, we can use either the direct or the EM approach for DBN parameter and structure learning under incomplete data. Specifically, for the EM method, the E-step must estimate the expected counts for the prior and transition network jointly. In the M-step however, the parameters and structures for the prior and transition network can be computed separately from the expected counts.

3.7.2.2 DBN inference

Like BN inference, DBN inference includes the posterior probability inference, the MAP inference, and the likelihood inference. Posterior probability inference can be further divided into filtering and prediction. For a DBN consisting of variables $\mathbf{X}^{1:t}$ and $\mathbf{Y}^{1:t}$, where $\mathbf{X}^{1:t} = \{\mathbf{X}^1, \mathbf{X}^2, \ldots, \mathbf{X}^t\}$ represents the unknown variables from time 1 to the current time t, and $\mathbf{Y}^{1:t} = \{\mathbf{Y}^1, \mathbf{Y}^2, \ldots, \mathbf{Y}^t\}$ stands for the corresponding observed variables from time 1 to the current time t. Filtering is computing $p(\mathbf{x}^t|\mathbf{y}^{1:t})$, which can be done recursively as follows:

$$\begin{aligned}
p(\mathbf{x}^t|\mathbf{y}^{1:t}) &= p(\mathbf{x}^t|\mathbf{y}^{1:t-1}, \mathbf{y}^t) \\
&\propto p(\mathbf{x}^t|\mathbf{y}^{1:t-1})p(\mathbf{y}^t|\mathbf{x}^t) \\
&= p(\mathbf{y}^t|\mathbf{x}^t)\int_{\mathbf{x}^{t-1}} p(\mathbf{x}^t|\mathbf{x}^{t-1})p(\mathbf{x}^{t-1}|\mathbf{y}^{1:t-1})d\mathbf{x}^{t-1}.
\end{aligned} \tag{3.135}$$

Note that this equation assumes that $\mathbf{Y}^{1:t-1}$ are independent of $\mathbf{X}^t$ given $\mathbf{X}^{t-1}$ and that $\mathbf{X}^t$ are independent of $\mathbf{Y}^{1:t-1}$ given $\mathbf{X}^{t-1}$. The former is called observation independence assumption, while the latter is called state independence assumption. Within the integration term, $p(\mathbf{x}^t|\mathbf{x}^{t-1})$ captures the state transition, and $p(\mathbf{x}^{t-1}|\mathbf{y}^{1:t-1})$ represents the filtering probability at time $t-1$, thus allowing recursive computation of $p(\mathbf{x}^t|\mathbf{y}^{1:t})$. Except for at $t=0$, both $p(\mathbf{y}^t|\mathbf{x}^t)$ and $p(\mathbf{x}^t|\mathbf{x}^{t-1})$ can be computed through BN inference on the transition network. At time $t=0$, BN inference can be performed on the prior network to compute $p(\mathbf{y}^t|\mathbf{x}^t)$. In CV, filtering is widely used for object tracking. The famous Kalman filtering and particle filtering methods are special DBNs widely used for object tracking. Particle filtering results from replacing the integral in Eq. (3.135) with the average of samples acquired from sampling x^{t-1} from the filtering probability at time $t-1$.

The second posterior probability inference is the prediction, that is, computing $p(\mathbf{x}^{t+h}|\mathbf{y}^{1:t})$, where h represents the prediction-ahead step. Starting from $h=1$, $p(\mathbf{x}^{t+h}|\mathbf{y}^{1:t})$ can be recursively computed as follows:

$$p(\mathbf{x}^{t+h}|\mathbf{y}^{1:t}) = \int_{\mathbf{x}^{t+h-1}} p(\mathbf{x}^{t+h}|\mathbf{x}^{t+h-1})p(\mathbf{x}^{t+h-1}|\mathbf{y}^{1:t})d\mathbf{x}^{t+h-1}. \tag{3.136}$$

Prediction inference becomes filtering inference when $h=0$. Similarly, with the same independence assumptions as for DBN filtering, DBN prediction inference can be implemented by recursively performing BN inference to compute $p(\mathbf{x}^{t+h}|\mathbf{x}^{t+h-1})$ on the transition network.

For MAP inference, the goal is to estimate the best configuration for $\mathbf{x}^{1:t}$ given $\mathbf{y}^{1:t}$,

$$\begin{aligned}
\mathbf{x}^{*1:t} &= \operatorname*{argmax}_{\mathbf{x}^{1:t}} p(\mathbf{x}^{1:t}|\mathbf{y}^{1:t}) \\
&\propto \operatorname*{argmax}_{\mathbf{x}^{1:t}} p(\mathbf{x}^{1:t}, \mathbf{y}^{1:t}).
\end{aligned} \tag{3.137}$$

Also called decoding, DBN MAP inference is computationally expensive since it requires the enumeration of all possible configurations for $\mathbf{x}^{1:t}$. Finally, DBN likelihood inference involves computing

$$p(\mathbf{y}^{1:t}|\mathcal{G}, \mathbf{\Theta}) = \sum_{\mathbf{x}^{1:t}} p(\mathbf{x}^{1:t}, \mathbf{y}^{1:t}|\mathcal{G}, \mathbf{\Theta}). \tag{3.138}$$

DBN likelihood inference remains expensive as it requires the summation over $\mathbf{x}^{1:t}$.

Besides the recursive methods discussed above, we can also employ the direct method. The direct method simply unrolls the DBN to T time slices to form an expanded BN and then performs DBN inferences using the exact or approximate BN inference methods.

Exact methods include the junction tree method and the approximate methods include MCMC sampling and variational methods. However, the size of unrolled DBN tends to be large. Efficient exact inference methods such as the forward and backward procedures are available for DBNs with chain structures such as HMM. They typically exploit the chain structure and develop recursive procedures to achieve efficient learning and inference. For approximate inference, the sampling methods or the particle filtering methods can be applied to both the direct and recursive methods. In general, the main challenge with DBN inference is its complexity, in particular, for exact inference. DBN inferences are at least T times computationally more expensive than BN inference, where T is the number of slices. There exist various exact and approximate inference methods. For a comprehensive review of different types of dynamic Bayesian networks, we refer the readers to [66]. Detailed information on DBN representation, learning, and inference can also be found in [67,68].

3.7.3 Special DBNs

DBNs represent generic dynamic graphical models. Many of the existing well-known dynamic models such as the hidden Markov models and the linear dynamic system models can be treated as special cases of DBNs. In this section, we show how DBNs can be used to represent these models, which usually have restricted topology and/or parameterization compared to DBNs.

3.7.3.1 Hidden Markov model (HMM)

The hidden Markov model (HMM) is a widely used dynamic model for modeling a dynamic process. It is a special kind of DBN, with two layers of chained structure.

3.7.3.1.1 HMM topology and parameterization

The nodes of an HMM are divided into two layers of chain structure. The bottom layer models the observed quantity; its nodes can be discrete or continuous random variables. The nodes in the top layer are discrete and latent (hidden), and they thus do not have observations during both training and testing. They represent the underlying states for the corresponding observational nodes in the bottom layer. The latent nodes' cardinality is the number of states, which is also unknown. Fig. 3.32 shows the typical topology for an HMM, where the hidden nodes are represented by Xs in the top layer, whereas the observations are represented by Ys in the bottom layer. The HMM is a special kind of state-observation model where the state is hidden.

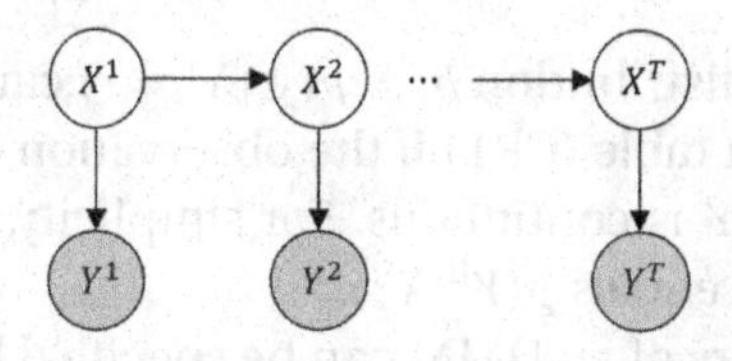

FIGURE 3.32 A hidden Markov model.

As a special kind of DBN, an HMM can be decomposed into a prior network and a transition network as shown in Figs. 3.33A and B, respectively. Specifically, the prior network

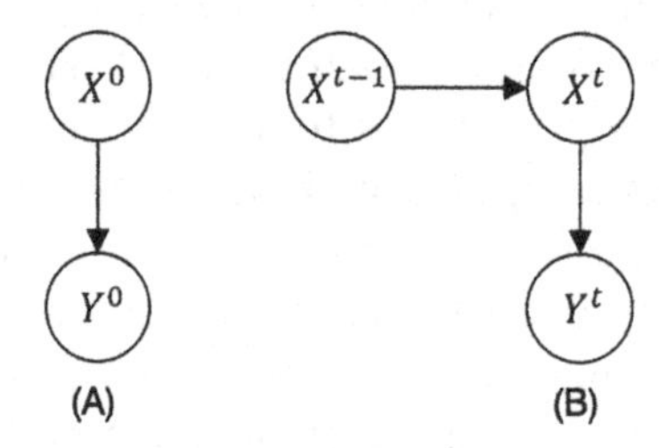

FIGURE 3.33 The prior network (A) and the transition network (B) for an HMM.

in Fig. 3.33A consists of one hidden state node $X^0 \in \{1, 2, \ldots, K\}$ and one observation node Y^0, which can be either continuous or discrete. Its distribution can be specified by a discrete prior probability distribution $p(X^0)$ for X^0 and a conditional probability distribution $p(Y^0|X^0)$ for Y^0; $p(X^0)$ can be written as

$$p(X^0 = i) = \pi_i, \ 1 \leq i \leq K, \tag{3.139}$$

where K is the number of hidden states. To make a valid probability distribution, we constrain $\sum_{i=1}^{K} \pi_i = 1$.

The transition network consists of nodes X^{t-1}, X^t, and Y^t as shown in Fig. 3.33B. The probability distribution for the transition network can be specified by the transition probability distribution $p(X^t|X^{t-1})$ and by the emission probability $p(Y^t|X^t)$. The transition probability captures the dynamics of the model and can be written as

$$p(X^t = j|X^{t-1} = i) = a_{ij}, \ 1 \leq i, j \leq K. \tag{3.140}$$

Similarly, to make a valid probability distribution, we constrain $\sum_{j=1}^{K} a_{ij} = 1$ for all i $(a_{ij} \geq 0)$. Like the DBN, the HMM also makes the first-order Markov and stationarity assumptions for the transition probability:

$$p(X^t|X^{t-1}, X^{t-2}, \ldots, X^1, Y^{t-1}, Y^{t-2}, ..) = p(X^t|X^{t-1})$$

and

$$p(X^{t_1}|X^{t_1-1}) = p(X^{t_2}|X^{t_2-1}), \ \forall t_1 \neq t_2.$$

The emission probability distribution $b_i = p(y^t|X^t = i)$ can be represented by a conditional probability distribution table (CPT) if the observation node Y is discrete or usually by a Gaussian distribution if Y is continuous. For simplicity, it is often assumed that the emission probability $p(Y^t|X^t)$ equals $p(Y^0|X^0)$.

In summary, the parameters of an HMM can be specified by $\boldsymbol{\Theta} = \{\pi_i, a_{ij}, b_i\}$. Given the specification of the prior and transition model and their parameterizations, the joint probability distribution of an HMM can be written as

$$p(X^0, \ldots, X^T, Y^0, \ldots, Y^T) = p(X^0)p(Y^0|X^0)\prod_{t=1}^{T} p(Y^t|X^t)p(X^t|X^{t-1}). \tag{3.141}$$

3.7.3.1.2 HMM inference

Similar to DBNs, HMM inference consists of mainly two types of inference: likelihood inference and decoding inference. Given a sequence of observations $\mathbf{y}^{1:T} = \{y^1, y^2, \ldots, y^T\}$, the likelihood inference is computing the joint likelihood $p(\mathbf{y}^{1:T}|\boldsymbol{\Theta})$ of the model parameter $\boldsymbol{\Theta}$. The decoding inference finds the most likely configuration for the latent variables i.e., the most likely state sequence, over time $\mathbf{x}^{1:T} = \{x^1, x^2, \ldots, x^T\}$, given the observation sequence $\mathbf{y}^{1:T}$, that is,

$$\mathbf{x}^{*1:T} = \underset{\mathbf{x}^{1:T}}{\operatorname{argmax}}\, p(\mathbf{x}^{1:T}|\mathbf{y}^{1:T}). \tag{3.142}$$

The decoding inference belongs to the MAP inference category. Like the DBN inference, HMM inference methods include the direct approach and the recursive approach. The direct approach unrolls the HMM into an expended BN of T time slices. Exact or approximate BN inference methods can then be applied to perform both likelihood and MAP inferences. For example, max-product belief propagation can be used to perform HMM decoding inference. Due to the presence of the latent variables $\mathbf{X}^{1:T}$, direct methods require marginalization over $\mathbf{X}^{1:T}$, which can be computationally expensive when the dimension of $\mathbf{X}^{1:T}$ is large. Because of the special chain structure of the HMMs, efficient algorithms based on recursive computation have been developed for both likelihood and decoding inference. They both exploit the HMM's underlying state and observation independence assumptions to recursively perform the inference.

Specifically, for likelihood inference, the forward-backward algorithm is often used to recursively compute the likelihood. According to the forward and backward algorithm, the likelihood inference can be written as

$$\begin{aligned}
p(\mathbf{y}^{1:T}|\boldsymbol{\Theta}) &= p(y^1, y^2, \ldots, y^T|\boldsymbol{\Theta}) \\
&= \sum_{k=1}^{K} p(y^1, y^2, \ldots, y^T, x^t = k|\boldsymbol{\Theta}) \\
&= \sum_{k=1}^{K} p(y^1, y^2, \ldots, y^t, x^t = k|\boldsymbol{\Theta})p(y^{t+1}, y^{t+2}, \ldots, y^T|x^t = k, \boldsymbol{\Theta}) \\
&= \sum_{k=1}^{K} \alpha_t(k)\beta_t(k),
\end{aligned} \tag{3.143}$$

where $\alpha_t(k) = p(y^1, y^2, \ldots, y^t, x^t = k)$ and $\beta_t(k) = p(y^{t+1}, y^{t+2}, \ldots, y^T|x^t = k)$ are referred to as the forward and backward probabilities, respectively. After some derivations, we can show that they can be recursively computed as follows:

$$\begin{aligned}
\alpha_t(k) &= p(y^1, y^2, \ldots, y^t, x^t = k) \\
&= \sum_{j=1}^{K} p(x^{t-1} = j, y^1, y^2, \ldots, y^{t-1}, y^t, x^t = k) \\
&= \sum_{j=1}^{K} p(x^{t-1} = j, y^1, y^2, \ldots, y^{t-1}) p(x^t = k | x^{t-1} = j) p(y^t | x^t = k) \\
&= \sum_{j=1}^{K} \alpha_{t-1}(j) p(x^t = k | x^{t-1} = j) p(y^t | x^t = k) \\
&= \sum_{j=1}^{K} \alpha_{t-1}(j) a_{jk} b_k. \qquad (3.144)
\end{aligned}$$

Similarly, $\beta_t(k)$ can be recursively computed as follows:

$$\begin{aligned}
\beta_t(k) &= p(y^{t+1}, y^{t+2}, \ldots, y^T | x^t = k) \\
&= \sum_{j=1}^{K} p(x^{t+1} = j, y^{t+1}, y^{t+2}, \ldots, y^T | x^t = k) \\
&= \sum_{j=1}^{K} p(x^{t+1} = j | x^t = k) p(y^{t+1} | x^{t+1} = j) p(y^{t+2}, \ldots, y^T | x^{t+1} = j) \\
&= \sum_{j=1}^{K} p(x^{t+1} = j | x^t = k) p(y^{t+1} | x^{t+1} = j) \beta_{t+1}(j) \\
&= \sum_{j=1}^{K} a_{kj} b_j \beta_{t+1}(j). \qquad (3.145)
\end{aligned}$$

Recursively computing the forward and backward probabilities using Eqs. (3.144) and (3.145), we can efficiently compute the likelihood using Eq. (3.143). Note that forward and backward procedures can be used separately to compute the likelihood, that is, $p(\mathbf{y}^{1:T}|\mathbf{\Theta}) = \sum_{k=1}^{K} \alpha_T(k)$ or $p(\mathbf{y}^{1:T}|\mathbf{\Theta}) = \sum_{k=1}^{K} \beta_1(k)$. Algorithm 3.12 provides the pseudocode for the forward–backward algorithm.

Algorithm 3.12 Forward–backward inference algorithm.

Initialize $\alpha_1(k) = \pi_k b_k(y^1)$ for $1 \le k \le K$
Initialize $\beta_T(k) = 1$ for $1 \le k \le K$
Arbitrarily choose a time t
Recursively compute $\alpha_t(k)$ and $\beta_t(k)$ using Eq. (3.144) or Eq. (3.145)
Termination: $p(\mathbf{y}^{1:T}|\mathbf{\Theta}) = \sum_{k=1}^{K} \alpha_t(k)\beta_t(k)$

Similar efficient algorithms have been developed for decoding inference, among which the most well-known is the Viterbi algorithm [69]. The decoding inference aims to find the most likely configuration for the latent variables over time $\mathbf{x}^{1:T}$ given the observation sequence $\mathbf{y}^{1:T}$:

$$\begin{aligned} \mathbf{x}^{*1:T} &= \underset{\mathbf{x}^{1:T}}{\operatorname{argmax}}\, p(\mathbf{x}^{1:T}|\mathbf{y}^{1:T}) \\ &= \underset{\mathbf{x}^{1:T}}{\operatorname{argmax}}\, p(\mathbf{x}^{1:T}, \mathbf{y}^{1:T}) \\ &= \underset{\mathbf{x}^{T}}{\operatorname{argmax}} \max_{\mathbf{x}^{1:T-1}} p(\mathbf{x}^{1:T}, \mathbf{y}^{1:T}). \end{aligned} \tag{3.146}$$

Let $\delta_t(\mathbf{x}^t) = \max_{\mathbf{x}^{1:t-1}} p(\mathbf{x}^{1:t}, \mathbf{y}^{1:t})$, which can be recursively computed as follows:

$$\begin{aligned} \delta_t(\mathbf{x}^t) &= \max_{\mathbf{x}^{1:t-1}} p(\mathbf{x}^{1:t}, \mathbf{y}^{1:t}) \\ &= \max_{\mathbf{x}^{1:t-1}} p(\mathbf{x}^{1:t-1}, \mathbf{x}^t, \mathbf{y}^{1:t-1}, \mathbf{y}^t) \\ &= \max_{\mathbf{x}^{1:t-1}} p(\mathbf{x}^{1:t-1}, \mathbf{y}^{1:t-1}) p(\mathbf{x}^t|\mathbf{x}^{t-1}) p(\mathbf{y}^t|\mathbf{x}^t) \\ &= \max_{\mathbf{x}^{t-1}} \{p(\mathbf{x}^t|\mathbf{x}^{t-1}) p(\mathbf{y}^t|\mathbf{x}^t) \max_{\mathbf{x}^{1:t-2}} p(\mathbf{x}^{1:t-1}, \mathbf{y}^{1:t-1})\} \\ &= \max_{\mathbf{x}^{t-1}} p(\mathbf{x}^t|\mathbf{x}^{t-1}) p(\mathbf{y}^t|\mathbf{x}^t) \delta_{t-1}(\mathbf{x}^{t-1}). \end{aligned} \tag{3.147}$$

Substituting Eq. (3.147) into Eq. (3.146) yields

$$\begin{aligned} \mathbf{x}^{*1:T} &= \underset{\mathbf{x}^{1:T}}{\operatorname{argmax}}\, p(\mathbf{x}^{1:T}|\mathbf{y}^{1:T}) \\ &= \underset{\mathbf{x}^{T}}{\operatorname{argmax}}\, \delta_T(\mathbf{x}^T) \\ &= \underset{\mathbf{x}^{T}}{\operatorname{argmax}} \max_{\mathbf{x}^{T-1}} \{p(\mathbf{x}^T|\mathbf{x}^{T-1}) p(\mathbf{y}^T|\mathbf{x}^T) \delta_{T-1}(\mathbf{x}^{T-1})\}, \end{aligned} \tag{3.148}$$

from which we have

$$x^{*t} = \underset{i}{\operatorname{argmax}}\, \delta_t(i)$$

for $t = 1, 2, \ldots, T$. The Viterbi algorithm is similar to the variable elimination algorithm for max-product (MAP) inference in Algorithm 3.2.

3.7.3.1.3 HMM learning

Like DBN learning with latent variables, variants of EM algorithms are used for HMM learning. Appendix 3.9.7 introduces a standard EM algorithm for HMM learning. Because of the special topology with HMM, efficient methods have been developed specifically for HMM learning. The most well-known HMM parameter learning method is the Baum–Welch algorithm [70].

Given M observation sequences $\mathbf{y} = \{\mathbf{y}_m\}_{m=1}^M$, we want to estimate the HMM parameters $\boldsymbol{\Theta}$, which consist of the prior probability π_i, transition probability a_{ij}, and the observation probability b_i. This is achieved by maximizing the log marginal likelihood:

$$\boldsymbol{\Theta}^* = \underset{\boldsymbol{\Theta}}{\text{argmax}} \log p(\mathbf{y}|\boldsymbol{\Theta}),$$

where $\log p(\mathbf{y}|\boldsymbol{\Theta})$ can be further written as

$$\begin{aligned} \log p(\mathbf{y}|\boldsymbol{\Theta}) &= \sum_{m=1}^{M} \log p(\mathbf{y}_m|\boldsymbol{\Theta}) \\ &= \sum_{m=1}^{M} \log \sum_{\mathbf{x}_m} p(\mathbf{x}_m, \mathbf{y}_m|\boldsymbol{\Theta}). \end{aligned} \tag{3.149}$$

Following the EM algorithm, maximizing the marginal log-likelihood is equivalent to maximizing the expected log-likelihood, that is,

$$\boldsymbol{\Theta}^* = \underset{\boldsymbol{\Theta}}{\text{argmax}} \sum_{m=1}^{M} \sum_{\mathbf{x}_m} q(\mathbf{x}_m|\mathbf{y}_m, \boldsymbol{\theta}_q) \log p(\mathbf{x}_m, \mathbf{y}_m|\boldsymbol{\Theta}), \tag{3.150}$$

where $q(\mathbf{x}_m|\mathbf{y}_m, \boldsymbol{\theta}_q)$ is chosen to be $p(\mathbf{x}_m|\mathbf{y}_m, \boldsymbol{\Theta}^-)$ with $\boldsymbol{\Theta}^-$ representing the parameters at the last iteration of the EM algorithm. Using the BN chain rule, $\log p(\mathbf{x}_m, \mathbf{y}_m|\boldsymbol{\Theta})$ can be further written as

$$\begin{aligned} \log p(\mathbf{x}_m, \mathbf{y}_m|\boldsymbol{\Theta}) &= \log p(x_m^0, \pi_i) + \log p(y_m^0|x_m^0, b_0) \\ &+ \sum_{t=1}^{T} [\log p(x_m^t|x_m^{t-1}, a_{ij}) + \log p(y_m^t|x_m^t, b_i)]. \end{aligned} \tag{3.151}$$

Following the EM procedure in Algorithm 3.8, in the E-step, $p(\mathbf{x}_m|\mathbf{y}_m, \boldsymbol{\Theta}^-)$ is estimated to obtain the probability (weight) for each configuration of $\mathbf{x}_m$. In the M-step, parameter estimation becomes expected counts of certain configurations in the training data. These updates iterate until convergence. The process can be initialized with a random $\boldsymbol{\Theta}^{\mathbf{0}}$.

The Baum–Welch algorithm follows a similar iterative procedure. According to the algorithm, given the current estimate of the parameters $\boldsymbol{\Theta}$ (started with a random initialization at $t = 0$). For each training sequence $\mathbf{y}_m$, we perform the E-step and M-step.

E-step:

$$\xi_t(i, j) = p(x^t = i, x^{t+1} = j|\mathbf{y}_m, \boldsymbol{\Theta}), \tag{3.152}$$

which can be computed using the forward–backward algorithm

$$\xi_t(i, j) = \frac{\alpha_t(i) a_{i,j} b_j(y^{t+1}) \beta_{t+1}(j)}{\sum_{i'=1}^{K} \sum_{j'=1}^{K} \alpha_t(i') a_{i',j'} b_{j'}(y^{t+1}) \beta_{t+1}(j')}, \tag{3.153}$$

where $\alpha_t(i)$ and $\beta_{t+1}(j)$ are respectively the forward and backward probabilities defined in Eqs. (3.144) and (3.145), and a_{ij} and $\beta_{t+1}(j')$ are the current parameters. Given $\xi_t(i, j)$, we can further define

$$\gamma_t(i) = p(x^t = i) = \sum_{j=1}^{K} \xi_t(i, j),$$

the probability of X^t being in state i at time t.

M-step:
Use $\xi_t(i, j)$ to update the parameters π_i, a_{ij}, and b_j of $\mathbf{\Theta}$ as follows:

$$\begin{aligned} \pi_i &= \gamma_1(i), \\ a_{i,j} &= \frac{\sum_{t=1}^{T-1} \xi_t(i, j)}{\sum_{t=1}^{T-1} \gamma_t(i)}, \end{aligned} \tag{3.154}$$

where $\gamma_1(i)$ is the probability that the state variable takes value i at time 1, $\sum_{t=1}^{T-1} \xi_t(i, j)$ is the total weighted count that the state variable transitions from state i to state j for the entire sequence, and $\sum_{t=1}^{T-1} \gamma_t(i)$ is the total weighted count that the state variable takes value i over the entire sequence.

For the emission probability b_i, if the observation y is discrete and $y \in \{1, 2, \ldots, J\}$, then

$$b_i(j) = \frac{\sum_{t=1, y^t=j}^{T} \gamma_t(i)}{\sum_{t=1}^{T} \gamma_t(i)}, \tag{3.155}$$

where $\sum_{t=1, y^t=j}^{T} \gamma_t(i)$ is the total count over the entire sequence that the state variable is at state i and observation assumes a value of j. For continuous observations y, for each state value i, collect data $\mathbf{y}_i = \{y^{t, x^t=i}\}_{t=1}^{T}$ and use $\mathbf{y}_i$ to estimate b_i. The E-step and M-step iterate until convergence.

Given that the individual parameter updates by each training sequence $\mathbf{y}_m$, the final estimated $\mathbf{\Theta}$ are the averages of the updated parameters for each $\mathbf{y}_m$. Using the updated parameters, the process then repeats until convergence. Like the EM algorithm, the method is sensitive to initialization and it is not guaranteed finding the global optimum. In addition, cross-validation is often used to determine the optimal cardinality of the hidden state variable X. Besides generative learning, discriminative learning can also be performed for HMM by maximizing the conditional log-likelihood. Further information on HMM learning and inference methods can be found in [71,67].

3.7.3.1.4 Variants of HMMs

Conventional HMMs are limited to modeling simple unary dynamics. Hence, many variants of HMMs have been developed to extend the standard HMMs. As shown in Fig. 3.34, they include the mixtures of Gaussian HMM, auto-regressive HMM (AR-HMM), input-output HMM (IO-HMM) [72], coupled HMM (CHMM) [73], factorial HMM [74], hierarchical HMM [75], and the hidden semi-Markov models (HSMMs). We further briefly summarize each variant of HMMs. Additional details on these models and their potential applications can be found in Section 17.6 of [11], [67], and [76].

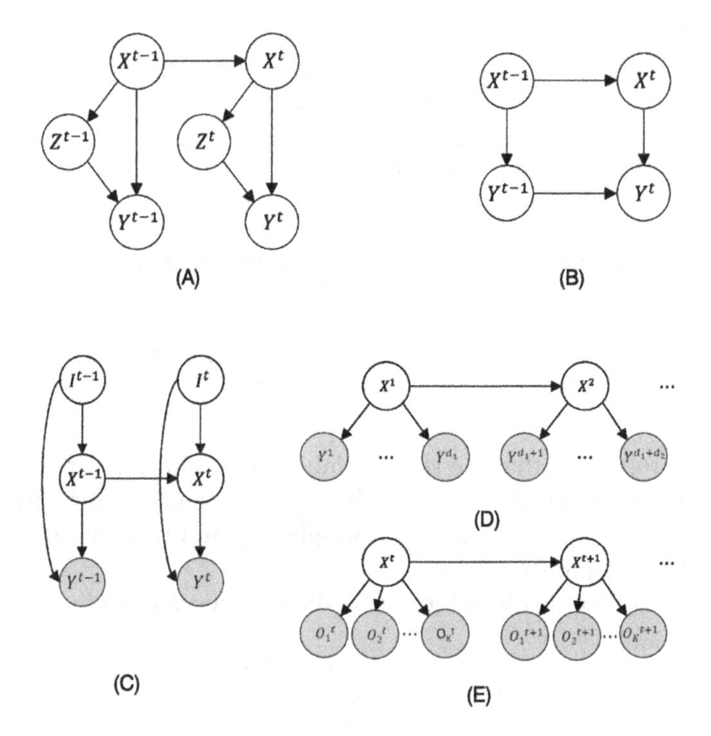

FIGURE 3.34 Variants of HMM. (A) Mixture of Gaussian HMM. (B) AR-HMM. (C) IO-HMM. (D) HSMM. (E) MOHMM.

For HMMs with continuous value observations, instead of using a single Gaussian to model the emission probability, the mixture of Gaussian HMM (Fig. 3.34A) models the emission probability as a mixture of Gaussians by introducing another discrete node to represent the mixture. The standard HMM assumes that observations are independent of each other given the hidden nodes. The auto-regressive HMM (AR-HMM) (Fig. 3.34B) relaxes this assumption by allowing a temporal link between consecutive observations. The input–output HMM (Fig. 3.34C) allows the hidden states to have both inputs and outputs. Finally, the hidden semi-Markov models (HSMM) (Fig. 3.34D) relaxes the HMM Markov state transition assumption by allowing state change as a function of time that has elapsed since entry into the current state. It explicitly models state duration, with each state emitting multiple observations. When the dimension of the feature vector is high, one drawback of HMM is that too many parameters of the observation node need to be estimated, so it tends to suffer from overfitting in training. To alleviate the problem of high dimensionality with observation space, the multiobservation hidden Markov models (MOHMMs, Fig. 3.34E) [77] tries to factorize the observation space to accommodate multiple types of observations, with the assumption that given the hidden state, different observation factors are independent of each other.

To model complex dynamic processes, composite models consisting of multiple HMMs have been introduced such as the coupled HMMs shown in Fig. 3.35. The coupled HMMs

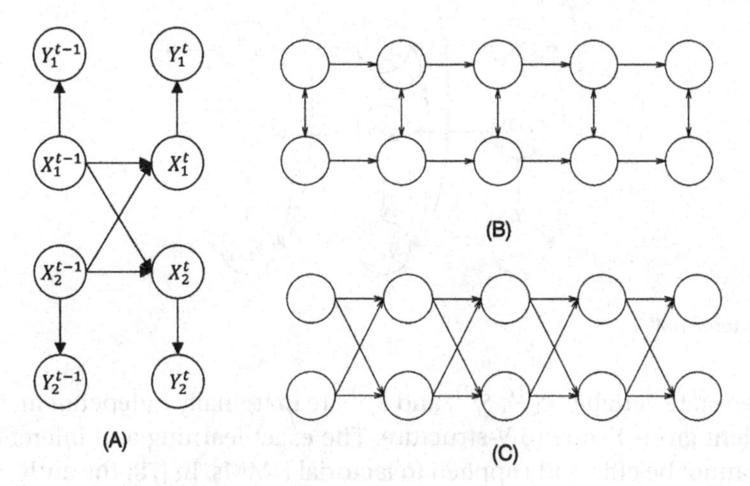

FIGURE 3.35 Coupled HMMs. (A) Coupled HMM. (B) Symmetric HMM. (C) Nonsymmetric HMM.

are used to model two interacting dynamic processes. Each process is captured by a separate HMM, and their interaction is captured by the links between their hidden states. CHMM can be further divided into symmetric CHMMs and nonsymmetric CHMMs as shown in Figs. 3.35B and C [73]. The symmetric CHMM in (A) captures the mutual interactions between the two entities. An undirected link (or double arrow) is used to connect the corresponding states of two entities. The nonsymmetric CHMM in (B) captures the temporal dependencies between the action of the two entities. The directed links between the corresponding states represent the temporal causality.

Another type of composite HMM is the factorial HMM, which expands the modeling capability of HMMs by assuming that the dynamic state can be represented by multiple chains of factorable hidden state sources, as shown in Fig. 3.36, where three chains of hidden states ($S^{(1)}$, $S^{(2)}$, and $S^{(3)}$) drive the same observations Y. The factorial HMM can factorize a complex hidden metastate into multiple independent simpler states such that they share the same observations. Each simple state can be represented by a single HMM, and each HMM can capture a different aspect of the complex dynamics. The outputs from each HMM are combined into a single output signal Y_t such that the output probabilities depend on the metastate. This HMM variant is good for modeling complex dynamic patterns that a single hidden state variable cannot capture. As there are no direct connections between the state variables in different chains, they are marginally independent.

Therefore, the metastate transition is factorizable:

$$p(S_t^{(1)}, S_t^{(2)}, S_t^{(3)} | S_{t-1}^{(1)}, S_{t-1}^{(2)}, S_{t-1}^{(3)}) = p(S_t^{(1)} | S_{t-1}^{(1)}) p(S_t^{(2)} | S_{t-1}^{(2)}) p(S_t^{(3)} | S_{t-1}^{(3)}).$$

The emission probability is characterized by $p(Y_t | S_t^{(1)}, S_t^{(2)}, S_t^{(3)})$. It can be parameterized as a linear Gaussian if Y_t is continuous or as a multinominal CPT if Y_t is discrete. Note

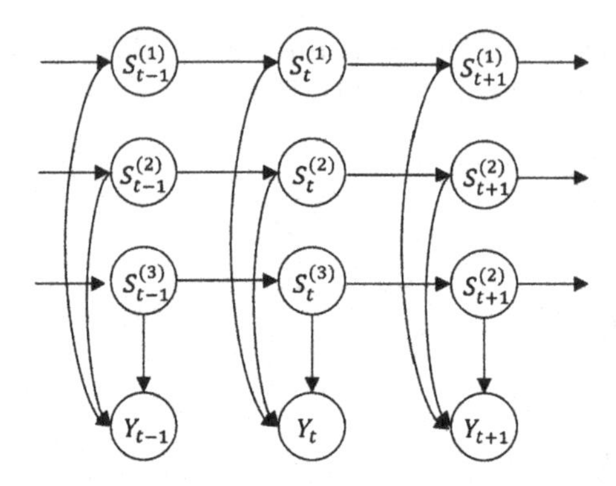

FIGURE 3.36 A factorial HMM.

that the hidden state variables $S_t^{(1)}$, $S_t^{(2)}$, and $S_t^{(3)}$ are marginally independent, but they become dependent given Y_t due to V-structure. The exact learning and inference methods with HMMs cannot be efficiently applied to factorial HMMs. In [78] the authors introduce efficient approximate learning and inference methods for factorial HMMs. It is clear that factorial HMMs can hence be treated as a particular case of coupled HMMs where different hidden state sources depend on each other through intrachannel state links. Brown and Hinton [79] introduced the product of HMM (PoHMM) shown in Fig. 3.37 to model the dynamic process as a product of several HMMs. PoHMM generalizes FHMMs, where the directed links among hidden states from different chains are replaced by undirected links to capture noncausal dependencies (undirected edges).

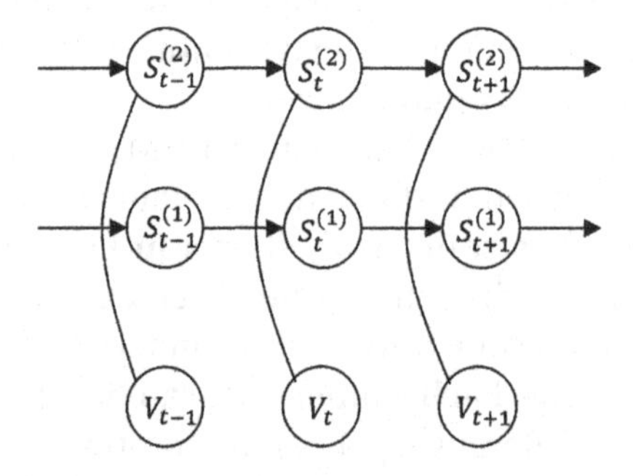

FIGURE 3.37 The product HMM.

Besides two-level HMMs, the hierarchical HMMs have also been introduced to model hidden states and their interactions at different levels. First introduced in [75], the hierarchical HMM (Fig. 3.38) allows for modeling domains with hierarchical state structures. It represents an extension to the factorial HMM to allow interactions among different hidden state variables. Furthermore, it extends the traditional HMM in a hierarchic manner to include a hierarchy of hidden states. Each state in the normal HMM is generalized recursively as another sub-HMM with special end states included to signal when the control of the activation is returned to the parent HMM. In HHMM, each state is considered to be a self-contained probabilistic model. More precisely, each state of the HHMM is itself an HHMM (or HMM). This implies that the states of the HHMM emit sequences of observation symbols rather than single observation symbols as is the case for the standard HMM states.

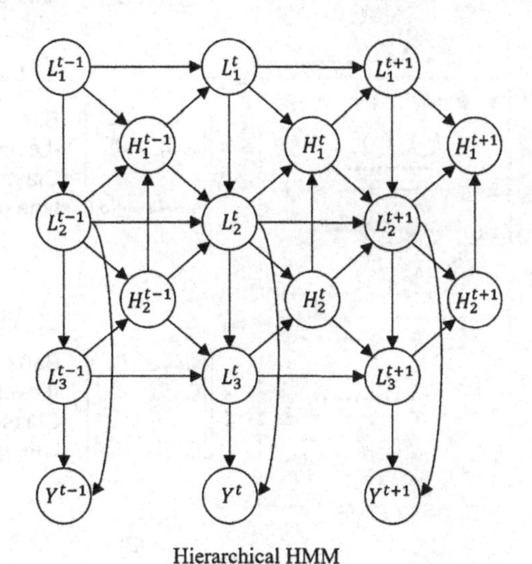

FIGURE 3.38 A hierarchical HMM.

Layered HMMs (LHMMs) are another type of hierarchical HMMs. As shown in Fig. 3.39, all the layers of the LHMM run in parallel; the lower-level layer generates the observation of higher level and the lowest level is the image observation. LHMMs allow efficient temporal patterns encoding at different levels of temporal granularity. Furthermore, LHMMs decompose the parameter space in a way that can enhance the robustness of the system by reducing training and tuning requirements. LHMMs can be regarded as a cascade of HMMs. At each level the HMMs perform classifications, and their outputs feed as input to the HMMs at the next level. Each HMM at each level is learned independently using the Baum–Welch algorithm.

Other extensions to HMMs include long-range HMMs, where the first-order Markov assumption is relaxed, switching HHMs [80], which introduce an additional switching variable to allow the HMM model to capture different types of dynamics, and the hierarchical HMMs [75].

3.7.3.2 Linear Dynamic System (LDS)

Another special case of DBN is the linear dynamic system (LDS), where the nodes form a two-level chain structure as shown in Fig. 3.40, much like as in HMM. Different from HMM, whose top layer is discrete and hidden, the top layer of an LDS can be either continuous or discrete, and it is often not assumed to be hidden.

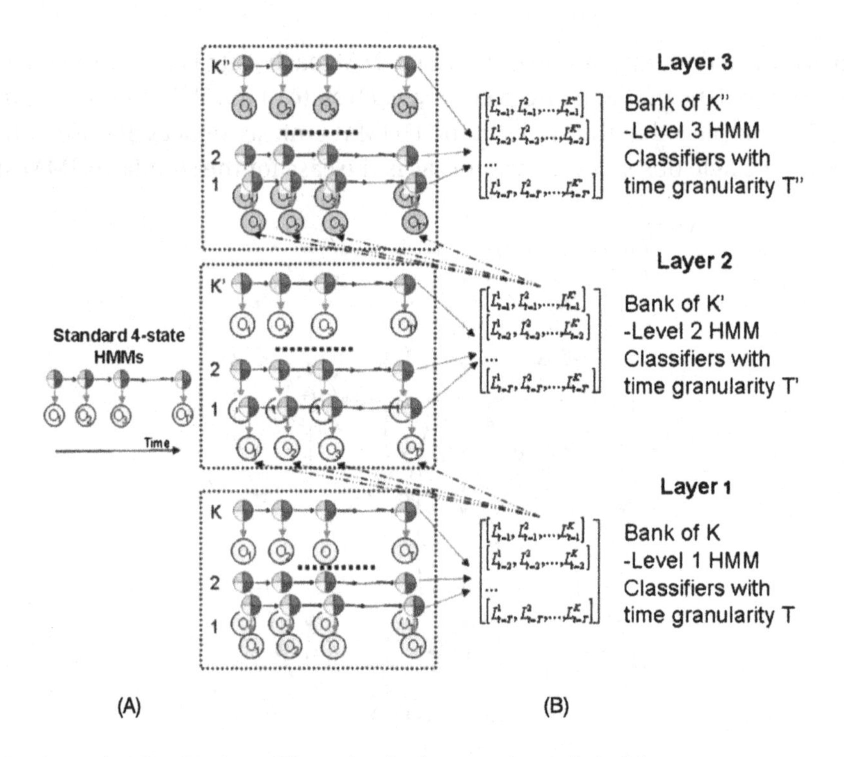

FIGURE 3.39 A layered HMM with three different levels of temporal granularity [4].

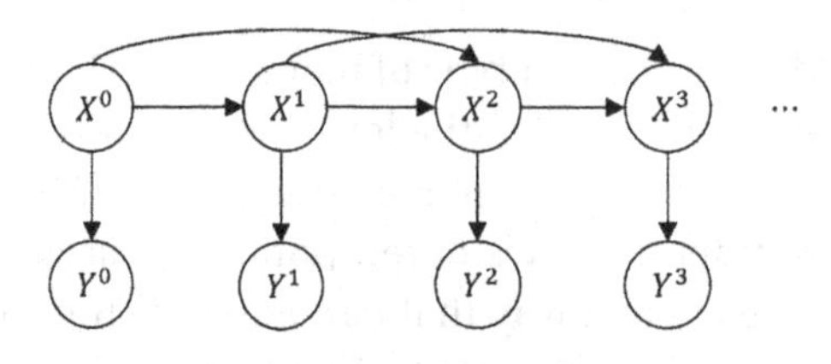

FIGURE 3.40 Graphical representation of the second-order linear dynamic system.

Furthermore, unlike traditional DBNs and HMMs, which typically assume first-order Markov state transition, an LDS can be of order p in state transition, as shown in Fig. 3.40, which illustrates a second-order LDS model. Following the parameterization of DBN, an LDS can be specified by a prior network, a transition network, and an observation model. The prior network consists of only one node, node X^0 in Fig. 3.40. The prior probability distribution can be specified as

$$p(X^0) = \mathcal{N}(\mu_0, \Sigma_0). \tag{3.156}$$

The pth-order transition model can be specified by $p(X^t|X^{t-1}, X^{t-2}, \ldots, X^{t-p})$. It is often specified through a regression function, whereby the current node value X^t is generated as a linear combination of the previous values (linear Gaussian),

$$X^t = A_1 X^{t-1} + A_2 X^{t-2} + \cdots A_p X^{t-p} + \epsilon, \tag{3.157}$$

where $\epsilon \sim \mathcal{N}(0, I^d)$ is a white noise vector, p is the order of the model, regression matrices $A_i \in \mathbb{R}^{d\times d}$, $i = 1, \ldots, p$, are the parameters,

$$p(X^t | X^{t-1}, X^{t-2}, \ldots, X^{t-p}) \sim \mathcal{N}(\mu_t, \Sigma_t), \tag{3.158}$$

$\mu_t = A_1 X_{t-1} + A_2 X_{t-2} + \ldots, A_p X_{t-p}$ is the mean of X^t, and Σ_t is its covariance matrix.

Finally, the observation probability distribution can be specified as

$$p(Y_t|X_t) \sim \mathcal{N}(\mu_t, \Sigma_t),\ t \geq 1. \tag{3.159}$$

Given this parameterization, the joint distribution of all the nodes of an LDS can be derived as follows:

$$\begin{aligned} & p(X^0, X^1, \ldots, X^T, Y^0, Y^1, \ldots, Y^T) \\ & = p(X^0)p(Y^0|X^0)p(X^1|X^0)p(Y^1|X^1)\ldots p(X^p|X^{p-1}, \ldots, X^{0)})p(Y^p|X^p) \\ & \prod_{t=p+1}^{T} p(X^t|X^{t-1}, X^{t-2}, \ldots, X^{t-p})p(Y^t|X^t). \end{aligned} \tag{3.160}$$

It can be proven easily that the joint probability distribution follows a Gaussian distribution. By assuming that the transition is first-order linear and that both the system and observation perturbations follow Gaussian distributions, an LDS becomes the Kalman filtering [81], which models a dynamic process as a unimodal linear-Gaussian system. By assuming that the nodes in the top layer are latent LDS becomes a latent LDS. By removing the bottom layer an LDS becomes an autoregressive dynamic model, which describes a time-varying process by assuming that the current state variable linearly depends on its own previous values. Other state-space dynamic models, such as the switching state-space model, for modeling multiple dynamic processes can also be treated as an extension of the LDS. DBNs generalize the LDS by allowing an arbitrary structure instead of limiting the structure to a two-layer chain.

3.8 Hierarchical Bayesian networks

Hierarchical models refer to graphical models with multiple layers of nodes. Similar to the organization of the human visual cortex, hierarchical models have demonstrated superior performance to the existing flat models in many applications. Despite their increasing significance, there is no unified definition of hierarchical graphical models. Based on our understanding, hierarchical graphical models can be classified into those with hyperparameter nodes and those with layers of hidden nodes. The first type of hierarchical models consists of random variables, their parameters, and hyperparameters. They are often called hierarchical Bayes models (HBMs). The second type consists of multiple layers of hidden random nodes. They are generically called deep models. In addition, there exist hybrid hierarchical models based on combining hidden layers with hyperparameters. In this section, we provide a brief introduction to each type of hierarchical models.

3.8.1 Hierarchical Bayes models

For many real-world applications, there often exists a large intraclass variation. For example, for human action/gesture recognition, each person may perform the same action differently in terms of its timing and spatial scope. Even the same person may perform the same action differently at different times. A single model may not have the capacity to adequately capture the large intraclass variations across all subjects/groups. Alternatively, we can develop a separate model for each group. However, this approach requires prior identification of the number of groups and the association of the data with a group. More importantly, it ignores the inherent dependencies among the data from different groups and requires much more data for training since it needs to train a separate model for each group. HBMs were introduced to overcome these limitations.

As a generalization of BNs, HBMs treat the parameters of a BN as additional RVs, and the distribution of the parameters is controlled by the hyperparameters. Random variables, parameters, and hyperparameters together form a hierarchical structure. Specifically, let $\mathbf{X}$ be the nodes for a BN $\mathcal{G}$, and let $\boldsymbol{\Theta}$ be the parameters of $\mathcal{G}$ that capture the joint probability distribution of $\mathbf{X}$. A hierarchical BN captures the joint distribution $p(\mathbf{X}, \boldsymbol{\Theta}|\boldsymbol{\alpha})$ of $\mathbf{X}$ and $\boldsymbol{\Theta}$, where $\boldsymbol{\alpha}$ represents the hyperparameters and $\boldsymbol{\Theta}$ are latent variables. According to the conditional chain rule, the joint variable and parameter distribution can be further factorized into the product of $p(\mathbf{X}|\boldsymbol{\Theta})$ and $p(\boldsymbol{\Theta}|\boldsymbol{\alpha})$ i.e. $p(\mathbf{X}, \boldsymbol{\Theta}|\boldsymbol{\alpha}) = p(\mathbf{X}|\boldsymbol{\Theta})p(\boldsymbol{\Theta}|\boldsymbol{\alpha})$; $p(\mathbf{X}|\boldsymbol{\Theta})$ captures the joint distribution of $\mathbf{X}$ given its parameters $\boldsymbol{\Theta}$, whereas $p(\boldsymbol{\Theta}|\boldsymbol{\alpha})$ captures the prior distribution of $\boldsymbol{\Theta}$, where $\boldsymbol{\alpha}$ stands for the hyperparameters that control the prior distribution. These hyperparameters are typically specified manually or learned from data. Compared to conventional BNs, hierarchical BNs are truly Bayesian in the sense that they capture the posterior distribution of the parameters $\boldsymbol{\Theta}$ if we interpret $p(\mathbf{X}|\boldsymbol{\Theta})$ as the likelihood of $\boldsymbol{\Theta}$. Moreover, they allow BN parameters to vary, and can hence better model the variability in the data.

An HBM can be concisely captured in a plate representation as shown in Fig. 3.41, where the plates represent the replicates (groups), that is, the quantities inside the rectangle should be repeated the given number of times. Specifically, $\mathbf{X}$ and $\mathbf{Y}$ represent the input and output variables, respectively. They are located in the inner most plate, which captures their relationships for all samples (N) from all groups, and vary with each sample. The parameters $\boldsymbol{\Theta}$ control the joint distribution of $\mathbf{X}$ and $\mathbf{Y}$, which is located within the second plate to capture the variation of $\boldsymbol{\Theta}$ as a result of $\mathbf{X}$ and $\mathbf{Y}$ belonging to different groups (K). They vary with each group. The hyperparameters $\boldsymbol{\alpha}$ determine the prior probability distribution $p(\boldsymbol{\Theta}|\boldsymbol{\alpha})$ of $\boldsymbol{\Theta}$. The hyper-hyperparameters $\boldsymbol{\gamma}$ control the prior distribution of $\boldsymbol{\alpha}$. Both hyperparameters and hyper-hyperparameters are outside the plates and hence are fixed and shared by **X** and **Y** in all groups. The hyper-hyperparameters are usually made noninformative or even ignored.

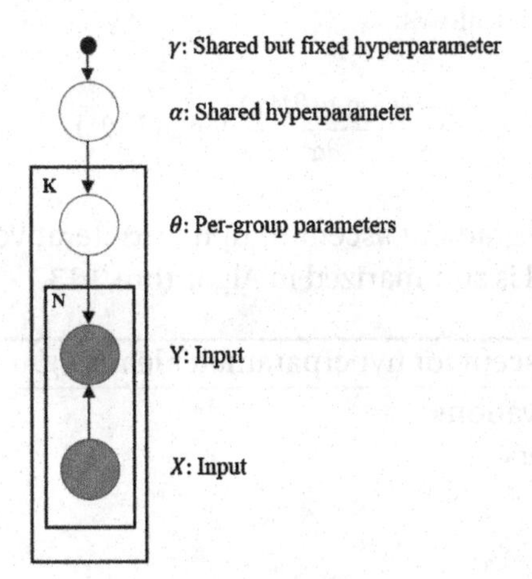

FIGURE 3.41 A hierarchical BN.

Given the hierarchical Bayes model $p(\mathbf{X}, \mathbf{Y}, \boldsymbol{\theta}|\boldsymbol{\alpha}) = p(\mathbf{X}, \mathbf{Y}|\boldsymbol{\theta})p(\boldsymbol{\theta}|\boldsymbol{\alpha})$, a top down ancestral sampling procedure can be applied to generate data by the following steps.

Step 1: obtain a sample $\hat{\alpha}$ from $p(\boldsymbol{\alpha}|\gamma)$.
Step 2: obtain a sample $\hat{\boldsymbol{\theta}}$ from $p(\boldsymbol{\theta}|\hat{\alpha})$.
Step 3: obtain a sample $\hat{x}$ and $\hat{y}$ from $p(\mathbf{X}, \mathbf{Y}|\hat{\boldsymbol{\theta}})$.
Step 4: repeat steps 1–3 until enough samples are collected.

The learning of HBMs is more challenging due to the presence of the latent variables $\boldsymbol{\theta}$. It involves learning the hyperparameters. Given training data $\mathcal{D} = \{\mathbf{x}^1, \mathbf{x}^2, ..., \mathbf{x}^N, \mathbf{y}^1, \mathbf{y}^2, ..., \mathbf{y}^N\}$, the hyperparameters α can be learned by maximizing the log marginal likelihood:

$$\begin{aligned}\boldsymbol{\alpha}^* &= \underset{\boldsymbol{\alpha}}{\operatorname{argmax}} \log p(\mathcal{D}|\boldsymbol{\alpha}) \\ &= \underset{\boldsymbol{\alpha}}{\operatorname{argmax}} \log \int_{\boldsymbol{\theta}} p(\mathcal{D}|\boldsymbol{\theta})p(\boldsymbol{\theta}|\boldsymbol{\alpha})d\boldsymbol{\theta}. \end{aligned} \tag{3.161}$$

Direct optimization of Eq. (3.161) is challenging due to the log integral term. Applying Jensen's inequality, we can instead maximize the lower bound of $\log p(\mathcal{D}|\alpha)$:

$$\begin{aligned}\boldsymbol{\alpha}^* &= \underset{\boldsymbol{\alpha}}{\operatorname{argmax}} \log p(\mathcal{D}|\boldsymbol{\alpha}) \\ &\approx \underset{\boldsymbol{\alpha}}{\operatorname{argmax}} \int_{\boldsymbol{\theta}} p(\boldsymbol{\theta}|\boldsymbol{\alpha}) \log p(\mathcal{D}|\boldsymbol{\theta})d\boldsymbol{\theta}. \end{aligned} \tag{3.162}$$

Eq. (3.162) can be solved via gradient ascent. The gradient of $\boldsymbol{\alpha}$ can be computed as

$$\nabla\boldsymbol{\alpha} = \int_{\boldsymbol{\theta}} p(\boldsymbol{\theta}|\boldsymbol{\alpha}) \frac{\partial \log p(\boldsymbol{\theta}|\alpha)}{\partial \alpha} \log p(\mathcal{D}|\boldsymbol{\theta})d\boldsymbol{\theta}. \tag{3.163}$$

The integral in Eq. (3.163) is difficult to exactly calculate. One solution is to replace it with the sum over samples of $\boldsymbol{\theta}$. By sampling $\boldsymbol{\theta}$ from $p(\boldsymbol{\theta}|\boldsymbol{\alpha})$ to obtain samples of $\boldsymbol{\theta}^s$ the gradient of α can be approximated as follows:

$$\nabla\boldsymbol{\alpha} \approx \sum_{s=1}^{S} \frac{\partial \log p(\boldsymbol{\theta}^s|\boldsymbol{\alpha})}{\partial\boldsymbol{\alpha}} \log p(\mathcal{D}|\boldsymbol{\theta}^s). \tag{3.164}$$

Given $\nabla\boldsymbol{\alpha}$, we can then use gradient ascent to update $\boldsymbol{\alpha}$ iteratively until convergence. The pseudocode of this method is summarized in Algorithm 3.13.

Algorithm 3.13 Gradient ascent for hyperparameter learning.

Input: $\mathcal{D} = \{\mathbf{x}^m\}_{m=1}^{M}$: observations
Output: $\boldsymbol{\alpha}$: hyperparameters
1: Initialization of $\boldsymbol{\alpha}^{(0)}$
2: $t \leftarrow 0$
3: **repeat**
4: Obtain samples of $\boldsymbol{\theta}^s$ from $p(\boldsymbol{\theta}|\boldsymbol{\alpha}^t)$
5: Compute $\nabla\boldsymbol{\alpha}$ using Eq. (3.164)
6: $\boldsymbol{\alpha}^{t+1} = \boldsymbol{\alpha}^t + \eta\nabla\boldsymbol{\alpha}$ //update α
7: t ← t+1
8: **until** convergence or reach maximum iteration number
9: **return** $\boldsymbol{\alpha}^{(t)}$

Alternatively, we can also replace the integral operation in Eq. (3.161) with the maximization operation to approximately estimate α:

$$\begin{aligned} \alpha^* &\approx \operatorname*{argmax}_{\boldsymbol{\alpha}} \log \max_{\boldsymbol{\theta}} p(\mathcal{D}|\boldsymbol{\theta}) p(\boldsymbol{\theta}|\boldsymbol{\alpha}) \\ &= \operatorname*{argmax}_{\boldsymbol{\alpha}} \log p(\boldsymbol{\theta}^*|\boldsymbol{\alpha}, \mathcal{D}), \end{aligned} \tag{3.165}$$

where $\boldsymbol{\theta}^*$ is the MAP estimate of $\boldsymbol{\theta}$ given the current $\boldsymbol{\alpha}$,

$$\boldsymbol{\theta}^* = \operatorname*{argmax}_{\boldsymbol{\theta}} \log p(\boldsymbol{\theta}|\boldsymbol{\alpha}, \mathcal{D}). \tag{3.166}$$

Using current $\boldsymbol{\alpha}$, $\boldsymbol{\theta}^*$ can also be estimated by maximizing the log posterior probability of $\boldsymbol{\theta}$:

$$\boldsymbol{\theta}^* = \operatorname*{argmax}_{\boldsymbol{\theta}} \left\{ \log p(\mathcal{D}|\boldsymbol{\theta}) + \log p(\boldsymbol{\theta}|\alpha) \right\}. \tag{3.167}$$

The pseudocode of this method is summarized in Algorithm 3.14. Finally, we can also use the EM method to solve empirical Bayesian learning by treating $\boldsymbol{\theta}$ as latent variables.

The inference for HBMs can be divided into empirical Bayesian inference and full Bayesian inference. For empirical Bayesian inference, the goal is to infer $\mathbf{y}^*$ given a query

---

Algorithm 3.14 Max-out for hyperparameter estimation.

Input: $\mathcal{D} = \{\mathbf{x}^m\}_{m=1}^M$: observations
Output: $\boldsymbol{\alpha}$: hyperparameters
1: Initialization of $\boldsymbol{\alpha}^{(0)}$
2: $t \leftarrow 0$
3: **repeat**
4: Obtain $\boldsymbol{\theta}^{*(t)}$ by solving Eq. (3.167)
5: Obtain $\boldsymbol{\alpha}^{(t+1)}$ by solving Eq. (3.165) given $\boldsymbol{\theta}^{(*t)}$
6: $t \leftarrow t+1$
7: **until** convergence or reach maximum iteration number
8: **return** $\boldsymbol{\alpha}^{(t)}$

---

input $\mathbf{x}'$, the training data $\mathcal{D}$, and the hyperparameters $\boldsymbol{\alpha}$:

$$\mathbf{y}^* = \max_{\mathbf{y}} p(\mathbf{y}|\mathbf{x}', \mathcal{D}, \boldsymbol{\alpha}), \tag{3.168}$$

where $p(\mathbf{y}|\mathbf{x}', \mathcal{D}, \boldsymbol{\alpha})$ can be rewritten as

$$p(\mathbf{y}|\mathbf{x}', \mathcal{D}, \boldsymbol{\alpha}) = \int_{\boldsymbol{\theta}} p(\mathbf{y}|\mathbf{x}', \boldsymbol{\theta}) p(\boldsymbol{\theta}|\mathcal{D}, \boldsymbol{\alpha}) d\boldsymbol{\theta}. \tag{3.169}$$

Eq. (3.169) requires integration over the parameters. It is difficult to compute, and no analytical solution exists. It can only be done numerically via sampling - in other words - by obtaining samples of $\boldsymbol{\theta}^s$ from $p(\boldsymbol{\theta}|\mathcal{D}, \boldsymbol{\alpha})$ and then approximating the integral over $\boldsymbol{\theta}^s$ by the sum of $\boldsymbol{\theta}^s$:

$$p(\mathbf{y}|\mathbf{x}', \mathcal{D}, \boldsymbol{\alpha}) = \frac{1}{S} \sum_{s=1}^{S} p(\mathbf{y}|\mathbf{x}', \boldsymbol{\theta}^s). \tag{3.170}$$

As $\boldsymbol{\theta}$ may be high-dimensional, MCMC or importance sampling may be used to perform the sampling of $p(\boldsymbol{\theta}|\mathcal{D}, \boldsymbol{\alpha})$. Alternatively, variational Bayes can be applied to approximate $p(\boldsymbol{\theta}|\mathcal{D}, \boldsymbol{\alpha})$ with a simple and factorized distribution to solve the integral problem in Eq. (3.169).

Full Bayesian inference infers $\mathbf{y}^*$ given only the query input $\mathbf{x}'$ and the training data $\mathcal{D}$, that is,

$$\mathbf{y}^* = \max_{\mathbf{y}} p(\mathbf{y}|\mathbf{x}', \mathcal{D}), \tag{3.171}$$

where $p(\mathbf{y}|\mathbf{x}', \mathcal{D})$ can be rewritten as

$$\begin{aligned} p(\mathbf{y}|\mathbf{x}', \mathcal{D}) &= \int_{\boldsymbol{\alpha}} \int_{\boldsymbol{\theta}} p(\mathbf{y}|\mathbf{x}', \boldsymbol{\theta}) p(\boldsymbol{\theta}|\mathcal{D}, \boldsymbol{\alpha}) p(\boldsymbol{\alpha}|\mathcal{D}) d\boldsymbol{\theta} d\boldsymbol{\alpha} \\ &= \int_{\boldsymbol{\theta}} p(\mathbf{y}|\mathbf{x}', \boldsymbol{\theta}) [\int_{\boldsymbol{\alpha}} p(\boldsymbol{\theta}|\mathcal{D}, \boldsymbol{\alpha}) p(\boldsymbol{\alpha}|\mathcal{D}) d\boldsymbol{\alpha}] d\boldsymbol{\theta}. \end{aligned} \tag{3.172}$$

The double integral in Eq. (3.172) is even more difficult to calculate. It can be only approximated by double sampling- that is, first sampling $\boldsymbol{\alpha}^{s_a}$ from $p(\boldsymbol{\alpha}|\mathcal{D})$ to approximate the $\boldsymbol{\alpha}$ integral with the $\boldsymbol{\alpha}$ sample average, then sampling $\boldsymbol{\theta}^{s_t}$ from $p(\theta|\mathcal{D}, \boldsymbol{\alpha}^{s_a})$ to approximate the $\boldsymbol{\theta}$ integral with $\boldsymbol{\theta}$ sample average, and finally approximating Eq. (3.172) using sample averages:

$$
\begin{aligned}
p(\boldsymbol{\theta}|\mathcal{D}) &= \frac{1}{S^a}\sum_{s_a=1}^{S_a} p(\boldsymbol{\theta}|\mathcal{D}, \boldsymbol{\alpha}^{s_a}), \text{ where } \boldsymbol{\alpha}^{s_a} \sim p(\boldsymbol{\alpha}|\mathcal{D}), \\
p(\mathbf{y}|\mathbf{x}', \mathcal{D}) &= \frac{1}{S^t}\sum_{s_t=1}^{S_t} p(\mathbf{y}|\mathbf{x}', \boldsymbol{\theta}^{s_t}), \text{ where } \boldsymbol{\theta}^{s_t} \sim p(\theta|\mathcal{D}, \boldsymbol{\alpha}^{s_a}).
\end{aligned}
\tag{3.173}
$$

Empirical Bayes inference is an approximation to the full Bayesian inference. It's a good approximation when the posterior probability of $\boldsymbol{\theta}$ is sharply peaked. When training data is insufficient or imbalanced, full Bayesian inference produces better inference results.

HBMs are effective in capturing the data variations. They assume that data are generated by different groups (clusters), that each group may be modeled by a different set of parameters, and that the parameter sets for different groups are samples from the parameter distribution controlled by the hyperparameters. Hierarchical modeling is similar to mixture models such as the mixture of Gaussians (MoG), but it differs from MoG in that it needs no a priori knowledge of the number of groups (number of Gaussian components). However, it requires that the parameters for each group follow the same distribution. Further information on HBMs can be found in [82,83].

3.8.2 Hierarchical deep models

Hierarchical deep models (HDMs) are multilayer graphical models with an input at the bottom layer, an output at the top layer, and multiple intermediate layers of hidden nodes. Each hidden layer represents input data at a certain level of abstraction. Models with latent layers or variables, such as the HMM, the MoG, and the latent Dirichlet allocation (LDA), have achieved better performance than models without latent variables. In addition, deep models with multiple layers of latent nodes have been proven to be significantly superior to the conventional "shallow" models. Fig. 3.42 contrasts a traditional BN (A) with a hierarchical deep BN (B), where $\mathbf{X}$ represents input variables, $\mathbf{Y}$ output variables, and $\mathbf{Z}^1, \mathbf{Z}^2, \ldots, \mathbf{Z}^n$ the intermediate hidden layers.

With multiple hidden layers, HDMs can represent the data at multiple levels of abstraction. They can be treated as hidden BNs, with latent variables $Z_1, Z_2, \ldots, Z_n$, can better capture the relationships between the input and output through the intermediate hidden layers. Fig. 3.43 shows the use of a deep model to represent an input image by geometric entities at different levels, that is, edges, parts, and objects.

Specifically, we can construct a deep regression BN [84] as shown in Fig. 3.44B. It consists of multiple layers, with the bottom layer representing the visible variables. The connections are directed from the upper layer to the lower layer, and no connections among

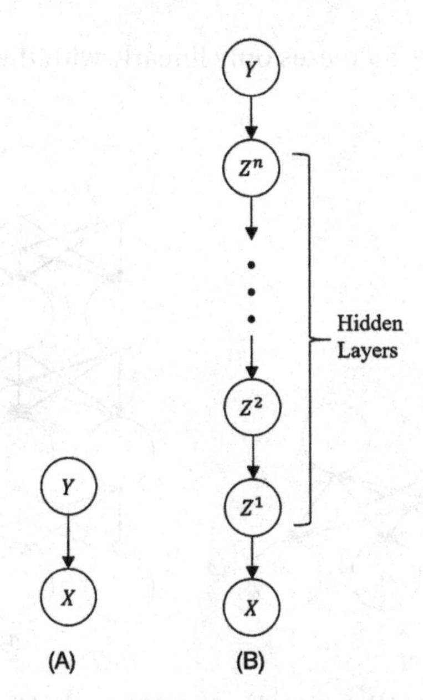

FIGURE 3.42 Comparison of a BN with a deep BN. (A) A conventional BN that captures $p(X, Y)$; (B) a hierarchical deep BN with multiple hidden layers that captures $p(X, Z_1, .., Z_n, Y)$, where $Z_1, Z_2.., Z_n$ are latent variables.

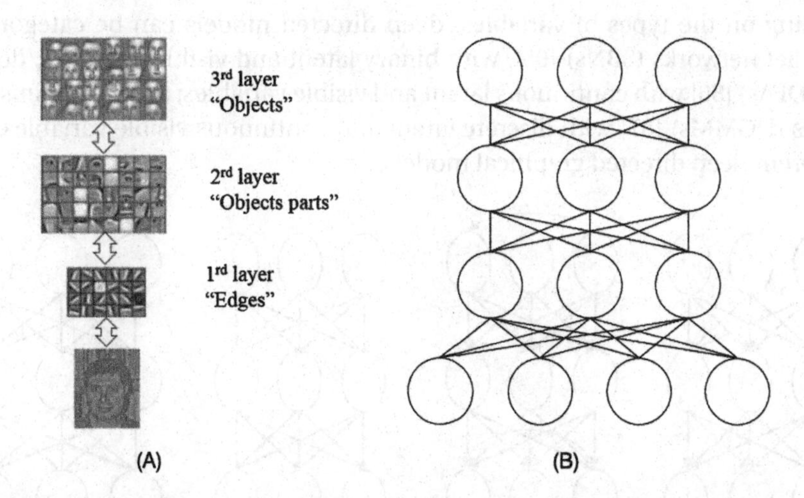

FIGURE 3.43 An illustration of the hierarchical representation of the input data by different hidden layers. (A) adapted from [5].

nodes within each layer are allowed. Its construction involves first determining a building block, the regression BN in Fig. 3.44A, and then stacking the building blocks on top of each other layer by layer, as shown in Fig. 3.44B. As discussed in Section 3.2.3.5, a regression BN is a BN, whose CPDs are specified by a linear regression of link weights. As a result, the total number of CPD parameters increases only linearly with the number of parameters for each node.

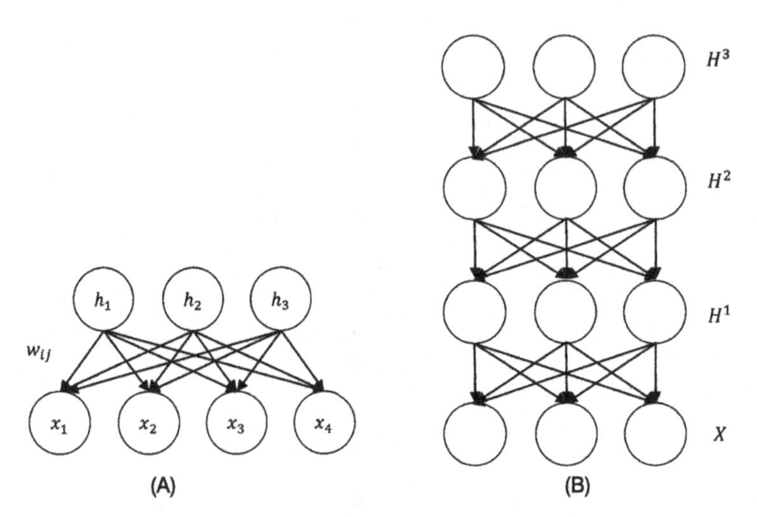

FIGURE 3.44 A deep Bayesian network. (A) A regression BN (RBN) as a building block; (B) a deep regression BN (DRBN) produced by stacking RBNs layer by layer.

Depending on the types of variables, deep directed models can be categorized into sigmoid belief networks (SBNs) [85], with binary latent and visible variables; deep factor analyzers (DFAs) [86] with continuous latent and visible variables; and deep Gaussian mixture models (DGMMs) [87] with discrete latent and continuous visible variables. Fig. 3.45 shows different deep directed graphical models.

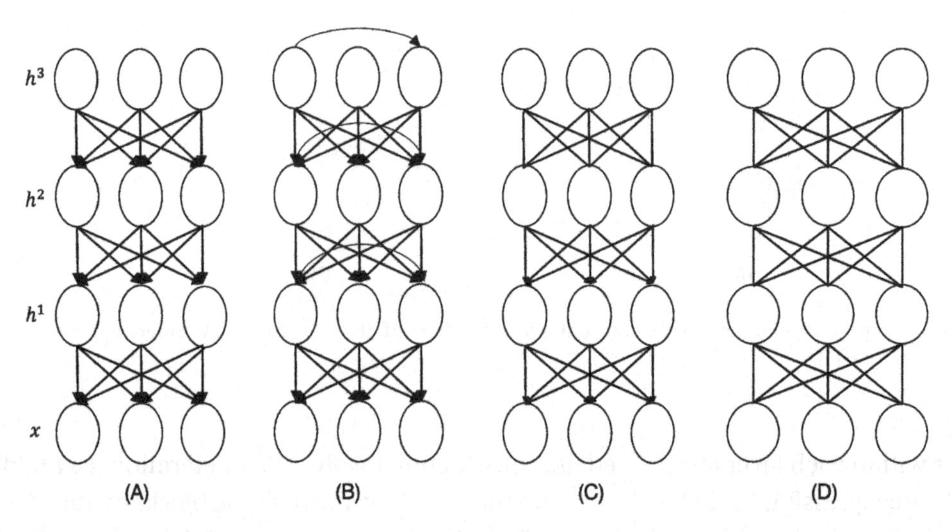

FIGURE 3.45 Different deep graphical models. (A) A deep BN, (B) a deep autoregressive model; (C) a deep belief network; (D) a deep Boltzmann machine.

Besides directed HDMs, we can also construct undirected HDMs such as the deep Boltzmann machine (DBM) in Fig. 3.45D, which is constructed by stacking layers of the restricted Boltzmann machine (RBM) on top of each other.[6] The DBM was once a major deep learning architecture. Compared to the undirected HDMs, directed HDMs enjoy several advantages. First, samples can be easily obtained by straightforward ancestral sampling. Second, there is no partition function issue since the joint distribution is obtained by multiplying all local conditional probabilities, which requires no further normalization. Finally but most importantly, directed models can naturally capture the dependencies among the latent variables given observations through the "explaining away" principle (i.e. the V-structure); thus, latent variables coordinate with each other to better explain the patterns in the data. The dependencies among the latent nodes, on the other hand, cause computational challenges in learning and inference. As a result, the DBM's inference is computationally less expensive as the hidden nodes are independent of each layer given the observation nodes.

Besides the directed and undirected HDMs, there are also the hybrid HDMs such as the deep belief networks as shown in Fig. 3.45C. Deep belief networks consist of directed layers, except for the top layer, which is undirected. This undirected top layer is introduced to alleviate the intractable posterior inference with the directed deep model by designing a special prior to make the latent variables conditionally independent such as the complementary prior [88], wherein the posterior probability for each latent variable can be individually computed. Compared with the deterministic deep models such as Convolutional Neural Networks (CNNs), deep probabilistic graphical models have the following advantages: 1) for classification/regression tasks, they can not only produce a prediction but can also produce a probability distribution in a principle manner such that the prediction uncertainty/confidence can be quantified. CNNs cannot effectively capture their prediction uncertainty. Recent developments in Bayesian neural networks are aimed at addressing this problem with CNNs. Second, as generative models, they can perform both discrimnative task (e.g. classification) and generative tasks such as data generation, while CNNs can only perform classifications, and the Generative Adversarial Networks (GANs) are used to perform data generation. Thirdly, they can perform both data generation and classification under incomplete input data, which both CNNs/GANs cannot. The main limitation with deep probabilistic graphical models are their computational complexity in both learning and inference, which prevents them from scaling up to large data/models. A detailed comparison of different types of HDMs can be found in [84].

The structure of a deep model is typically fixed. As a result, deep model learning involves learning the parameters for each observable and hidden node. Deep model learning typically consists of two stages, pretraining and refining. During the pretraining stage, parameters for each layer are separately learned using the outputs from the previous layer as inputs. Then, during refining stage, parameters for each layer are refined by jointly learning all parameters. The refining stage can be performed in an unsupervised or a supervised

[6] Readers may refer to Chapter 4 for a discussion of undirected graphical models; Section 4.2.4 is specifically about RBMs.

manner. Supervised learning can be done either generatively or discriminatively. Because of the presence of the latent nodes, learning for both pretraining and refining can be performed using either the direct method by directly maximizing the marginal likelihood or the EM method by maximizing the expected marginal likelihood. Furthermore, due to the large number of hidden nodes in each layer, it becomes intractable to exactly compute the gradient in the direct method or exactly compute the expectation in the EM method.

Given a learned deep model, the inference often involves estimating the values of the hidden nodes at the top layer for a given input observation. This can be done via MAP inference. Because of the dependencies among the latent nodes, exact MAP inference for a directed deep model is not feasible. Approximate inferences such as coordinate ascent or variational inference can be used instead. In general, learning and inference with HDMs are much more challenging than with the corresponding deterministic deep models such as the deep neural networks. They do not scale up well, which explains why HDMs, despite their powerful probabilistic representations, have not been widely adopted for deep learning. Further information on the learning and inference for deep BNs can be found in [84].

3.8.3 Hybrid hierarchical models

Finally, we briefly discuss the hybrid hierarchical models (HHMs), which combine HBN with HDM to exploit their respective strengths. As shown in Fig. 3.46, an HHM typically consists of an the observation layer, one or more hidden layers, parameter layers, and one hyperparameter layer. The hidden layers make it possible to represent input data at different levels of abstraction. The parameter layers allow the parameters at each hidden layer to vary to capture the variation in each hidden layer. The hyperparameters capture the distributions of hidden layer parameters.

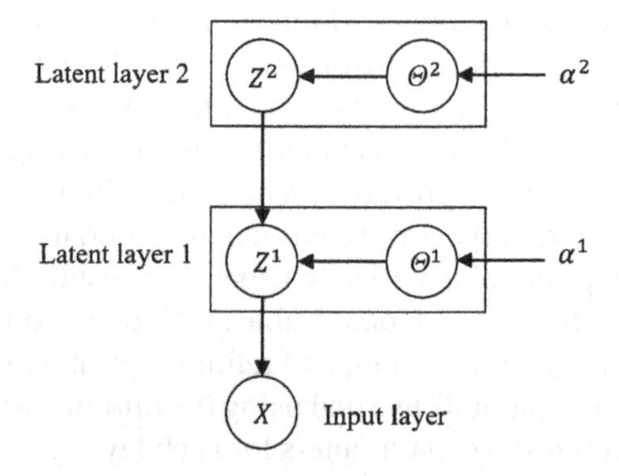

FIGURE 3.46 An example of a hybrid hierarchical graphical model consisting of an input observation layer X, two hidden layers Z^1 and Z^2, parameters $\boldsymbol{\theta}^1$ and $\boldsymbol{\theta}^2$ for hidden layers, and hyperparameters $\boldsymbol{\alpha}^1$ and $\boldsymbol{\alpha}^2$ that control hidden layer parameters. The model captures $p(X, Z^1, Z^2, \Theta^1, \Theta^2|\alpha^1, \alpha^2)$, where Z^1, Z^2, Θ^1, and Θ^2 are all latent variables.

Compared to HDMs, HHMs can use fewer hidden nodes to represent the variations in the input data. In addition, instead of assuming that the parameters for each node are independent, as is often done by HDMs, HHMs capture the dependencies among the parameters of hidden nodes through the hyperparameters. Finally, HHMs also have a much smaller number of parameters to learn since they only need to learn the hyperparameters, which are typically much fewer in number than the parameters.

A good example of HHMs is the latent Dirichlet allocation (LDA) [6]. It consists of an observation layer, one latent variable layer, the parameter layer, and the hyperparameter layer. Whereas the latent variable captures the latent topics in the data, the hyperparameters accommodate the variability with the parameters of the latent variable. Specifically, as shown in Fig. 3.47, an LDA model in plate notation consists of the observation variable w_{ij}, the j-th word in i-th document, the latent variable z_{ij}, the latent topic that word w_{ij} belongs to, the parameters θ_i that specify the topic distribution for document i, and the hyperparameters α and β. The latent variable z_{ij} is discrete and typically represents the clusters (or topics) in the data. The cardinality of z_{ij} is typically determined empirically through cross-validation. The hyperparameters α and β are for the Dirichlet prior of θ_i, respectively, and of the word for each topic. LDA assumes that a document (data) is composed of a mixture of latent topics (clusters) z_{ij}, and that each topic, in turn, consists of a mixture of words (features) w_{ij}. The model captures the joint distribution of w_{ij}, z_{ij}, θ_i, and following the BN chain rule, $p(w_{ij}, z_{ij}, \theta_i|\alpha, \beta) = p(w_{ij}|z_{ij}, \beta)p(z_{ij}|\theta_i)p(\theta_i|\alpha)$, where θ_i and z_{ij} are latent variables. LDA can hence be treated as a hierarchical mixture model.

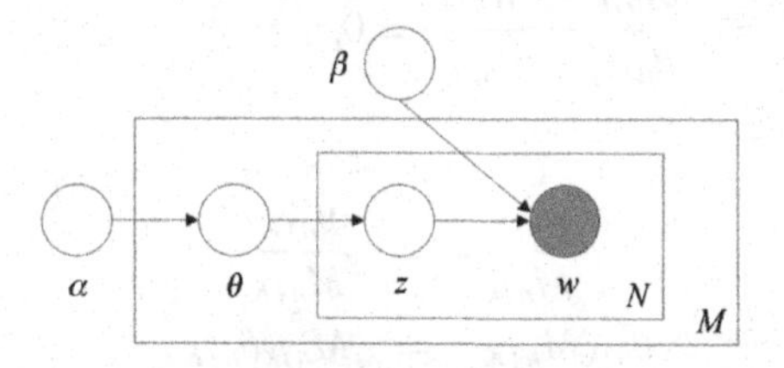

FIGURE 3.47 An LDA model, where z_{ij} represents the latent topics for word w_{ij}, θ_i parameterizes the topic distribution for the i the document, α is the hyperparameters to specify the prior distribution of θ_i, β parameterizes the distribution of the words w_{ij} for each topic, M denotes the number of documents, and N is the number of words in a document. Figure from [6].

Given training data, we can use the methods discussed in Section 3.8.1 to learn the hyperparameters. With the learned hyperparameters α and β, LDA inference computes z, that is, $z^* = \text{argmax}_z\, p(z|w, \alpha, \beta)$, which is the most likely topic for an observed word w. It may also be used to infer the most likely topic distribution for a given document i, i.e., $\theta_i^* = \arg\max_{\theta_i} p(\theta_i|\mathbf{w}, \alpha, \beta)$, based on which document i may be classified into one of the document categories. Further information on LDA can be found in [6].

LDA represents a generalization of probabilistic latent semantic analysis (pLSA) [89]. pLSA is a latent topic model, widely used in CV and in natural language processing. Like LDA, it assumes that a document consists of a mixture of latent topics, which, in turn, consists of a mixture of features (words). Unlike LDA, which assumes that the topic parameters follow a Dirichlet distribution, pLSA does not treat its parameters as random variables. Hence, pLSA is not strictly a hierarchical Bayes model. In addition, pLSA requires learning the mixture weights for each document. As a result, the number of parameters increases linearly with the number of training samples. Finally, there is no natural way for pLSA to assign a probability to a new testing observation.

3.9 Appendix

3.9.1 Proof of Eq. (3.64)

$$
\begin{aligned}
\frac{\partial LL(\boldsymbol{\theta}_{nj}:D)}{\partial \theta_{njk}} &= \frac{\partial \sum_{k=1}^{K} M_{njk} \log \theta_{njk}}{\partial \theta_{njk}} \\
&= \frac{\partial [\sum_{k=1}^{K-1} M_{njk} \log \theta_{njk} + M_{njK} \log \theta_{njK}]}{\partial \theta_{njk}} \\
&= \frac{\partial [\sum_{k=1}^{K-1} M_{njk} \log \theta_{njk} + M_{njK} \log (1 - \sum_{k=1}^{K-1} \theta_{njk})]}{\partial \theta_{njk}} \\
&= \frac{M_{njk}}{\theta_{njk}} - \frac{M_{njK}}{1 - \sum_{k=1}^{K-1} \theta_{njk}} \\
&= \frac{M_{njk}}{\theta_{njk}} - \frac{M_{njK}}{\theta_{njK}} = 0,
\end{aligned}
\tag{3.174}
$$

$$
\begin{aligned}
\frac{\theta_{njk}}{M_{njk}} &= \frac{\theta_{njK}}{M_{njK}}, \\
\theta_{njk} M_{njK} &= M_{njk} \theta_{njK}, \\
\sum_{k=1}^{K} \theta_{njk} M_{njK} &= \sum_{k=1}^{K} M_{njk} \theta_{njK}, \\
M_{njK} &= M_{nj} \theta_{njK}, \\
\theta_{njK} &= \frac{M_{njK}}{M_{nj}}.
\end{aligned}
\tag{3.175}
$$

3.9.2 Proof of Gaussian Bayesian network

Let X_n be a node in a Gaussian BN, and let $\pi(X_n)$ be the parent nodes of X_n. According to the definition of a linear Gaussian, we have

$$
X_n = \sum_{X_k \in \pi(X_n)} \alpha_n^k X_k + \beta_n + \epsilon_n, \tag{3.176}
$$

where $\epsilon_n \sim \mathcal{N}(0, \sigma^2_{n|\pi_n})$ is the Gaussian noise for node n. By Eq. (3.176) we have

$$\mu_n = E(X_n) = E(\sum_{X_k \in \pi(X_n)} \alpha_n^k X_k + \beta_n + \epsilon_n) = \sum_{X_k \in \pi(X_n)} \alpha_n^k \mu_k + \beta_n, \tag{3.177}$$

$$\begin{aligned}
\sigma_n^2 &= E((X_n - \mu_n)^2) \\
&= E[(\sum_{X_k \in \pi(X_n)} \alpha_n^k (X_k - \mu_k) + \epsilon_n)^2] \\
&= E[(\sum_{X_k \in \pi(X_n)} \alpha_n^k (X_k - \mu_k))^2] + \sigma^2_{n|\pi_n} \\
&= \sum_{X_k \in \pi(X_n)} (\alpha_n^k)^2 E((X_k - \mu_k))^2 + 2 \sum_{\substack{X_k \in \pi(X_n) \\ X_m \in \pi(X_n) \\ X_n \neq X_m}} \alpha_n^k \alpha_n^m E[(X_k - \mu_k)(X_m - \mu_m)] + \sigma^2_{n|\pi_n} \\
&= \sum_{X_k \in \pi(X_n)} (\alpha_n^k)^2 \sigma_k^2 + 2 \sum_{\substack{X_k \in \pi(X_n) \\ X_m \in \pi(X_n) \\ X_n \neq X_m}} \alpha_n^k \alpha_n^m \sigma^2_{X_k X_m} + \sigma^2_{n|\pi_n},
\end{aligned} \tag{3.178}$$

where $\sigma^2_{X_k X_m}$ is the covariance of two parents of X_n. Let $\mathbf{x} = (X_1, X_2, \ldots, X_N)^\top$ be the vector representing the N nodes in the BN, and let $\boldsymbol{\mu} = (\mu_1, \mu_2, \ldots, \mu_N)^\top$ be the mean vector. Following Section 10.2.5 of [11], we can compute the joint covariance matrix Σ for $\mathbf{x}$ as

$$\boldsymbol{\Sigma} = E[(\mathbf{x} - \boldsymbol{\mu})(\mathbf{x} - \boldsymbol{\mu})^\top] \tag{3.179}$$

with $(\mathbf{x} - \boldsymbol{\mu})$ computed as

$$\mathbf{x} - \boldsymbol{\mu} = \mathbf{A}(\mathbf{x} - \boldsymbol{\mu}) + \boldsymbol{\epsilon}, \tag{3.180}$$

where $\mathbf{A}$ is a matrix of weights α_n^m,[7] and $\boldsymbol{\epsilon} = (\epsilon_1, \epsilon_2, \ldots \epsilon_N)^\top$ is the Gaussian noise vector of all nodes. From Eq. (3.180) we have

$$\mathbf{x} - \boldsymbol{\mu} = (\mathbf{I} - \mathbf{A})^{-1} \boldsymbol{\epsilon}. \tag{3.181}$$

Plugging Eq. (3.181) into Eq. (3.179) yields

$$\begin{aligned}
\boldsymbol{\Sigma} &= E[(\mathbf{x} - \boldsymbol{\mu})(\mathbf{x} - \boldsymbol{\mu})^\top] \\
&= E[((\mathbf{I} - \mathbf{A})^{-1} \boldsymbol{\epsilon})((\mathbf{I} - \mathbf{A})^{-1} \boldsymbol{\epsilon})^\top] \\
&= ((\mathbf{I} - \mathbf{A})^{-1}) E(\boldsymbol{\epsilon} \boldsymbol{\epsilon}^z)((\mathbf{I} - \mathbf{A})^{-t}) \\
&= (\mathbf{I} - \mathbf{A})^{-1} \mathbf{S} (\mathbf{I} - \mathbf{A})^{-t},
\end{aligned} \tag{3.182}$$

where $\mathbf{S}$ is a diagonal matrix with conditional variances $\sigma^2_{i|\pi_n}$ on the diagonal.

[7] $\mathbf{A}$ is a lower triangular matrix if the nodes are arranged in the topological order.

Given the joint distribution captured by $\boldsymbol{\mu}$ and $\boldsymbol{\Sigma}$, it is easy to show that the marginal distribution of any subset of variables $\mathbf{x}_s \subset \mathbf{x}$ remains Gaussian, $p(\mathbf{x}_s) \sim \mathcal{N}(\boldsymbol{\mu}_s, \boldsymbol{\Sigma}_s)$. Moreover, $\boldsymbol{\mu}_s$ and $\boldsymbol{\Sigma}_s$ can be directly extracted from the corresponding elements in $\boldsymbol{\mu}$ and $\boldsymbol{\Sigma}$. In addition, this also applies to the conditional distribution. Let $\mathbf{x}_{s'} \subset \mathbf{x}$ be another subset in $\mathbf{x}$. Then $p(\mathbf{x}_s|\mathbf{x}_{s'})$ also follows a Gaussian distribution,

$$p(\mathbf{x}_s|\mathbf{x}_{s'}) = \mathcal{N}(\boldsymbol{\mu}_{s|s'}, \boldsymbol{\Sigma}_{s|s'}) \tag{3.183}$$

with

$$\begin{aligned} \boldsymbol{\mu}_{s|s'} &= \boldsymbol{\mu}_s + \boldsymbol{\Sigma}_{ss'}\boldsymbol{\Sigma}_{s'}^{-1}(\mathbf{x}_{s'} - \boldsymbol{\mu}_{s'}), \\ \boldsymbol{\Sigma}_{s|s'} &= \boldsymbol{\Sigma}_s - \boldsymbol{\Sigma}_{ss'}\boldsymbol{\Sigma}_{s'}^{-1}\boldsymbol{\Sigma}_{s's}, \end{aligned} \tag{3.184}$$

where $\boldsymbol{\Sigma}_{ss'} = E[(\mathbf{s}-\boldsymbol{\mu}_s)(\mathbf{s}'-\boldsymbol{\mu}_{s'})^\top]$ is the covariance matrix of subsets s and s'. The elements of $\boldsymbol{\Sigma}_{ss'}$ can be directly extracted from the corresponding elements of $\boldsymbol{\Sigma}$. Similarly, $\boldsymbol{\Sigma}_{s's}$ is the covariance matrix of subsets s' and s. It is the transpose of $\boldsymbol{\Sigma}_{ss'}$. Refer to Chapter 2.3.1 of [90] and Theorem 7.4 of [7] for the proof. Take Fig. 3.7, for example, the GBN where $\mathbf{x} = (X_1, X_2, X_3, X_4, X_5)^\top$. By the previous equations the joint mean is $\boldsymbol{\mu} = (\mu_1, \mu_2, \mu_3, \mu_4, \mu_5)^\top$, where $\mu_3 = \alpha_3^1\mu_1 + \alpha_3^2\mu_2 + \beta_3$, $\mu_4 = \alpha_4^3\mu_3 + \beta_4$, and $\mu_5 = \alpha_5^3\mu_3 + \beta_4$. The covariance matrix $\boldsymbol{\Sigma}$ can be computed as

$$\boldsymbol{\Sigma} = (\mathbf{I} - \mathbf{A})^{-1}\mathbf{S}(\mathbf{I} - \mathbf{A})^{-T},$$

where $\mathbf{A}$ and $\mathbf{S}$ can be computed as follows:

$$\mathbf{A} = \begin{pmatrix} 0 & 0 & 0 & 0 & 0 \\ 0 & 0 & 0 & 0 & 0 \\ \alpha_3^1 & \alpha_3^2 & 0 & 0 & 0 \\ 0 & 0 & \alpha_4^3 & 0 & 0 \\ 0 & 0 & \alpha_5^3 & 0 & 0 \end{pmatrix},$$

$$\mathbf{S} = \begin{pmatrix} \sigma_1^2 & 0 & 0 & 0 & 0 \\ 0 & \sigma_2^2 & 0 & 0 & 0 \\ 0 & 0 & \sigma_3^2 & 0 & 0 \\ 0 & 0 & 0 & \sigma_4^2 & 0 \\ 0 & 0 & 0 & 0 & \sigma_5^2 \end{pmatrix}.$$

3.9.3 Laplace approximation

The goal of the Laplace approximation is to approximate a complex probability distribution by a Gaussian distribution. Specifically, for a given joint distribution $p(\mathbf{x})$ over a set of random variables $\mathbf{X}$ of N dimensions, the Laplace approximation finds a Gaussian distribution $q(\mathbf{x})$ centered on a mode of $p(\mathbf{x})$ to approximate $p(\mathbf{x})$. Let $\mathbf{x}_0$ be a mode of $p(\mathbf{x})$.

Taking the second-order Taylor expansion of $\log p(\mathbf{x})$ around $\mathbf{x}_0$ produces

$$\begin{aligned}\log p(\mathbf{x}) &\approx \log p(\mathbf{x}_0) + (\mathbf{x}-\mathbf{x}_0)^\top \frac{\partial \log p(\mathbf{x}_0)}{\partial \mathbf{x}} \\ &\quad + \frac{(\mathbf{x}-\mathbf{x}_0)^\top \frac{\partial^2 \log p(\mathbf{x}_0)}{\partial^2 \mathbf{x}} (\mathbf{x}-\mathbf{x}_0)}{2} \\ &= \log p(\mathbf{x}_0) + \frac{(\mathbf{x}-\mathbf{x}_0)^\top \frac{\partial^2 \log p(\mathbf{x}_0)}{\partial^2 \mathbf{x}} (\mathbf{x}-\mathbf{x}_0)}{2}. \end{aligned} \tag{3.185}$$

Taking the exponential of both sides of Eq. (3.185) yields

$$p(\mathbf{x}) \approx p(\mathbf{x}_0) \exp \frac{(\mathbf{x}-\mathbf{x}_0)^\top \frac{\partial^2 \log p(\mathbf{x}_0)}{\partial^2 \mathbf{x}} (\mathbf{x}-\mathbf{x}_0)}{2} \tag{3.186}$$

Letting $A = -[\frac{\partial^2 \log p(\mathbf{x}_0)}{\partial^2 \mathbf{x}}]$, where $\frac{\partial^2 \log p(\mathbf{x}_0)}{\partial^2 \mathbf{x}}$ is called the Hessian matrix of $\log p(\mathbf{x})$, and substituting it into Eq. (3.186) yield

$$p(\mathbf{x}) \approx p(\mathbf{x}_0) \exp - \frac{(\mathbf{x}-\mathbf{x}_0)^\top A (\mathbf{x}-\mathbf{x}_0)}{2}, \tag{3.187}$$

where $p(\mathbf{x}_0)$ is the normalization constant equal to $\frac{|A|^{1/2}}{(2\pi)^{N/2}}$. Hence we have

$$q(\mathbf{x}) = \frac{|A|^{1/2}}{(2\pi)^{N/2}} \exp - \frac{(\mathbf{x}-\mathbf{x}_0)^\top A (\mathbf{x}-\mathbf{x}_0)}{2} = \mathcal{N}(\mathbf{x}_0, A^{-1}). \tag{3.188}$$

3.9.4 Inference under uncertain evidence

For a Bayesian Network (BN), we can perform probability inference for a query node X_Q given an evidence $X_E = x_e$ by computing $P(X_Q|X_E = x_E)$. However, the conventional inference method assumes there exists no uncertainty in the evidence. For many real world applications, the evidence we observe contains uncertainty, which may originate from noise in the data or from the imprecision with the evidence measuring device. We refer such evidence as *uncertain evidence.* Based on how the uncertainty of the evidence is interpreted and represented, we can classify the uncertain evidence into two categories: soft evidence and virtual evidence.

Let X_E be the uncertain evidence variable. Soft evidence captures the uncertainty of X_E by a probability $q(X_E)$, while the virtual evidence encodes the uncertainty in X by it's likelihood ratio with respect to a virtual binary variable Z. Inference with soft evidence can be performed using Jeffery's rule [94] as in Eq. (3.57), that is,

$$p(\mathbf{x}_Q|q(\mathbf{x}_E)) = \sum_{\mathbf{x}_E} p(\mathbf{x}_Q|\mathbf{x}_E) q(\mathbf{x}_E) \tag{3.189}$$

For virtual evidence, according to Judea Pearl [99], we can introduce a virtual node Z as a child of X_E. Z has the same states as node X_E. The CPT for node Z is partially specified by the likelihood ratio of X_E, i.e., $\frac{p(Z=z|X_E=x_E)}{p(Z=z|X_E\neq x_E)}$. Given this specification, we can then perform inference of $p(X_Q|Z=z)$. Further details about these two types of uncertain evidences can be found in [100–105].

3.9.5 Hamiltonian Monte-Carlo (HMC) sampling

The benefit of the MH algorithm is that it allows sampling from an un-normalized probability. Despite it's strengths, traditional MH algorithm uses a simple proposal distribution, largely based on random walk, such as a Gaussian distribution; it is hence inefficient and cannot scale up well to high dimensional space. HMC [106] was introduced to address this limitation by employing Hamiltonian dynamics to speed up proposal generation. In a Hamilton dynamical system, denote $\boldsymbol{x}$ as the position, r as the momentum, t as the time, $H=H(\boldsymbol{x},r,t)$ is the Hamiltonian function that captures the total energy of the system and it satisfies the Hamiltonian equations:

$$\frac{dr}{dt}=-\frac{\partial\mathcal{H}}{\partial\boldsymbol{x}},\quad \frac{d\boldsymbol{x}}{dt}=+\frac{\partial\mathcal{H}}{\partial r} \tag{3.190}$$

For a closed system, the total energy equals to sum of the potential energy and the kinetic energy. Denote the random variables that we want to sample as $\mathbf{x}$; the position $\boldsymbol{x}$ in Eq. (3.190) is now replaced by $\mathbf{x}$. Define the potential energy function as $U(\mathbf{x})=-\log p(\mathbf{x})$ and the kinetic energy function as $K(r)=\frac{1}{2}r^T\Sigma^{-1}r$, the Hamiltonian function can be written as

$$H(\mathbf{x},r)=U(\mathbf{x})+K(r)=-\log p(\mathbf{x})+\frac{1}{2}r^T\Sigma^{-1}r \tag{3.191}$$

where we assume r follows a Gaussian distribution $\mathcal{N}(0,\Sigma)$. Combining Eq. (3.190) and Eq. (3.191), we can derive the update equations for the proposal in HMC. In a discretized version, denote the step size as ϵ, the iteration number as i, and we have the update equations:

$$\begin{aligned} r_i &= r_{i-1}-\epsilon\nabla U(\mathbf{x}_{i-1}) \\ \mathbf{x}_i &= \mathbf{x}_{i-1}+\epsilon\nabla K(r_i)=\mathbf{x}_{i-1}+\epsilon\Sigma^{-1}r_i \end{aligned} \tag{3.192}$$

The update equation guarantees next sample has a higher probability than the current sample. Algorithm 3.15 provides the pseudo-code for the HMC sampling method. Further information about the HMC method may be found in [107].

The benefit of HMC algorithm compared to traditional MH algorithm is that the dynamics speeds up inference because the momentum r of the system prevents the random walk behavior. Distances between successively generated proposal points from HMC are typically large, so we need fewer iterations to obtain representative samples. And since the proposal distribution moves towards the direction that maximizes the target distribution, HMC in most cases accepts new states. To summarize, by explicitly exploiting the Hamiltonian dynamics, HMC is significantly more efficient than traditional MH algorithm.

Algorithm 3.15 Hamiltonian Monte Carlo algorithm.

Input: Unnormalized target distribution $\tilde{p}(\mathbf{x})$, momentum distribution $p(r) = \mathcal{N}(0, \Sigma)$, potential energy function $U(\mathbf{x}) = -\log \tilde{p}(\mathbf{x})$, the total energy function $U(\mathbf{x}) + \frac{1}{2}r^T\Sigma^{-1}r$, and step size ϵ.
Initialize: Starting position $\mathbf{x}_0$ at $t = 0$.
for $t = 0, 1, 2, \ldots$ **do**
  $r^t \sim \mathcal{N}(0, \Sigma)$; ▷ Sample momentum r^t
  $r^t \leftarrow r^t - \frac{\epsilon}{2}\nabla U(\mathbf{x}^t)$;
  for $i = 1, 2, \ldots m$ **do** ▷ Simulate discretized Hamiltonian dynamics
    $\mathbf{x}_i \leftarrow \mathbf{x}_{i-1} + \epsilon\Sigma^{-1}r_{i-1}$;
    $r_i \leftarrow r_{i-1} - \epsilon\nabla U(\mathbf{x}_i)$;
  end for
  $r_m \leftarrow r_m - \frac{\epsilon}{2}\nabla U(\mathbf{x}_m)$;
  set $(\mathbf{x}', r') = (\mathbf{x}_m, r_m)$;
  $A(\mathbf{x}', \mathbf{x}^t) = \min\left(1, e^{H(\mathbf{x}^t, r^t) - H(\mathbf{x}', r')}\right)$; ▷ Compute the acceptance probability
  $u \sim \mathcal{U}(0, 1)$; ▷ Generate a uniform random number
  if $u \leq A(\mathbf{x}', \mathbf{x}^t)$ **then** $\mathbf{x}^{t+1} = \mathbf{x}'$; ▷ Accept $\mathbf{x}'$ based on $A(\mathbf{x}', \mathbf{x}^t)$
  else $\mathbf{x}^{t+1} = \mathbf{x}^t$; ▷ Reject $\mathbf{x}'$, and use the old state
  end if
end for

3.9.6 BN belief propagation examples

Given the structure and parameters of a Bayesian Network as shown in Fig. 3.48, we perform detailed step-by-step calculations to perform the sum-product and max-product inference.

3.9.6.1 Sum-product BP examples

We first show the process for sum-product inference given a boundary evidence. This is then followed by sum-product inference given an non-boundary evidence.

Sum-product inference given boundary evidence

We perform the sum-product inference given the evidence: `M`(Marry Calls) = '1'. Particularly, we follow the order of nodes below for message passing. Please note this is not necessary the optimal order.

1. Node A (Alarm) collects all messages from its children M and J and its parents E and B, updates its belief, and normalizes
2. Node B (Burglary) collects its message from child A, updates its belief, and normalizes
3. Node E (Earthquake) collects its message from child A, updates its belief, and normalizes

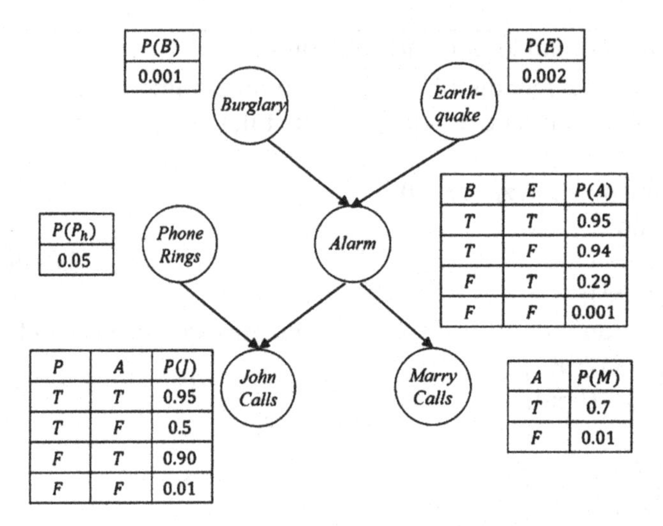

FIGURE 3.48 An example Bayesian Network.

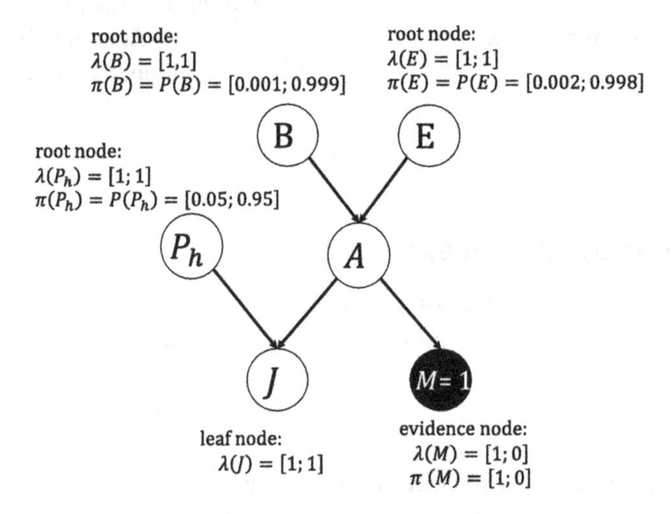

FIGURE 3.49 Initialization of messages for sum of product inference with a boundary evidence.

4. Node J (John Calls) collects its messages from it's parents Ph and A, updates its belief, and normalizes
5. Node Ph (Phone Rings) collects its message from child J, updates its belief, and normalizes
6. Node M (Marry Calls) collects its message from parent A, updates its belief, and normalizes (optional)

We first initialize the messages for the boundary nodes and the evidence node as shown in Fig. 3.49. The messages for the remaining nodes are initialized to ones. In addition, the incoming messages to each node (including the evidence node) are initialized to ones. For the evidence node M, we need revise the entries of the CPTs involving node M as follows: the CPT entries corresponding to $m \neq 1$ are set to zeros, i.e., $p(m=0|a)=0$ and the entries corresponding to $m=1$, i.e., $p(m=1|a)$, remain unchanged. Below, we show the detailed calculation of messages and belief updating of each node using Eqs. (3.23)–(3.26) for the first iteration.

Message computation and belief updating for node A

The message node A receives from its parent E is

$$\pi_e(a) = \pi(e) \prod_{c \in \text{Child}(e) \backslash a} \lambda_c(e) = \pi(e) = [0.002; 0.998] \tag{3.193}$$

The message node A receives from its parent B is

$$\pi_b(a) = \pi(b) \prod_{c \in \text{Child}(b) \backslash a} \lambda_c(b) = \pi(b) = [0.001; 0.999] \tag{3.194}$$

The total message node A receives from its parents nodes E and B is

$$\begin{aligned} \pi(a) &= \sum_{b,e} p(a|b,e)\pi_e(a)\pi_b(a) \\ &= \begin{bmatrix} \sum_{b,e} p(a=1|b,e)\pi_e(a)\pi_b(a) \\ \sum_{b,e} p(a=0|b,e)\pi_e(a)\pi_b(a) \end{bmatrix} = \begin{bmatrix} 0.002516 \\ 0.997484 \end{bmatrix} \end{aligned} \tag{3.195}$$

The message node A receives from its child J is

$$\begin{aligned} \lambda_j(a) &= \sum_j \lambda(j) \sum_{p_h} p(j|a, p_h)\pi_{p_h}(j) \\ &= \begin{bmatrix} \sum_j \lambda(j) \sum_{p_h} p(j|a=1, p_h)\pi_{p_h}(j) \\ \sum_j \lambda(j) \sum_{p_h} p(j|a=0, p_h)\pi_{p_h}(j) \end{bmatrix} = \begin{bmatrix} 2 \\ 2 \end{bmatrix} \end{aligned} \tag{3.196}$$

where $\pi_{p_h}(j) = [1; 1]$ as initialized.
The message node A receives from its child M is

$$\begin{aligned} \lambda_m(a) &= \sum_m \lambda(m) p(m|a) \\ &= \begin{bmatrix} \sum_m \lambda(m) p(m|a=1) \\ \sum_m \lambda(m) p(m|a=0) \end{bmatrix} = \begin{bmatrix} 0.7 \\ 0.01 \end{bmatrix} \end{aligned} \tag{3.197}$$

The total message node A receives from its children nodes J and M is

$$\lambda(a) = \lambda_j(a)\lambda_m(a) = \begin{bmatrix} 1.4 \\ 0.02 \end{bmatrix} \tag{3.198}$$

Given current messages, we obtain the normalized belief of node A:

$$Bel(a) = \alpha\pi(a)\lambda(a) = \begin{bmatrix} 0.150068 \\ 0.849932 \end{bmatrix} \quad (3.199)$$

Message computation and belief updating for node B

The message node B receives from its child A is

$$\begin{aligned} \lambda_a(b) &= \sum_a \lambda(a) \sum_e p(a|b,e)\pi_e(a) \\ &= \begin{bmatrix} \sum_a \lambda(a) \sum_e p(a|b=1,e)\pi_e(a) \\ \sum_a \lambda(a) \sum_e p(a|b=0,e)\pi_e(a) \end{bmatrix} = \begin{bmatrix} 1.317228 \\ 0.022178 \end{bmatrix} \end{aligned} \quad (3.200)$$

where $\pi_e(a) = [0.002; 0.998]$ is calculated in Eq. (3.193). The total message node B receives from its child A is

$$\lambda(b) = \lambda_a(b) = \begin{bmatrix} 1.317228 \\ 0.022178 \end{bmatrix} \quad (3.201)$$

Given current messages, we obtain the normalized belief of node B:

$$Bel(b) = \alpha\pi(b)\lambda(b) = \begin{bmatrix} 0.056117 \\ 0.943883 \end{bmatrix} \quad (3.202)$$

Message computation and belief updating for node E

The message node E receives from its child A is

$$\begin{aligned} \lambda_a(e) &= \sum_a \lambda(a) \sum_b p(a|b,e)\pi_b(a) \\ &= \begin{bmatrix} \sum_a \lambda(a) \sum_b p(a|b,e=1)\pi_b(a) \\ \sum_a \lambda(a) \sum_b p(a|b,e=0)\pi_b(a) \end{bmatrix} = \begin{bmatrix} 0.421111 \\ 0.022676 \end{bmatrix} \end{aligned} \quad (3.203)$$

where $\pi_b(a) = [0.001; 0.999]$ is calculated in Eq. (3.194).
The total message node E receives from its child A is

$$\lambda(e) = \lambda_a(e) = \begin{bmatrix} 0.421111 \\ 0.022676 \end{bmatrix} \quad (3.204)$$

Given current messages, we obtain the normalized belief of node E

$$Bel(e) = \alpha\pi(e)\lambda(e) = \begin{bmatrix} 0.035881 \\ 0.964119 \end{bmatrix} \quad (3.205)$$

Message computation and belief updating for node J

The message node J receives from its parent A is

$$\pi_a(j) = \pi(a) \prod_{c \in \text{Child}(a) \backslash j} \lambda_c(a) = \pi(a)\lambda_m(a) = \begin{bmatrix} 0.001761 \\ 0.009975 \end{bmatrix} \tag{3.206}$$

where $\lambda_m(a) = [0.7; 0.01]$ is calculated in Eq. (3.197).
The message node J receives from its parent Ph is

$$\pi_{p_h}(j) = \pi(p_h) \prod_{c \in \text{Child}(p_h) \backslash j} \lambda_c(p_h) = \pi(p_h) = \begin{bmatrix} 0.05 \\ 0.95 \end{bmatrix} \tag{3.207}$$

The total message node J receives from its parent nodes Ph and A is

$$\begin{aligned} \pi(j) &= \sum_{p_h, a} p(j|p_h, a)\pi_{p_h}(j)\pi_a(j) \\ &= \begin{bmatrix} \sum_{p_h,a} p(j=1|p_h, a)\pi_{p_h}(j)\pi_a(j) \\ \sum_{p_h,a} p(j=0|p_h, a)\pi_{p_h}(j)\pi_a(j) \end{bmatrix} = \begin{bmatrix} 0.001933 \\ 0.009803 \end{bmatrix} \end{aligned} \tag{3.208}$$

where $\pi_a(j) = [0.001761; 0.009975]$ is calculated in Eq. (3.206) and $\pi_{p_h}(j) = [0.05; 0.95]$ is calculated in Eq. (3.207).
Given current messages, we obtain the normalized belief of node J

$$Bel(j) = \alpha\pi(j)\lambda(j) = \begin{bmatrix} 0.164707 \\ 0.835293 \end{bmatrix} \tag{3.209}$$

Message computation and belief updating for node Ph

The message node Ph receives from its child J is

$$\begin{aligned} \lambda_j(p_h) &= \sum_j \lambda(j) \sum_a p(j|p_h, a)\pi_a(j) \\ &= \begin{bmatrix} \sum_j \lambda(j) \sum_a p(j|p_h=1, a)\pi_a(j) \\ \sum_j \lambda(j) \sum_a p(j|p_h=0, a)\pi_a(j) \end{bmatrix} = \begin{bmatrix} 0.011736 \\ 0.011736 \end{bmatrix} \end{aligned} \tag{3.210}$$

where $\pi_a(j) = [0.001761; 0.009975]$ is calculated in Eq. (3.206).
The total message node Ph receives from its child J is

$$\lambda(p_h) = \lambda_j(p_h) = \begin{bmatrix} 0.011736 \\ 0.011736 \end{bmatrix} \tag{3.211}$$

Given current messages, we obtain the normalized belief of node Ph

$$Bel(p_h) = \alpha\pi(p_h)\lambda(p_h) = \begin{bmatrix} 0.0500 \\ 0.9500 \end{bmatrix} \tag{3.212}$$

Message computation and belief updating for node M

$$\pi_a(m) = \pi(a) \prod_{c \in \text{Child}(a) \setminus m} \lambda_c(a) = \pi(a)\lambda_j(a) = \begin{bmatrix} 0.005032 \\ 1.994968 \end{bmatrix} \tag{3.213}$$

where $\lambda_j(a) = [2; 2]$ is calculated in Eq. (3.196). The total message node M receives from its parent A is

$$\pi(m) = \sum_a p(m|a)\pi_a(m) = \begin{bmatrix} 0.023472 \\ 0 \end{bmatrix} \tag{3.214}$$

where $p(m = 0|a) = 0$ as we revise given the evidence $m = 1$. Given current messages, we obtain the normalized belief of node M

$$Bel(m) = \alpha\pi(m)\lambda(m) = \begin{bmatrix} 1 \\ 0 \end{bmatrix} \tag{3.215}$$

where $\lambda(m) = [1; 0]$ as initialized. Since node M is the evidence node, its belief doesn't change over iterations. Note updating belief for node M is optional as it is a boundary evidence node.

We now finish the first iteration. We repeat the above process for the second and third iteration. During each iteration, for each node, we update its messages based on the current messages the node receives from its parents and children. Comparing the messages and beliefs from the second iteration and the third iteration, we can observe that there is no change, i.e., the belief propagation converges after the second iteration. In the end, we obtain the belief of each node given the evidence as

$$\begin{aligned} Bel(a) &= p(a|m = 1) = [0.150068; 0.849932] \\ Bel(b) &= p(b|m = 1) = [0.056117; 0.943883] \\ Bel(e) &= p(e|m = 1) = [0.035881; 0.964119] \\ Bel(j) &= p(j|m = 1) = [0.164707; 0.835293] \\ Bel(p_h) &= p(p_h|m = 1) = [0.050000; 0.950000] \end{aligned} \tag{3.216}$$

Sum-product inference given non-boundary evidence

We now consider another example where we are given a non-boundary evidence `Alarm = '1'`. The initialization of the boundary nodes, the non-boundary nodes, and evidence nodes remain the same as we did for the boundary evidence case as shown in Fig. 3.50. In addition, the incoming messages to each node are initialized to ones. For the evidence A, the entries of CPTs involving node A are revised as follows: the entries of the CPTs corresponding to $a \neq 1$ are set to zeros, i.e., $p(a = 0|b, e) = 0$, $p(j|p_h, a = 0) = 0$ and $p(m|a = 0) = 0$, while the entries of CPTs corresponding to $a = 1$, i.e., $p(a = 1|b, e)$, $p(j|p_h, a = 1)$ and $p(m|a = 1)$, remain unchanged as the original conditional probabilities. Given the initialization, the belief propagation process remains the same. In the following,

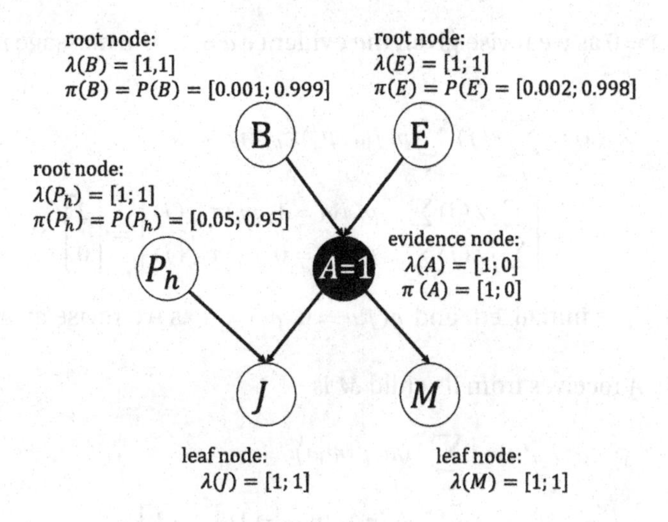

FIGURE 3.50 Initialization of messages for sum of product inference with a non-boundary evidence node.

we perform the sum-product inference given an non-boundary evidence `Alarm = 1` for the same example shown in Fig. 3.48 and following the same node order.

Message computation and belief updating for node A[8]

The message node A receives from its parent E is

$$\pi_e(a) = \pi(e) \prod_{c \in \text{Child}(e) \backslash a} \lambda_c(e) = \pi(e) = [0.002; 0.998] \tag{3.217}$$

The message node A receives from its parent B is

$$\pi_b(a) = \pi(b) \prod_{c \in \text{Child}(b) \backslash a} \lambda_c(b) = \pi(b) = [0.001; 0.999] \tag{3.218}$$

The total message node A receives from its parents nodes E and B is

$$\begin{aligned} \pi(a) &= \sum_{b,e} p(a|b,e)\pi_e(a)\pi_b(a) \\ &= \begin{bmatrix} \sum_{b,e} p(a=1|b,e)\pi_e(a)\pi_b(a) \\ \sum_{b,e} p(a=0|b,e)\pi_e(a)\pi_b(a) \end{bmatrix} = \begin{bmatrix} 0.002516 \\ 0 \end{bmatrix} \end{aligned} \tag{3.219}$$

[8] Updating the messages and belief for node A is necessary as it is not a boundary evidence node and it's messages are needed to update the belief of other nodes.

where $p(a=0|b,e)=0$ as we revise given the evidence $a=1$. The message node A receives from its child J is

$$\begin{aligned}\lambda_j(a) &= \sum_j \lambda(j) \sum_{p_h} p(j|a,p_h)\pi_{p_h}(j) \\ &= \begin{bmatrix} \sum_j \lambda(j) \sum_{p_h} p(j|a=1,p_h)\pi_{p_h}(j) \\ \sum_j \lambda(j) \sum_{p_h} p(j|a=0,p_h)\pi_{p_h}(j) \end{bmatrix} = \begin{bmatrix} 2 \\ 0 \end{bmatrix}\end{aligned} \tag{3.220}$$

where $\pi_{p_h}(j)=[1;1]$ as initialized, and $p(j|a=0,p_h)=0$ as we revise given the evidence $a=1$.

The message node A receives from its child M is

$$\begin{aligned}\lambda_m(a) &= \sum_m \lambda(m)p(m|a) \\ &= \begin{bmatrix} \sum_m \lambda(m)p(m|a=1) \\ \sum_m \lambda(m)p(m|a=0) \end{bmatrix} = \begin{bmatrix} 1 \\ 0 \end{bmatrix}\end{aligned} \tag{3.221}$$

where $p(m|a=0)=0$ as we revise given the evidence $a=1$.

The total message node A receives from its children nodes J and M is

$$\lambda(a) = \lambda_j(a)\lambda_m(a) = \begin{bmatrix} 2 \\ 0 \end{bmatrix} \tag{3.222}$$

Given current messages, we obtain the normalized belief of node A:

$$Bel(a) = \alpha\pi(a)\lambda(a) = \begin{bmatrix} 1 \\ 0 \end{bmatrix} \tag{3.223}$$

Since node A is the evidence node, its belief doesn't change over iterations.

Messages computation and belief updating for node B

The message node B receives from its child A is

$$\begin{aligned}\lambda_a(b) &= \sum_a \lambda(a) \sum_e p(a|b,e)\pi_e(a) \\ &= \begin{bmatrix} \sum_a \lambda(a) \sum_e p(a|b=1,e)\pi_e(a) \\ \sum_a \lambda(a) \sum_e p(a|b=0,e)\pi_e(a) \end{bmatrix} = \begin{bmatrix} 1.880040 \\ 0.003156 \end{bmatrix}\end{aligned} \tag{3.224}$$

where $\pi_e(a)=[0.002; 0.998]$ is calculated in Eq. (3.217).

The total message node B receives from its child A is

$$\lambda(b) = \lambda_a(b) = \begin{bmatrix} 1.880040 \\ 0.003156 \end{bmatrix} \tag{3.225}$$

Given current messages, we obtain the normalized belief for node B

$$Bel(b) = \alpha\pi(b)\lambda(b) = \begin{bmatrix} 0.373551 \\ 0.626449 \end{bmatrix} \tag{3.226}$$

Message computation and belief updating for node E

The message node E receives from its child A is

$$\begin{aligned} \lambda_a(e) &= \sum_a \lambda(a) \sum_b p(a|b,e)\pi_b(a) \\ &= \begin{bmatrix} \sum_a \lambda(a) \sum_b p(a|b,e=1)\pi_b(a) \\ \sum_a \lambda(a) \sum_b p(a|b,e=0)\pi_b(a) \end{bmatrix} = \begin{bmatrix} 0.581320 \\ 0.003878 \end{bmatrix} \end{aligned} \tag{3.227}$$

where $\pi_b(a) = [0.001; 0.999]$ is calculated in Eq. (3.218).
The total message node E receives from its child A is

$$\lambda(e) = \lambda_a(e) = \begin{bmatrix} 0.581320 \\ 0.003878 \end{bmatrix} \tag{3.228}$$

Given current messages, we obtain the normalized belief of node E

$$Bel(e) = \alpha\pi(e)\lambda(e) = \begin{bmatrix} 0.231009 \\ 0.768991 \end{bmatrix} \tag{3.229}$$

Message computation and belief updating for node J

The message node J receives from its parent A is

$$\pi_a(j) = \pi(a) \prod_{c \in \text{Child}(a)\setminus j} \lambda_c(a) = \pi(a)\lambda_m(a) = \begin{bmatrix} 0.002516 \\ 0 \end{bmatrix} \tag{3.230}$$

where $\lambda_m(a) = [1; 0]$ is calculated in Eq. (3.221).
The message node J receives from its parent Ph is

$$\pi_{p_h}(j) = \pi(p_h) \prod_{c \in \text{Child}(p_h)\setminus j} \lambda_c(p_h) = \pi(p_h) = \begin{bmatrix} 0.05 \\ 0.95 \end{bmatrix} \tag{3.231}$$

The total message node J receives from its parent nodes Ph and A is

$$\begin{aligned} \pi(j) &= \sum_{p_h,a} p(j|p_h,a)\pi_{p_h}(j)\pi_a(j) \\ &= \begin{bmatrix} \sum_{p_h,a} p(j=1|p_h,a)\pi_{p_h}(j)\pi_a(j) \\ \sum_{p_h,a} p(j=0|p_h,a)\pi_{p_h}(j)\pi_a(j) \end{bmatrix} = \begin{bmatrix} 0.002271 \\ 0.000245 \end{bmatrix} \end{aligned} \tag{3.232}$$

where $\pi_a(j) = [0.002516; 0]$ is calculated in Eq. (3.230) and $\pi_{p_h}(j) = [0.05; 0.95]$ is calculated in Eq. (3.231).
Given current messages, we obtain the normalized belief for node J

$$Bel(j) = \alpha\pi(j)\lambda(j) = \begin{bmatrix} 0.902623 \\ 0.097377 \end{bmatrix} \tag{3.233}$$

Message computation and belief updating for node Ph

The message node Ph receives from its child J is

$$\begin{aligned} \lambda_j(p_h) &= \sum_j \lambda(j) \sum_a p(j|p_h, a)\pi_a(j) \\ &= \begin{bmatrix} \sum_j \lambda(j) \sum_a p(j|p_h = 1, a)\pi_a(j) \\ \sum_j \lambda(j) \sum_a p(j|p_h = 0, a)\pi_a(j) \end{bmatrix} = \begin{bmatrix} 0.002516 \\ 0.002516 \end{bmatrix} \end{aligned} \tag{3.234}$$

where $\pi_a(j) = [0.002516; 0]$ is calculated in Eq. (3.230).
The total message node Ph receives from child J is

$$\lambda(p_h) = \lambda_j(p_h) = \begin{bmatrix} 0.002516 \\ 0.002516 \end{bmatrix} \tag{3.235}$$

Given current messages, we obtain the normalized belief for node Ph

$$b_{p_h}(p_h) = \alpha\pi(p_h)\lambda(p_h) = \begin{bmatrix} 0.0500 \\ 0.9500 \end{bmatrix} \tag{3.236}$$

Message computation and belief updating for node M

The message node M receives from its parent A is

$$\pi_a(m) = \pi(a) \prod_{c \in \text{Child}(a) \backslash m} \lambda_c(a) = \pi(a)\lambda_j(a) = \begin{bmatrix} 0.005032 \\ 0 \end{bmatrix} \tag{3.237}$$

where $\lambda_j(a) = [2; 0]$ is calculated in Eq. (3.220).
The total message node M receives from its parent A is

$$\pi(m) = \sum_a p(m|a)\pi_a(m) = \begin{bmatrix} 0.003522 \\ 0.001510 \end{bmatrix} \tag{3.238}$$

Given current messages, we obtain the normalized belief of node M

$$Bel(m) = \alpha\pi(m)\lambda(m) = \begin{bmatrix} 0.7000 \\ 0.3000 \end{bmatrix} \tag{3.239}$$

We now finish the first iteration. We repeat the above process for the second and third iteration. Comparing the messages and beliefs from the second iteration and the third iteration, we can observe that there is no change. i.e., the belief propagation converges after the second iteration. In the end, we obtain the marginal distribution of each node given the evidence as

$$\begin{aligned}
Bel(b) &= p(b|a=1) = [0.373551; 0.626449] \\
Bel(e) &= p(e|a=1) = [0.231009; 0.768991] \\
Bel(j) &= p(j|a=1) = [0.902623; 0.097377] \\
Bel(p_h) &= p(p_h|a=1) = [0.050000; 0.950000] \\
Bel(m) &= p(m|a=1) = [0.700000; 0.300000]
\end{aligned} \tag{3.240}$$

3.9.6.2 Max-product inference given non-boundary evidence

To perform max-product inference, we only need to replace the summation operation with the maximization operation. The initialization remains the same. To illustrate the process, we perform the max-product inference given non-boundary evidence: `Alarm = 1`. The initialization, the revision of CPTs involving node A, and the order of nodes for updating remain the same as we did for sum-product inference given non-boundary evidence. In the following, we show the detailed calculation of messages and belief updating of each node for the first iteration.

Message computation and belief updating for node A

The message node A receives from its parent E is

$$\pi_e(a) = \pi(e) \prod_{c \in \text{Child}(e) \backslash a} \lambda_c(e) = \pi(e) = [0.002; 0.998] \tag{3.241}$$

The message node A receives from its parent B is

$$\pi_b(a) = \pi(b) \prod_{c \in \text{Child}(b) \backslash a} \lambda_c(b) = \pi(b) = [0.001; 0.999] \tag{3.242}$$

The total message node A receives from its parents nodes E and B is

$$\begin{aligned}
\pi(a) &= \max_{b,e} p(a|b,e)\pi_e(a)\pi_b(a) \\
&= \begin{bmatrix} \max_{b,e} p(a=1|b,e)\pi_e(a)\pi_b(a) \\ \max_{b,e} p(a=0|b,e)\pi_e(a)\pi_b(a) \end{bmatrix} = \begin{bmatrix} 0.000997 \\ 0 \end{bmatrix}
\end{aligned} \tag{3.243}$$

where $p(a=0|b,e)=0$ as we revise given the evidence $a=1$. The message node A receives from its child J is

$$\begin{aligned}
\lambda_j(a) &= \max_j \lambda(j) \max_{p_h} p(j|a, p_h)\pi_{p_h}(j) \\
&= \begin{bmatrix} \max_j \lambda(j) \max_{p_h} p(j|a=1, p_h)\pi_{p_h}(j) \\ \max_j \lambda(j) \max_{p_h} p(j|a=0, p_h)\pi_{p_h}(j) \end{bmatrix} = \begin{bmatrix} 0.9500 \\ 0 \end{bmatrix}
\end{aligned} \tag{3.244}$$

where $\pi_{p_h}(j) = [1; 1]$ as initialized and $p(j|a=0, p_h) = 0$ as we revise given the evidence $a = 1$.
The message node A receives from its child M is

$$\lambda_m(a) = \max_m \lambda(m) p(m|a) = \begin{bmatrix} \max_m \lambda(m) p(m|a=1) \\ \max_m \lambda(m) p(m|a=0) \end{bmatrix} = \begin{bmatrix} 0.7000 \\ 0 \end{bmatrix} \tag{3.245}$$

where $p(m|a=0) = 0$ as we revise given the evidence $a = 1$.
The total message node A receives from its children nodes J and M is

$$\lambda(a) = \lambda_j(a)\lambda_m(a) = \begin{bmatrix} 0.665000 \\ 0 \end{bmatrix} \tag{3.246}$$

Given current messages, we obtain the normalized belief of node A:

$$Bel(a) = \alpha\pi(a)\lambda(a) = \begin{bmatrix} 1 \\ 0 \end{bmatrix} \tag{3.247}$$

Since node A is the evidence node, its belief doesn't change over iterations.

Message computation and belief updating for node B

The message node B receives from its child A is

$$\lambda_a(b) = \max_a \lambda(a) \max_e p(a|b, e)\pi_e(a) = \begin{bmatrix} \max_a \lambda(a) \max_e p(a|b=1, e)\pi_e(a) \\ \max_a \lambda(a) \max_e p(a|b=0, e)\pi_e(a) \end{bmatrix} = \begin{bmatrix} 0.623850 \\ 0.000664 \end{bmatrix} \tag{3.248}$$

where $\pi_e(a) = [0.002; 0.998]$ is calculated in Eq. (3.241).
The total message node B receives from its child A is then

$$\lambda(b) = \lambda_a(b) = \begin{bmatrix} 0.623850 \\ 0.000664 \end{bmatrix} \tag{3.249}$$

Given current messages, we obtain the normalized belief of node B

$$Bel(b) = \alpha\pi(b)\lambda(b) = \begin{bmatrix} 0.484662 \\ 0.515338 \end{bmatrix} \tag{3.250}$$

Message computation and belief updating for node E

The message node E receives from its child A is

$$\lambda_a(e) = \max_a \lambda(a) \max_b p(a|b,e)\pi_b(a)$$
$$= \begin{bmatrix} \max_a \lambda(a) \max_b p(a|b,e=1)\pi_b(a) \\ \max_a \lambda(a) \max_b p(a|b,e=0)\pi_b(a) \end{bmatrix} = \begin{bmatrix} 0.192657 \\ 0.000664 \end{bmatrix} \tag{3.251}$$

where $\pi_b(a) = [0.001; 0.999]$ is calculated in Eq. (3.242).
The total message node E receives from its child A is then

$$\lambda(e) = \lambda_a(e) = \begin{bmatrix} 0.192657 \\ 0.000664 \end{bmatrix} \tag{3.252}$$

Given current messages, we obtain the normalized belief of node E

$$Bel(e) = \alpha\pi(e)\lambda(e) = \begin{bmatrix} 0.367671 \\ 0.632329 \end{bmatrix} \tag{3.253}$$

Message computation and belief updating for node J

The message node J receives from its parent A is

$$\pi_a(j) = \pi(a) \prod_{c \in \text{Child}(a) \setminus j} \lambda_c(a) = \pi(a)\lambda_m(a) = \begin{bmatrix} 0.000698 \\ 0 \end{bmatrix} \tag{3.254}$$

where $\lambda_m(a) = [0.7000; 0]$ is calculated in Eq. (3.245).
The message node J receives from its parent Ph is

$$\pi_{p_h}(j) = \pi(p_h) \prod_{c \in \text{Child}(p_h) \setminus j} \lambda_c(p_h) = \pi(p_h) = \begin{bmatrix} 0.05 \\ 0.95 \end{bmatrix} \tag{3.255}$$

The total message node J receives from its parent nodes Ph and A is then

$$\pi(j) = \max_{p_h,a} p(j|p_h,a)\pi_{p_h}(j)\pi_a(j)$$
$$= \begin{bmatrix} \max_{p_h,a} p(j=1|p_h,a)\pi_{p_h}(j)\pi_a(j) \\ \max_{p_h,a} p(j=0|p_h,a)\pi_{p_h}(j)\pi_a(j) \end{bmatrix} = \begin{bmatrix} 0.000597 \\ 0.000066 \end{bmatrix} \tag{3.256}$$

Given current messages, we obtain the normalized belief of node J

$$Bel(j) = \alpha\pi(j)\lambda(j) = \begin{bmatrix} 0.900452 \\ 0.099548 \end{bmatrix} \tag{3.257}$$

Message computation and belief updating for node Ph

The message node Ph receives from its child J is

$$\lambda_j(p_h) = \max_j \lambda(j) \max_a p(j|p_h, a)\pi_a(j)$$
$$= \begin{bmatrix} \max_j \lambda(j) \max_a p(j|p_h = 1, a)\pi_a(j) \\ \max_j \lambda(j) \max_a p(j|p_h = 0, a)\pi_a(j) \end{bmatrix} = \begin{bmatrix} 0.000663 \\ 0.000628 \end{bmatrix} \tag{3.258}$$

where $\pi_a(j) = [0.000698; 0]$ is calculated in Eq. (3.254).
The total message node Ph receives from child J is then

$$\lambda(p_h) = \lambda_j(p_h) = \begin{bmatrix} 0.000663 \\ 0.000628 \end{bmatrix} \tag{3.259}$$

Given current messages, we obtain the normalized belief of node Ph

$$Bel(p_h) = \alpha\pi(p_h)\lambda(p_h) = \begin{bmatrix} 0.052640 \\ 0.947360 \end{bmatrix} \tag{3.260}$$

Message computation and belief updating for node M

The message node M receives from its parent A is

$$\pi_a(m) = \pi(a) \prod_{c \in \text{Child}(a) \backslash m} \lambda_c(a) = \pi(a)\lambda_j(a) = \begin{bmatrix} 0.000947 \\ 0 \end{bmatrix} \tag{3.261}$$

where $\lambda_j(a) = [0.9500; 0]$ is calculated in Eq. (3.244).
The total message node M receives from its parent A is then

$$\pi(m) = \max_a p(m|a)\pi_a(m) = \begin{bmatrix} 0.000663 \\ 0.000284 \end{bmatrix} \tag{3.262}$$

Given current messages, we obtain the normalized belief of node M

$$Bel(m) = \alpha\pi(m)\lambda(m) = \begin{bmatrix} 0.7000 \\ 0.3000 \end{bmatrix} \tag{3.263}$$

We now finish the first iteration. We repeat the above process for two more iterations, and observe that there is no change in the messages and beliefs at third iteration, i.e., the belief propagation converges after the second iteration. We can then find the unique MAP assignment by performing max marginal for each node independently if there are no ties in any of the updated node beliefs, yielding the MAP configuration for the example as

$$[0, 0, 1, 0, 1] = \arg \max_{b,e,j,p_h,m} p(b, e, j, p_h, m|a = 1) \tag{3.264}$$

via $x^* = \arg\max_x Bel(x)$ for $x \in \{B, E, J, Ph, M\}$.

3.9.7 EM learning for HMM

In this section, we introduce HMM learning using the standard EM method. Let an HMM be defined over the state variables $\mathbf{Q}$, observation variables $\mathbf{O}$, and parameters $\boldsymbol{\lambda} = \{\Lambda, A, B\}$. Given the training sequences $\mathbf{O} = \{O(m)\}_{m=1}^{M}$, where $O(m) = \{O^t(m)\}_{t=0}^{t_m}$, HMM learning is to find $\boldsymbol{\lambda}$ by maximizing it's log-likelihood, i.e.,

$$\boldsymbol{\lambda}^* = \arg\max_{\boldsymbol{\lambda}} \log p(\mathbf{O}|\boldsymbol{\lambda})$$

Let $Q(m)$ be the unobserved state sequence corresponding to $\mathbf{O}(m)$, $\log p(\mathbf{O}|\boldsymbol{\lambda})$ can be computed as follows

$$\begin{aligned}
\log p(\mathbf{O}|\boldsymbol{\lambda}) &= \sum_{m=1}^{M} \log p(O(m)|\boldsymbol{\lambda}) \\
&= \sum_{m=1}^{M} \log \sum_{Q(m)} p(O(m), Q(m)|\boldsymbol{\lambda}), \\
&= \sum_{m=1}^{M} \log \sum_{Q(m)} q(Q(m)|O(m), \Theta_q) \frac{p(O(m), Q(m)|\boldsymbol{\lambda})}{q(Q(m)|O(m), \Theta_q)} \\
&\geq \sum_{m=1}^{M} \sum_{Q(m)} q(Q(m)|O(m), \Theta_q) \log p(O(m), Q(m)|\boldsymbol{\lambda}) \quad \text{Jensen's inequality} \\
&= \sum_{m=1}^{M} \sum_{Q(m)} q(Q(m)|O(m), \Theta_q) \log p(O^0(m)|\Lambda) \prod_{t=1}^{t_m} p(O^t(m)|Q^t(m)|\mathbf{B}) p(Q^t(m)|Q^{t-1}(m), \mathbf{A}) \\
&= \sum_{m=1}^{M} \sum_{Q(m)} q(Q(m)|O(m), \Theta_q) \log p(O^0(m)|\Lambda) + \\
&\quad \sum_{m=1}^{M} \sum_{Q(m)} q(Q(m)|O(m), \Theta_q) \sum_{t=1}^{t_m} \log(O^t(m)|Q^t(m), \mathbf{B}) + \\
&\quad \sum_{m=1}^{M} \sum_{Q(m)} q(Q(m)|O(m), \Theta_q) \sum_{t=1}^{t_m} \log p(Q^t(m)|Q^{t-1}(m), \mathbf{A})
\end{aligned} \tag{3.265}$$

It is clear from Eq. (3.265) that given $q(Q(m)|O(m), \Theta_q)$, parameters Λ, $\mathbf{A}$, and $\mathbf{B}$ can be computed separately by maximizing their expected likelihoods, corresponding to the three terms. Based on this, we can introduce the EM method as follows.

Set $q(Q(m)|O(m), \Theta_q) = p(Q(m)|O(m), \boldsymbol{\lambda}^{t-1})$ and initialize $\boldsymbol{\lambda}$ to $\boldsymbol{\lambda}^0$.
E-step:
Compute $p(Q(m)|O(m), \boldsymbol{\lambda}^{t-1})$ for all possible configurations of $Q(m)$ and for all sequences
M-step:
Find Λ^t, $\mathbf{A}^t$, and $\mathbf{B}^t$ by maximizing the their expected loglikelihood, i.e.,

$$\Lambda^t = \arg\max_{\Lambda} \sum_{m=1}^{M} \sum_{Q(m)} p(Q(m)|O(m), \boldsymbol{\lambda}^{t-1}) \log p(O^0(m)|\Lambda)$$

$$\mathbf{B}^t = \arg\max_{\mathbf{B}} \sum_{m=1}^{M} \sum_{Q(m)} p(Q(m)|O(m), \boldsymbol{\lambda}^{t-1}) \sum_{t=1}^{t_m} \log(O^t(m)|Q^t(m), \mathbf{B})$$

$$\mathbf{A}^t = \arg\max_{\mathbf{A}} \sum_{m=1}^{M} \sum_{Q(m)} p(Q(m)|O(m), \boldsymbol{\lambda}^{t-1}) \sum_{t=1}^{t_m} \log p(Q^t(m)|Q^{t-1}(m), \mathbf{A})$$

Repeat the E and M steps until convergence

Algorithm 3.16 HMM EM learning pseudo-code.

Initialize $\boldsymbol{\lambda}$ to $\boldsymbol{\lambda}^0$ and $w(c, m)$ and $S(m, t, i, j)$ to zeros
E-step:
for $m = 1$ *to* M **do**
  for $c = 1$ *to* C_m **do** ▷ C_m is the number of configurations of $Q(m)$
    $w(c, m) = p(Q_c(m), O(m)|\boldsymbol{\lambda}^{t-1})$ ▷ $Q_c(m)$ is the c-th configuration of $Q(m)$
  end for
end for
Compute the expected state transition counts
for $m = 1$ *to* M **do**
  for $c = 1$ *to* C_m **do**
    for $i = 1$ *to* N **do**
      for $j = 1$ *to* N **do**
        $S(m, t, i, j) = S(m, t, i, j) + I(Q_c^{t-1}(m) = i \wedge Q_c^t(m) = j) w(c, m)$
      end for
    end for
  end for
end for
M-step:

$$a_{ij} = \frac{\sum_{m=1}^{M} \sum_{t=0}^{t_m} S(m, t, i, j)}{\sum_{m=1}^{M} \sum_{t=1}^{t_m} \sum_{j=1}^{N} S(m, t, i, j)}$$

$$b_i(k) = \frac{\sum_{m=1}^{M} \sum_{t=0}^{t_m} \sum_{j=1}^{N} S(m, t, i, j) I(O^t(m) = k)}{\sum_{m=1}^{M} \sum_{t=0}^{t_m} \sum_{j=1}^{N} S(m, t, i, j)}$$

$$\pi_i = \frac{\sum_{m=1}^{M} \sum_{j=1}^{N} S(m, 0, i, j)}{\sum_{m=1}^{M} C_m}$$

Repeat E and M step until convergence

References

[1] N. Friedman, D. Koller, Tutorial on learning Bayesian networks from data, in: NIPS, 2001 [online], available: http://www.cs.huji.ac.il/~nirf/NIPS01-Tutorial/Tutorial.pps.

[2] R.E. Neapolitan, et al., Learning Bayesian Networks, 2004.

[3] Variational inference, http://www.cs.cmu.edu/~guestrin/Class/10708/recitations/r9/VI.ppt.

[4] N. Oliver, E. Horvitz, A. Garg, Layered representation for human activity recognition, Computer Vision and Image Understanding (2004).

[5] H. Lee, R. Grosse, R. Ranganath, A.Y. Ng, Convolutional deep belief networks for scalable unsupervised learning of hierarchical representations, in: Proceedings of the 26th Annual International Conference on Machine Learning, ACM, 2009, pp. 609–616.

[6] D.M. Blei, A.Y. Ng, M.I. Jordan, Latent Dirichlet allocation, The Journal of Machine Learning Research 3 (2003) 993–1022.

[7] D. Koller, N. Friedman, Probabilistic Graphical Models: Principles and Techniques, MIT Press, 2009.

[8] H. Hu, Z. Li, A.R. Vetta, Randomized experimental design for causal graph discovery, in: Advances in Neural Information Processing Systems, 2014, pp. 2339–2347.

[9] C. Meek, Strong completeness and faithfulness in Bayesian networks, in: Proceedings of the Eleventh Conference on Uncertainty in Artificial Intelligence, Morgan Kaufmann Publishers Inc., 1995, pp. 411–418.

[10] J. Pearl, T.S. Vermal, Equivalence and synthesis of causal models, in: Proceedings of Sixth Conference on Uncertainty in Artificial Intelligence, 1991, pp. 220–227.

[11] K.P. Murphy, Machine Learning: A Probabilistic Perspective, MIT Press, 2012.

[12] P.P. Shenoy, Inference in hybrid Bayesian networks using mixtures of Gaussians, arXiv preprint, arXiv: 1206.6877, 2012.

[13] N. Friedman, D. Geiger, M. Goldszmidt, Bayesian network classifiers, Machine Learning 29 (2–3) (1997) 131–163.

[14] C. Bielza, P. Larrañaga, Discrete Bayesian network classifiers: a survey, ACM Computing Surveys (CSUR) 47 (1) (2014) 5.

[15] R.M. Neal, Connectionist learning of belief networks, Artificial Intelligence 56 (1) (1992) 71–113.

[16] S. Srinivas, A generalization of the noisy-or model, in: Proceedings of the Ninth International Conference on Uncertainty in Artificial Intelligence, Morgan Kaufmann Publishers Inc., 1993, pp. 208–215.

[17] G.F. Cooper, The computational complexity of probabilistic inference using Bayesian belief networks, Artificial Intelligence 42 (2) (1990) 393–405.

[18] C. Cannings, E. Thompson, H. Skolnick, The recursive derivation of likelihoods on complex pedigrees, Advances in Applied Probability 8 (4) (1976) 622–625.

[19] J. Pearl, Reverend Bayes on inference engines: a distributed hierarchical approach, in: AAAI Conference on Artificial Intelligence, 1982, pp. 133–136.

[20] J. Pearl, Probabilistic Reasoning in Intelligent Systems: Networks of Plausible Inference, Elsevier, 1998.

[21] Y. Weiss, W.T. Freeman, On the optimality of solutions of the max-product belief-propagation algorithm in arbitrary graphs, IEEE Transactions on Information Theory 47 (2) (2001) 736–744.

[22] S.L. Lauritzen, D.J. Spiegelhalter, Local computations with probabilities on graphical structures and their application to expert systems, Journal of the Royal Statistical Society. Series B (Methodological) (1988) 157–224.

[23] P.P. Shenoy, G. Shafer, Axioms for probability and belief-function propagation, in: Classic Works of the Dempster-Shafer Theory of Belief Functions, Springer, 2008, pp. 499–528.

[24] F.V. Jensen, An Introduction to Bayesian Networks, vol. 210, UCL Press, London, 1996.

[25] F.V. Jensen, S.L. Lauritzen, K.G. Olesen, Bayesian updating in causal probabilistic networks by local computations, Computational Statistics Quarterly (1990).

[26] V. Lepar, P.P. Shenoy, A comparison of Lauritzen–Spiegelhalter, Hugin, and Shenoy–Shafer architectures for computing marginals of probability distributions, in: Proceedings of the Fourteenth Conference on Uncertainty in Artificial Intelligence, Morgan Kaufmann Publishers Inc., 1998, pp. 328–337.

[27] A.P. Dawid, Applications of a general propagation algorithm for probabilistic expert systems, Statistics and Computing 2 (1) (1992) 25–36.

[28] P. Dagum, M. Luby, Approximating probabilistic inference in Bayesian belief networks is NP-hard, Artificial Intelligence 60 (1) (1993) 141–153.
[29] K.P. Murphy, Y. Weiss, M.I. Jordan, Loopy belief propagation for approximate inference: an empirical study, in: Proceedings of the Fifteenth Conference on Uncertainty in Artificial Intelligence, Morgan Kaufmann Publishers Inc., 1999, pp. 467–475.
[30] S.C. Tatikonda, M.I. Jordan, Loopy belief propagation and Gibbs measures, in: Proceedings of the Eighteenth Conference on Uncertainty in Artificial Intelligence, Morgan Kaufmann Publishers Inc., 2002, pp. 493–500.
[31] M. Henrion, Propagating uncertainty in Bayesian networks by probabilistic logic sampling, in: Uncertainty in Artificial Intelligence 2 Annual Conference on Uncertainty in Artificial Intelligence, UAI-86, Elsevier Science, Amsterdam, NL, 1986, pp. 149–163.
[32] R. Fung, K.C. Chang, Weighing and integrating evidence for stochastic simulation in Bayesian networks, in: Annual Conference on Uncertainty in Artificial Intelligence, UAI-89, Elsevier Science, New York, N. Y., 1989, pp. 209–219.
[33] A. Darwiche, Modeling and Reasoning With Bayesian Networks, Cambridge University Press, 2009.
[34] S. Geman, D. Geman, Stochastic relaxation, Gibbs distributions, and the Bayesian restoration of images, IEEE Transactions on Pattern Analysis and Machine Intelligence 6 (1984) 721–741.
[35] D. Lunn, D. Spiegelhalter, A. Thomas, N. Best, The bugs project: evolution, critique and future directions, Statistics in Medicine 28 (25) (2009) 3049–3067.
[36] B. Carpenter, A. Gelman, M. Hoffman, D. Lee, B. Goodrich, M. Betancourt, M.A. Brubaker, J. Guo, P. Li, A. Riddell, Stan: a probabilistic programming language, Journal of Statistical Software (2016).
[37] W.K. Hastings, Monte Carlo sampling methods using Markov chains and their applications, Biometrika 57 (1) (1970) 97–109.
[38] M.I. Jordan, Z. Ghahramani, T.S. Jaakkola, L.K. Saul, An introduction to variational methods for graphical models, Machine Learning 37 (2) (1999) 183–233.
[39] L.K. Saul, T. Jaakkola, M.I. Jordan, Mean field theory for sigmoid belief networks, Journal of Artificial Intelligence Research 4 (1) (1996) 61–76.
[40] G.E. Hinton, R.S. Zemel, Autoencoders, minimum description length, and Helmholtz free energy, in: Advances in Neural Information Processing Systems, 1994, p. 3.
[41] A. Mnih, K. Gregor, Neural variational inference and learning in belief networks, arXiv preprint, arXiv: 1402.0030, 2014.
[42] S. Nie, D.D. Maua, C.P. de Campos, Q. Ji, Advances in learning Bayesian networks of bounded treewidth, in: Advances in Neural Information Processing Systems 27, 2014.
[43] D. Heckerman, M.P. Wellman, Bayesian networks, Communications of the ACM 38 (3) (1995) 27–31.
[44] D. Heckerman, A tutorial on learning with Bayesian networks, in: Learning in Graphical Models, Springer, 1998, pp. 301–354.
[45] G. Schwarz, et al., Estimating the dimension of a model, The Annals of Statistics 6 (2) (1978) 461–464.
[46] R.E. Kass, A.E. Raftery, Bayes factors, Journal of the American Statistical Association 90 (430) (1995) 773–795.
[47] W. Buntine, Theory refinement on Bayesian networks, in: Proceedings of the Seventh Conference on Uncertainty in Artificial Intelligence, Morgan Kaufmann Publishers Inc., 1991, pp. 52–60.
[48] G.F. Cooper, E. Herskovits, A Bayesian method for the induction of probabilistic networks from data, Machine Learning 9 (4) (1992) 309–347.
[49] D. Heckerman, D. Geiger, D.M. Chickering, Learning Bayesian networks: the combination of knowledge and statistical data, Machine Learning 20 (3) (1995) 197–243.
[50] Bayesian structure learning scoring functions, http://www.lx.it.pt/~asmc/pub/talks/09-TA/ta_pres.pdf.
[51] D. Heckerman, A. Mamdani, M.P. Wellman, Real-world applications of Bayesian networks, Communications of the ACM 38 (3) (1995) 24–26.
[52] C. Chow, C. Liu, Approximating discrete probability distributions with dependence trees, IEEE Transactions on Information Theory 14 (3) (1968) 462–467.
[53] C.P. De Campos, Q. Ji, Efficient structure learning of Bayesian networks using constraints, The Journal of Machine Learning Research 12 (2011) 663–689.
[54] M. Bartlett, J. Cussens, Advances in Bayesian network learning using integer programming, arXiv preprint, arXiv:1309.6825, 2013.

[55] X. Zheng, B. Aragam, P.K. Ravikumar, E.P. Xing, DAGs with no tears: continuous optimization for structure learning, in: Advances in Neural Information Processing Systems, 2018, pp. 9491–9502.
[56] D.M. Chickering, Optimal structure identification with greedy search, The Journal of Machine Learning Research 3 (2003) 507–554.
[57] C. Yuan, B. Malone, X. Wu, Learning optimal Bayesian networks using A* search, in: Proceedings – International Joint Conference on Artificial Intelligence, vol. 22, no. 3, Citeseer, 2011, p. 2186.
[58] M. Schmidt, A. Niculescu-Mizil, K. Murphy, et al., Learning graphical model structure using L1-regularization paths, in: AAAI Conference on Artificial Intelligence, vol. 7, 2007, pp. 1278–1283.
[59] G.F. Cooper, A simple constraint-based algorithm for efficiently mining observational databases for causal relationships, Data Mining and Knowledge Discovery 1 (2) (1997) 203–224.
[60] M. Scutari, Bayesian network constraint-based structure learning algorithms: parallel and optimised implementations in the bnlearn R package, arXiv preprint, arXiv:1406.7648, 2014.
[61] D. Geiger, D. Heckerman, Learning Gaussian networks, in: Proceedings of the Tenth International Conference on Uncertainty in Artificial Intelligence, Morgan Kaufmann Publishers Inc., 1994, pp. 235–243.
[62] S. Huang, J. Li, J. Ye, A. Fleisher, K. Chen, T. Wu, E. Reiman, A.D.N. Initiative, et al., A sparse structure learning algorithm for Gaussian Bayesian network identification from high-dimensional data, IEEE Transactions on Pattern Analysis and Machine Intelligence 35 (6) (2013) 1328–1342.
[63] A.L. Yuille, A. Rangarajan, The concave–convex procedure (CCCP), in: Advances in Neural Information Processing Systems, 2002, pp. 1033–1040.
[64] A.P. Dempster, N.M. Laird, D.B. Rubin, Maximum likelihood from incomplete data via the EM algorithm, Journal of the Royal Statistical Society. Series B (Methodological) (1977) 1–38.
[65] P. Dagum, A. Galper, E. Horvitz, Dynamic network models for forecasting, in: Proceedings of the Eighth International Conference on Uncertainty in Artificial Intelligence, Morgan Kaufmann Publishers Inc., 1992, pp. 41–48.
[66] V. Mihajlovic, M. Petkovic, Dynamic Bayesian Networks: A State of the Art, 2001.
[67] K.P. Murphy, Dynamic Bayesian Networks: Representation, Inference and Learning, Ph.D. dissertation, University of California, Berkeley, 2002.
[68] Z. Ghahramani, Learning dynamic Bayesian networks, in: Adaptive Processing of Sequences and Data Structures, Springer, 1998, pp. 168–197.
[69] A.J. Viterbi, Error bounds for convolutional codes and an asymptotically optimum decoding algorithm, IEEE Transactions on Information Theory 13 (2) (1967) 260–269.
[70] L.E. Baum, T. Petrie, G. Soules, N. Weiss, A maximization technique occurring in the statistical analysis of probabilistic functions of Markov chains, The Annals of Mathematical Statistics 41 (1) (1970) 164–171.
[71] L.R. Rabiner, A tutorial on hidden Markov models and selected applications in speech recognition, Proceedings of the IEEE 77 (2) (1989) 257–286.
[72] Y. Bengio, P. Frasconi, An input output HMM architecture, in: Advances in Neural Information Processing Systems, 1995, pp. 427–434.
[73] M. Brand, N. Oliver, A. Pentland, Coupled hidden Markov models for complex action recognition, in: IEEE Computer Society Conference on Computer Vision and Pattern Recognition, 1997, pp. 994–999.
[74] Z. Ghahramani, M.I. Jordan, Factorial hidden Markov models, Machine Learning 29 (2–3) (1997) 245–273.
[75] S. Fine, Y. Singer, N. Tishby, The hierarchical hidden Markov model: analysis and applications, Machine Learning 32 (1) (1998) 41–62.
[76] Z. Ghahramani, An introduction to hidden Markov models and Bayesian networks, International Journal of Pattern Recognition and Artificial Intelligence 15 (01) (2001) 9–42.
[77] T. Xiang, S. Song, Video behavior profiling for anomaly detection, IEEE Transactions on Pattern Analysis and Machine Intelligence (2008).
[78] Z. Ghahramani, M.I. Jordan, Factorial hidden Markov models, in: Advances in Neural Information Processing Systems, 1996, pp. 472–478.
[79] A.D. Brown, G.E. Hinton, Products of hidden Markov models, in: AISTATS, 2001.
[80] V. Pavlovic, J.M. Rehg, J. MacCormick, Learning switching linear models of human motion, in: Advances in Neural Information Processing Systems, 2001, pp. 981–987.

[81] R.E. Kalman, A new approach to linear filtering and prediction problems, Journal of Basic Engineering 82 (1) (1960) 35–45.
[82] G.M. Allenby, P.E. Rossi, R.E. McCulloch, Hierarchical Bayes model: a practitioner's guide, Journal of Bayesian Applications in Marketing (2005) 1–4.
[83] Hbc: Hierarchical Bayes compiler, http://www.umiacs.umd.edu/~hal/HBC/.
[84] S. Nie, M. Zheng, Q. Ji, The deep regression Bayesian network and its applications: probabilistic deep learning for computer vision, IEEE Signal Processing Magazine 35 (1) (2018) 101–111.
[85] Z. Gan, R. Henao, D.E. Carlson, L. Carin, Learning deep sigmoid belief networks with data augmentation, in: AISTATS, 2015.
[86] Y. Tang, R. Salakhutdinov, G. Hinton, Deep mixtures of factor analysers, arXiv preprint, arXiv:1206.4635, 2012.
[87] A. van den Oord, B. Schrauwen, Factoring variations in natural images with deep Gaussian mixture models, in: Advances in Neural Information Processing Systems, 2014, pp. 3518–3526.
[88] G.E. Hinton, S. Osindero, Y.-W. Teh, A fast learning algorithm for deep belief nets, Neural Computation 18 (7) (2006) 1527–1554.
[89] T. Hofmann, Probabilistic latent semantic indexing, in: Proceedings of the 22nd Annual International ACM SIGIR Conference on Research and Development in Information Retrieval, ACM, 1999, pp. 50–57.
[90] C.M. Bishop, Pattern Recognition and Machine Learning, Springer, 2006.
[91] J.D. Park, MAP complexity results and approximation methods, arXiv preprint, arXiv:1301.0592, 2012.
[92] Q. Liu, A. Ihler, Variational algorithms for marginal MAP, The Journal of Machine Learning Research 14 (1) (2013) 3165–3200.
[93] D. Maua, C. De Campos, Anytime marginal MAP inference, arXiv preprint, arXiv:1206.6424, 2012.
[94] R.C. Jeffrey, The Logic of Decision, University of Chicago Press, 1990.
[95] H. Steck, T.S. Jaakkola, On the Dirichlet prior and Bayesian regularization, in: Advances in Neural Information Processing Systems, 2003, pp. 713–720.
[96] D. Heckerman, D. Geiger, D.M. Chickering, Learning Bayesian networks: the combination of knowledge and statistical data, Machine Learning 20 (3) (1995) 197–243.
[97] X. Zheng, B. Aragam, P.K. Ravikumar, E.P. Xing, DAGs with NO TEARS: continuous optimization for structure learning, Advances in Neural Information Processing Systems 31 (2018) 9472–9483.
[98] G.C. Wei, M.A. Tanner, A Monte Carlo implementation of the EM algorithm and the poor man's data augmentation algorithms, Journal of the American Statistical Association 85 (411) (1990) 699–704.
[99] J. Pearl, Probabilistic Reasoning in Intelligent Systems: Networks of Plausible Inference, Elsevier, 2014.
[100] H. Chan, A. Darwiche, On the revision of probabilistic beliefs using uncertain evidence, Artificial Intelligence 163 (1) (2005) 67–90.
[101] M. Valtorta, Y.-G. Kim, J. Vomlel, Soft evidential update for probabilistic multiagent systems, International Journal of Approximate Reasoning 29 (1) (2002) 71–106.
[102] Y. Peng, S. Zhang, R. Pan, Bayesian network reasoning with uncertain evidences, International Journal of Uncertainty, Fuzziness and Knowledge-Based Systems 18 (05) (2010) 539–564.
[103] J. Bilmes, On virtual evidence and soft evidence in Bayesian networks, 2004.
[104] F.J. Groen, A. Mosleh, Foundations of probabilistic inference with uncertain evidence, International Journal of Approximate Reasoning 39 (1) (2005) 49–83.
[105] R. Pan, Y. Peng, Z. Ding, Belief update in Bayesian networks using uncertain evidence, in: 2006 18th IEEE International Conference on Tools with Artificial Intelligence, ICTAI'06, IEEE, 2006, pp. 441–444.
[106] S. Duane, A.D. Kennedy, B.J. Pendleton, D. Roweth, Hybrid Monte Carlo, Physics Letters B 195 (2) (1987) 216–222.
[107] R.M. Neal, et al., MCMC using Hamiltonian dynamics, in: Handbook of Markov Chain Monte Carlo, vol. 2(11), 2011, p. 2.

4 Undirected probabilistic graphical models

4.1 Introduction

As discussed in Chapter 1, there are two types of graphical models, directed and undirected PGMs. Both types of PGMs are widely used in CV. In fact, undirected PGMs such as Markov random fields (MRF), and conditional random fields (CRF), have been applied to a wide range of vision tasks from image denoising, segmentation, motion estimation, stereo to object recognition and image editing. Following the format of Chapter 3, we will start by discussing the definitions and properties of undirected PGMs and then discussing the major learning and inference methods for them. In addition, we will first introduce basic undirected PGMs, that is, the Markov networks (MNs). We will then discuss their variants, including the CRF and the restricted Boltzmann machine. In our discussions, we will also draw a contrast between directed and undirected PGMs , and also discuss their commonalities.

4.1.1 Definitions and properties

Undirected PGMs are graphs consisting of nodes and undirected links, where nodes represent the RVs, and links capture their dependencies. Different from the links in the directed PGMs, the links in the undirected PGMs represent mutual interactions between the connected RVs. Like a directed PGM, an undirected PGM can compactly encode the joint probability distribution of a set of random variables.

4.1.1.1 Definitions

A Markov network (MN), also called an MRF, is an undirected graph satisfying the Markov property. Formally, an MN can be defined as follows: Let $\mathbf{X}$=$\{X_1, X_2, \ldots, X_N\}$ denote a set of N random variables. An MN is a graphical representation of the joint probability distribution $p(X_1, X_2, \ldots, X_N)$. An MN over $\mathbf{X}$ can be defined as a two-tuple $\mathcal{M} = \{\mathcal{G}, \mathbf{\Theta}\}$, where $\mathcal{G}$ defines the qualitative (structure) part of the MN, whereas $\mathbf{\Theta}$ defines the quantitative part of the MN; $\mathcal{G}$ can be further represented as $\mathcal{G} = \{\mathcal{E}, \mathcal{V}\}$, where $\mathcal{V}$ represents nodes of $\mathcal{G}$ corresponding to the variables in $\mathbf{X}$ and $\mathcal{E} = \{E_{ij}\}$ represents the undirected edges (links) between nodes i and j. They capture the probabilistic dependence between the variables represented by nodes i and j. Different from the directed links for BNs, the links in MN capture the mutual dependencies or mutual interactions between the two connected variables. The neighbors N_{X_i} of a node X_i are defined as the nodes directly connected to X_i. The parameters $\mathbf{\Theta}$ of the model collectively characterize the strengths of the links. An undirected graph $\mathcal{M}$ is an MN only if it satisfies the Markov condition, that is,

$$X_i \perp X_j \mid N_{X_i} \; \forall X_j \in \mathbf{X} \setminus N_{X_i} \setminus X_i. \tag{4.1}$$

The Markov condition states that a node is independent of all other nodes, given its neighboring nodes. Fig. 4.1 shows an MN with five nodes. According to the Markov condition, node A is independent of node E given its two neighbors B and C.

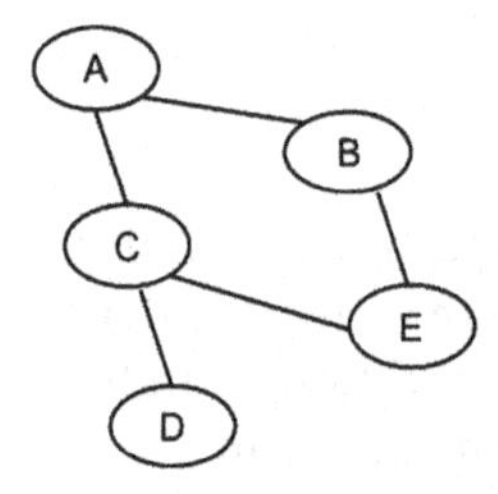

FIGURE 4.1 An example of a Markov network.

Following the graph theories, a clique is defined as a fully connected subgraph, that is, there is a link between every pair of nodes in the subgraph. A clique may vary in the number of nodes. Fig. 4.2 shows examples of cliques of different sizes. The maximal cliques

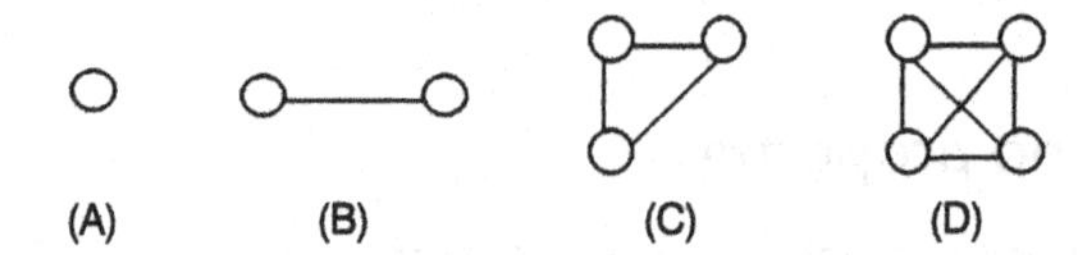

FIGURE 4.2 Examples of cliques of (A) one node, (B) two nodes, (C) three nodes, and (D) four nodes.

of a graph correspond to the smallest number of cliques that cover all nodes. They are unique. For the MN in Fig. 4.3, its maximal cliques are $\{X_1, X_2\}$, $\{X_2, X_3, X_4\}$, $\{X_4, X_5\}$, and $\{X_6\}$.

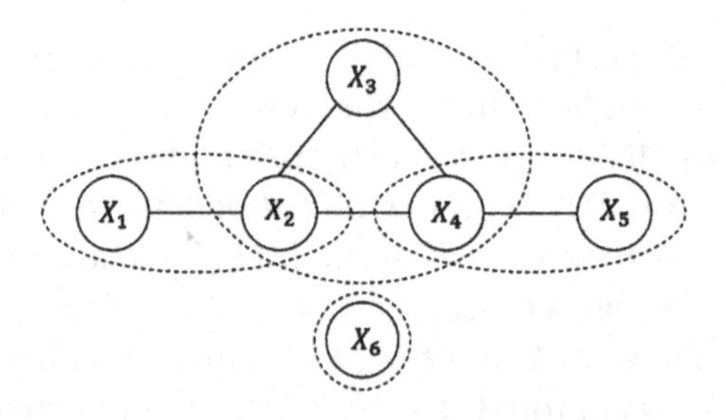

FIGURE 4.3 An example of maximum cliques.

According to the Hammersley–Clifford (HC) theorem [1], the joint probability distribution of all nodes of an MN equals the product of the normalized potential functions of its maximal cliques:

$$p(x_1, x_2, \ldots, x_N) = \frac{1}{Z} \prod_{c \in C} \psi_c(\mathbf{x}_c), \tag{4.2}$$

where C represents the set of maximal cliques, c is a maximal clique, $\mathbf{x}_c$ are nodes in clique c, $\psi_c(\mathbf{x}_c)$ is the potential function for clique c, and Z is the partition function that normalizes the right-hand side of Eq. (4.2) to be between 0 and 1. For discrete MNs, $Z = \sum_{x_1, x_2, \ldots, x_N} \prod_{c \in C} \psi_c(\mathbf{x}_c)$. Note that whereas the HC theorem specifies the joint distribution of an MN in terms of products of potential functions for maximal cliques, the joint distribution of an MN can in fact be parameterized by any set of cliques that cover all nodes though such parameterizations may not be unique.

For the example in Fig. 4.3, according to the HC theorem, the joint probability of its nodes in terms of maximal cliques can be written as

$$p(x_1, x_2, \ldots, x_6) = \frac{1}{Z} \psi_{12}(x_1, x_2) \psi_{234}(x_2, x_3, x_4) \psi_{45}(x_4, x_5) \psi_6(x_6), \tag{4.3}$$

where $Z = \sum_{x_1, x_2, \ldots, x_6} \psi_{12}(x_1, x_2) \psi_{234}(x_2, x_3, x_4) \psi_{45}(x_4, x_5) \psi_6(x_6)$.

Different from a BN, which is parameterized by the CPDs, an MN is parameterized by the potential function $\psi()$. It measures the compatibility among the variables; the higher the potential value, the stronger the compatibility among the variables. Compared with the CPDs, a potential function is a more symmetric parameterization. Furthermore, a potential function can be flexible, and, in fact, can be any nonnegative function. One issue with potential function specification is that it is unnormalized. In fact, it does not have to be a probability. One type of widely used potential functions is the log-linear function,

$$\psi_c(\mathbf{x}_c) = \exp(-\mathbf{w}_c E_c(\mathbf{x}_c)), \tag{4.4}$$

where $E_c(\mathbf{X}_c)$ is called the energy function for clique c, and $\mathbf{w}_c$ are its parameters. Due to the negative sign, a high value of the potential function corresponds to a low value of the energy function. Such a log-linear representation can be generalized to any potential functions. Given the log-linear potential function, the joint probability distribution for an MN can be written as a Gibbs distribution

$$p(x_1, x_2, \ldots, x_N) = \frac{1}{Z} \exp[-\sum_{c \in C} \mathbf{w}_c E_c(\mathbf{x}_c)] \tag{4.5}$$

4.1.1.2 Properties

Like a BN, an MN also embeds local and global independence properties because of the Markov condition. Local independence includes the Markov property and pairwise independence. As defined in Eq. (4.1), the Markov property states that a node X_i is independent of all other nodes, given its neighbors, that is, the neighbors of X_i completely shield X_i from all other nodes. For MNs, the Markov blanket (MB) for a node consists of its nearest (immediate) neighbors, that is, $MB_{X_i} = N_{X_i}$. For the example in Fig. 4.4B, the neighbors (or its

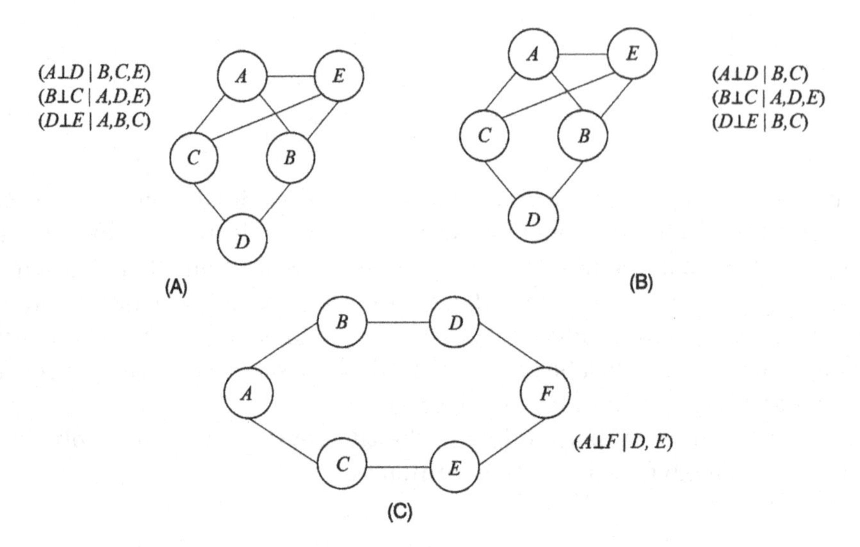

FIGURE 4.4 Examples of local and global independencies. (A) Pairwise. (B) Markov Blanket. (C) Global independence.

MB) of a node C consist of nodes $\{A, D, E\}$. Following the Markov property, we have

$$C \perp B|\{A, D, E\}.$$

The Markov property can be extended to local pairwise independence, which states that any two disconnected nodes are independent of each other given all other nodes. Formally, given two disconnected nodes X_i and X_j, the pairwise independence can be written as

$$X_i \perp X_j|\mathbf{X} \setminus X_i \setminus X_j.$$

This property and it's complement can be employed to perform constraint-based MN structure learning. For the MN example in Fig. 4.4A, nodes D and E are independent of each other given nodes A, B, and C.

Besides local independence properties, MN also has global independence through the D-separation principle. Any two nodes are independent of each other if every path between the two nodes is blocked. A path between two nodes X_i and X_j is a sequence of nodes between them such that any successive nodes are connected by an undirected edge and no node appears in the sequence twice. A path is blocked if one of the nodes in the path is given. For the example shown in Fig. 4.4C, nodes A and F are independent of each other given D and E. This is because two paths between A and F, path A–B–D–F and path A–C–E–F, are both blocked given D and E.

4.1.1.3 I-map

Like BNs, there is also an issue of faithfulness between an MN and a distribution p. An MN $\mathcal{M}$ is an I-map of a distribution p if $I(\mathcal{M}) \subseteq I(p)$, and it is a perfect I-map if $I(\mathcal{M}) = I(p)$, where $I()$ represents the independencies.

4.2 Pairwise Markov networks

The most commonly used MN is the pairwise MN, where the maximum clique size is two. Pairwise MNs are attractive because of their simplicity and efficient representation. The number of parameters of a discrete pairwise MN is quadratic with respect to the number of nodes. The potential function for each clique involves only two neighboring nodes, X_i and X_j, where X_j is one of the immediate neighbors of node X_i. Following the HC theorem, the joint probability for a pairwise MN can be written as

$$p(x_1, x_2, \ldots, x_n) = \frac{1}{Z} \prod_{(x_i, x_j) \in \mathcal{E}} \psi_{ij}(x_i, x_j), \tag{4.6}$$

where $\psi_{ij}(x_i, x_j)$ is the pairwise potential function between nodes X_i and X_j. Besides the pairwise potential function, a bias (prior) potential function $\phi_i(x_i)$ is often added in practice to represent the bias (prior probability) for each node. The joint probability for a pairwise MN with bias terms can be written as

$$p(x_1, x_2, \ldots, x_n) = \frac{1}{Z} \prod_{(x_i, x_j) \in \mathcal{E}} \psi_{ij}(x_i, x_j) \prod_{x_i \in \mathcal{V}} \phi_i(x_i). \tag{4.7}$$

With a log-linear potential function, the joint probability can be written as

$$p(x_1, x_2, \ldots, x_n) = \frac{1}{Z} \exp[- \sum_{(x_i, x_j) \in \mathcal{E}} w_{ij} E_{ij}(x_i, x_j) - \sum_{x_i \in \mathcal{V}} \alpha_i E_i(x_i)], \tag{4.8}$$

where $E_{ij}(x_i, x_j)$ and $E_i(x_i)$ are the pairwise and bias (prior) energy functions, respectively, and w_{ij} and α_i are their parameters. The pairwise MNs can be further divided into discrete and continuous pairwise MNs as discussed further.

4.2.1 Discrete pairwise Markov networks

For a discrete pairwise MN, all nodes represent discrete RVs. One of the most used pairwise MNs is the Potts model. According to this model, the pairwise energy function $E_{ij}(x_i, x_j)$ is defined as follows:

$$E_{ij}(x_i, x_j) = 1 - \delta(x_i, x_j) \tag{4.9}$$

or more generically as

$$E_{ij}(x_i, x_j) = \begin{cases} -\xi & \text{if } x_i = x_j, \\ \xi & \text{otherwise,} \end{cases} \tag{4.10}$$

where ξ is a positive constant, and δ is the Kronecker function, which equals 1 whenever its arguments are equal and 0 otherwise. The energy function measures the interaction strength between the two nodes and encourages local coherence. It becomes 0 whenever X_i and X_j are the same and 1 otherwise. The prior energy function is often parameterized as

$$E_i(x_i) = -x_i. \tag{4.11}$$

By these definitions the joint probability distribution of **X** can be written as:

$$\begin{aligned} p(x_1, x_2, \ldots, x_N) &= \frac{1}{Z}\exp\{-\sum_{(x_i,x_j)\in\mathcal{E}} w_{ij}[1-\delta(x_i,x_j)] \\ &+ \sum_{x_i\in\mathcal{V}} \alpha_i x_i\}. \end{aligned} \tag{4.12}$$

A particular case of discrete MNs is a binary MN, where each node represents a binary RV. A common binary pairwise MN model is the Ising model. Named after the physicist Ernst Ising for his work in statistical physics, the Ising model consists of binary variables that can be in one of two states +1 or -1, that is, $X_i \in \{+1, -1\}$. The pairwise energy function can be written as

$$E_{ij}(x_i, x_j) = -x_i x_j. \tag{4.13}$$

When X_i and X_j are the same, the energy contribution by them is the lowest at -1 and $+1$ otherwise. The joint probability distribution for a binary pairwise MN with the Ising model can be written as

$$p(x_1, x_2, \ldots, x_N) = \frac{1}{Z}\exp[\sum_{(x_i,x_j)\in\mathcal{E}} w_{ij} x_i x_j + \sum_{x_i\in\mathcal{V}} \alpha_i x_i]. \tag{4.14}$$

4.2.2 Label-observation Markov networks

A special kind of pairwise MNs is a label-observation MN. Also called metric MRF [2] in CV, it is used to specifically perform the labeling task. It divides the nodes into label nodes $\mathbf{X} = \{X_i\}$ and observation nodes $\mathbf{Y} = \{Y_i\}$, with one-to-one correspondence between a label node X_i and the corresponding observation node Y_i. Whereas the label nodes X_i must be discrete (or integer), the observation nodes Y_i can be either discrete or continuous. A label-observation MN is typically arranged in a grid (lattice) format as shown in Fig. 4.5. Each observation node Y_i connects to the corresponding label node X_i, and there are no connections among observation nodes. Hence a label-observation MN defines the joint distribution of **X** and **Y**, i.e., $p(\mathbf{X}, \mathbf{Y})$.

Given the structure of a label-observation MN, the energy function can be divided into the pairwise energy function $E_{ij}(x_i, x_j)$ between the two label nodes X_i and X_j and the unary energy function (also called the likelihood energy) $E_i(y_i|x_i)$ between a label node X_i and its observation node Y_i. Following the log-linear potential parameterization, the posterior label distribution can be written as

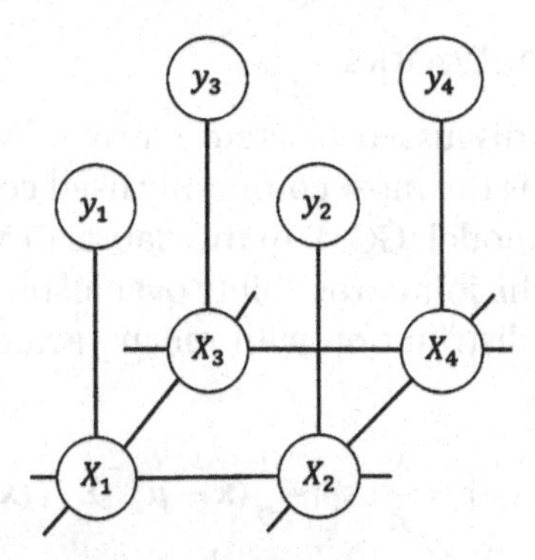

FIGURE 4.5 An example of a label-observation MN, where the links between X_i and Y_i are often represented by directed edges pointing from X_i to Y_i, leading to a hybrid model.

$$
\begin{aligned}
& p(x_1, x_2, ..., x_N, y_1, y_2, .., y_N) \\
= \; & \frac{1}{Z} p(x_1, x_2, \ldots, x_N) p(y_1, y_2, .., y_N | x_1, x_2, \ldots, x_N) \\
= \; & \frac{1}{Z} \exp[\sum_{i \in \mathcal{V}_x} -\alpha_i E_i(y_i|x_i) - \sum_{(i,j) \in \mathcal{E}_x} w_{i,j} E_{ij}(x_i, x_j)], \qquad (4.15)
\end{aligned}
$$

where $\mathcal{V}_x$ and $\mathcal{E}_x$ respectively represent all X nodes and all links among X nodes. The pairwise energy function $E_{ij}(x_i, x_j)$ between x_i and x_j measures the label compatibility for nearby X nodes. As the label compatibility defines the default prior relationships among the labels, it may vary from task to task. In general, label compatibility is often used to impose local smoothness between two neighboring nodes X_i and X_j. Its value should be small when their labels are compatible with each other and large otherwise. One such parameterization is the Potts model or the Ising model if the label nodes are binary. A simple variant of the Ising model is $E_{ij}(x_i, x_j) = 0$ when $x_i = x_j$ and 1 otherwise. The unary energy function $E_i(y_i|x_i)$ measures the compatibility between a label node and its observation in terms of label likelihood. Hence it is called the likelihood energy. In CV the unary energy function typically equals the (negative) log-likelihood of the label given its observation, that is, $E_i(y_i|x_i) = -\log p(y_i|x_i)$. If y_i is continuous, then $p(y_i|x_i)$ can follow a Gaussian distribution for unary Y_i or multivariate Gaussian for multivariate Y_i. If, on the other hand, Y_i is discrete (or even binary), then $p(y_i|x_i)$ follows a Bernoulli, categorical, or an integer distribution.

The label-observation MNs are widely used in CV and image processing for image denoising or segmentation. Given an MN, the goal of image denoising or segmentation is finding the values for $\mathbf{X}$, given values of $\mathbf{y}$, by maximizing the conditional probability of $p(\mathbf{x}|\mathbf{y})$ through a MAP inference, that is, $\mathbf{x}^* = \arg\max_{\mathbf{x}} p(\mathbf{x}|\mathbf{y})$. Details on MRF application to image labeling will be discussed in Section 5.2.3.

4.2.3 Gaussian Markov networks

In the previous sections, we discussed discrete pairwise MNs. In this section, we discuss the Gaussian MN, which is the most commonly used continuous pairwise MNs. Also called the Gaussian graphical model (GGM) or the Gaussian Markov random field (GMRF), a Gaussian MN assumes that the joint probability over all nodes $\mathbf{X} = (X_1, X_2, \dots, X_N)^\top$ follows a multivariate Gaussian distribution with mean vector $\boldsymbol{\mu}$ and covariance matrix $\boldsymbol{\Sigma}$, that is,

$$p(x_1, x_2, \dots, x_N) = \frac{1}{Z}\exp[-\frac{1}{2}(\mathbf{x}-\boldsymbol{\mu})^\top \boldsymbol{\Sigma}^{-1}(\mathbf{x}-\boldsymbol{\mu})], \tag{4.16}$$

where Z is the normalization constant. Define $\mathbf{W} = \boldsymbol{\Sigma}^{-1}$ as the precision matrix, the expression in the exponent in Eq. (4.16) can be rewritten as

$$(\mathbf{x}-\boldsymbol{\mu})^\top \mathbf{W}(\mathbf{x}-\boldsymbol{\mu}) = \mathbf{x}^\top \mathbf{W}\mathbf{x} - 2\mathbf{x}^\top \mathbf{W}\boldsymbol{\mu} + \boldsymbol{\mu}^\top \mathbf{W}\boldsymbol{\mu},$$

where the last term is constant and it can therefore be folded into the normalization constant. The joint probability can be rewritten as

$$p(x_1, x_2, \dots, x_N) = \frac{1}{Z}\exp[-(\frac{1}{2}\mathbf{x}^\top \mathbf{W}\mathbf{x} - \mathbf{x}^\top \mathbf{W}\boldsymbol{\mu})]. \tag{4.17}$$

By breaking up the expression in the exponent into bias and pairwise terms, we have

$$\begin{aligned} p(x_1, x_2, \dots, x_N) &= \frac{1}{Z}\exp[-\frac{1}{2}(\sum_{i\in\mathcal{V}} w_{ii}x_i^2 + 2\sum_{(i,j)\in\mathcal{E}} w_{ij}x_ix_j) \\ &\quad + (\sum_{i\in\mathcal{V}} w_{ii}x_i\mu_i + \sum_{(i,j)\in\mathcal{E}} w_{ij}x_i\mu_j)], \end{aligned} \tag{4.18}$$

where w_{ii} and w_{ij} are elements of $\mathbf{W}$. After some rearrangements of elements of Eq. (4.18), we have

$$\begin{aligned} p(x_1, x_2, \dots, x_N) &= \frac{1}{Z}\exp[-(\frac{1}{2}\sum_{i\in\mathcal{V}} w_{ii}x_i^2 - \sum_{i\in\mathcal{V}} w_{ii}x_i\mu_i - \sum_{(i,j)\in\mathcal{E}} w_{ij}x_i\mu_j) \\ &\quad -(\sum_{(i,j)\in\mathcal{E}} w_{ij}x_ix_j)], \end{aligned} \tag{4.19}$$

where the first three terms give rise to the bias (prior) energy term $E_i(x_i)$, and the fourth term produces the pairwise energy term $E_{ij}(x_i, x_j)$, that is,

$$\begin{aligned} E_i(x_i) &= \frac{1}{2}w_{ii}x_i^2 - w_{ii}x_i\mu_i - w_{ij}x_i\mu_j, \\ E_{ij}(x_i, x_j) &= w_{ij}x_ix_j. \end{aligned}$$

Hence a Gaussian MN has a linear pairwise energy function and a quadratic bias energy function.

Like the Gaussian Bayesian network, an important property of GMN is the independence property. Given $\mathbf{\Sigma}$ and $\mathbf{W}$ for a GMN, if $\sigma_{ij} = 0$ (σ_{ij} is an element of $\mathbf{\Sigma}$), this means that X_i and X_j are marginally independent, that is, $X_i \perp X_j$. If $w_{ij} = 0$ (w_{ij} is an element of $\mathbf{W}$), then X_i and X_j are conditionally independent given all other nodes, that is, $X_i \perp X_j | X_{-i,i\neq j}$. Moreover, $w_{ij} = 0$ is called structural zeros as this means that there is no link between nodes X_i and X_j. This property is often used during GMN structure learning by incorporating the ℓ_1 norm on W to yield a sparsely connected GMN.

4.2.4 Restricted Boltzmann machines

A Boltzmann machine (BM) is a particular kind of pairwise binary MNs with values 0 or 1 for each node. A general BM is fully connected. Its potential function is specified by the log-linear model with the energy function specified by the Ising model plus a bias term, that is,

$$\begin{aligned} E_{ij}(x_i, x_j) &= -x_i x_j, \\ E_i(x_i) &= -x_i. \end{aligned} \tag{4.20}$$

The energy function is the lowest only when $x_i = x_j = 1$. The joint probability distribution of a BM can be written as

$$p(x_1, x_2, \ldots, x_N) = \frac{1}{Z} \exp[\sum_{(x_i, x_j) \in \mathcal{E}} w_{ij} x_i x_j + \sum_{x_i \in \mathcal{V}} \alpha_i x_i]. \tag{4.21}$$

Despite the strong representation power of a general BM, its learning and inference are impractical. A special kind of BM is the Restricted Boltzmann Machine or RBM. Graphically, it consists of two layers, including the observation layer consisting of nodes X_i and the hidden layer consisting of the hidden nodes H_j as shown in Fig. 4.6A, where each visible node X_i is connected to each hidden node H_j and vice versa. There are no direct connections among visible nodes and among hidden nodes. Following the BM's parameterization, the energy functions of an RBM can be parameterized as follows:

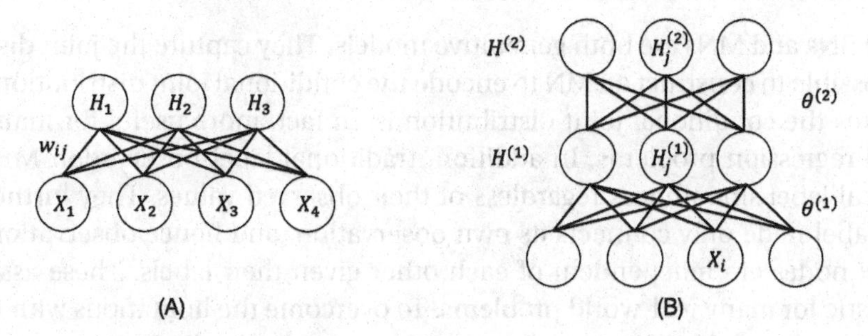

FIGURE 4.6 (A) An example of the restricted Boltzmann machine and (B) the deep Boltzmann machine.

$$E_{ij}(x_i, h_j) = -x_i h_j,$$

$$
\begin{aligned}
E_i(x_i) &= -x_i, \\
E_j(h_j) &= -h_j,
\end{aligned}
\tag{4.22}
$$

where the first equation is the pairwise energy term between visible and hidden nodes, and the second and third equations are the bias terms for the visible and hidden nodes, respectively. Given these energy functions, the joint probability distribution of an RBM can be written as

$$
\begin{aligned}
& p(x_1, x_2, \dots, x_N, h_1, h_2, \dots, h_M) \\
&= \frac{1}{Z} \exp[\sum_i \sum_j w_{ij} x_i h_j + \sum_i \alpha_i x_i + \sum_j \beta_j h_j].
\end{aligned}
\tag{4.23}
$$

Given the RBM topology in Fig. 4.6, we can show that both $p(\mathbf{h}|\mathbf{x})$ and $p(\mathbf{x}|\mathbf{h})$ can factorize, that is, $p(\mathbf{h}|\mathbf{x}) = \prod_{j=1}^{M} p(h_j|\mathbf{x})$ and $p(\mathbf{x}|\mathbf{h}) = \prod_{i=1}^{N} p(x_i|\mathbf{h})$. Moreover, using Eq. (4.23), we have

$$
\begin{aligned}
p(h_j = 1|\mathbf{x}) &= \sigma(\beta_j + \sum_i w_{ij} x_i), \\
p(x_i = 1|\mathbf{h}) &= \sigma(\alpha_i + \sum_j w_{ij} h_j),
\end{aligned}
$$

where σ is the sigmoid function, $\mathbf{h} = (h_1, h_2, \dots, h_M)$, and $\mathbf{x} = (x_1, x_2, \dots, x_N)$. An RBM can be extended with discrete X_i and continuous X_i. An RBM with continuous X_i is called a Gaussian Bernoulli RBM. An RBM is one of the major types of building blocks for constructing deep probabilistic models. A deep undirected generative model can be constructed by stacking layers of RBMs on top of each other to form the deep Boltzmann machine (DBM) [3] as shown in Fig. 4.6B.

4.3 Conditional random fields

Traditional BNs and MNs are both generative models. They capture the joint distribution. It is also possible to construct an MN to encode the conditional joint distribution. A model that captures the conditional joint distribution is, in fact, more useful for many classification and regression problems. In addition, traditional label-observation MRF models assume local label smoothness regardless of their observed values. They further assume that each label node only connects its own observation, and hence observations for different label nodes are independent of each other given their labels. These assumptions are unrealistic for many real-world problems. To overcome the limitations with MRFs, the conditional random field (CRF) model was introduced [27]. As a discriminative model, CRF directly models the posteriori probability distribution of the target variables $\mathbf{X}$ given their observations $\mathbf{Y}$. The CRF, therefore, encodes $p(\mathbf{X}|\mathbf{Y})$ instead of their joint probability as in MRF. Fig. 4.7A gives an example of a CRF model.

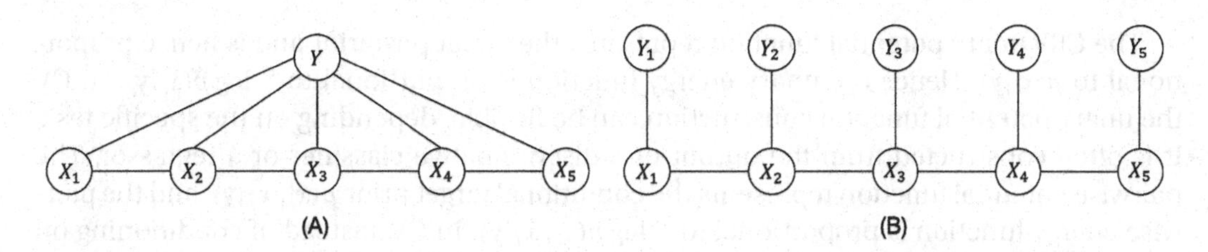

FIGURE 4.7 (A) An example CRF that captures $p(\mathbf{X}|\mathbf{Y})$, and (B) a similar label-observation MN that captures $p(\mathbf{X}, \mathbf{Y})$.

Compared to the label-observation MN in Fig. 4.7B, the target nodes **X** of a CRF are conditioned on all observations **Y**[1] instead of individual local observation Y_i. As a result, a CRF follows the local conditional Markov property. Let X_i and X_j be two nodes in **X** such that

$$X_i \perp X_j | N_{X_i}, \mathbf{Y} \; \forall X_j \in \mathbf{X} \setminus N_{X_i} \setminus X_i.$$

It states that given its neighbors, X_i is independent of X_j only given **Y**. For example, in Fig. 4.7A, according to the CRF, X_1 and X_3 are conditionally independent of each other, only given X_2 and **Y**. On the other hand, for the MN in Fig. 4.7B, X_1 and X_3 are conditionally independent of each other, only given X_2. Compared to the Markov property for an MN, the local conditional independencies for CRF are more restricted and hence weaker. Hence it can capture the dependencies among the local observations instead of assuming their independencies as by MRF.

A CRF model over random variables $\mathbf{X} = \{X_1, \ldots X_N\}$ conditioned on their observations **Y** can be defined as follows:

$$p(x_1, x_2, \ldots, x_n|\mathbf{y}) = \frac{1}{Z(\mathbf{y})} \prod_{(x_i,x_j)\in\mathcal{E}} \psi_{ij}(x_i, x_j|\mathbf{y}) \prod_{x_i\in\mathcal{V}} \phi_i(x_i|\mathbf{y}), \tag{4.24}$$

where $Z(\mathbf{y})$ is the partition function. Both the unary potential function $\phi_i(x_i|\mathbf{y})$ and pairwise potential function $\psi_{ij}(x_i, x_j|\mathbf{y})$ are conditioned on **Y**. With the log-linear potential function, the joint probability can be written as

$$p(x_1, x_2, \ldots, x_n|\mathbf{y}) = \frac{1}{Z(\mathbf{y})} \exp(-E(\mathbf{x}|\mathbf{y})). \tag{4.25}$$

For a pairwise CRF model, the posterior energy function $E(\mathbf{x}|\mathbf{y})$ can be defined as

$$E(\mathbf{x}|\mathbf{y}) = \sum_{x_i\in\mathcal{V}} -\alpha_i E_i(x_i|\mathbf{y}) - \sum_{x_i,x_j\in\mathcal{E}} w_{ij} E_{ij}(x_i, x_j|\mathbf{y}),$$

where $E_i(x_i|\mathbf{y})$ and $E_{ij}(x_i, x_j|\mathbf{y})$ respectively represent the conditional unary and pairwise energy functions, both of which are conditioned on the observations **y**.

[1] In practice, for computational efficiency, **Y** may not be the complete observations for all sites but rather the observations for only nearby sites.

The CRF unary potential function represents the target posterior and is hence proportional to $p(x_i|\mathbf{y})$. Hence the unary energy function is proportional to $-\log p(x_i|\mathbf{y})$. In CV the unary potential function construction can be flexible, depending on the specific task. It is often constructed from the output of a discriminative classifier or a regressor. The pairwise potential function represents the conditional target prior $p(x_i, x_j|\mathbf{y})$, and the pairwise energy function is proportional to $-\log p(x_i, x_j|\mathbf{y})$. In CV, instead of conditioning on the entire observations $\mathbf{Y}$, for computational simplicity, the pairwise energy function is usually conditioned on Y_i and Y_j, the observations of X_i and X_j. Moreover, the pairwise energy function $E_{ij}(x_i, x_j|y_i, y_j)$ is often formulated as the product of a label compatibility function and a penalty term. The label compatibility function measures the consistency between X_i and X_j, such as the Potts model for discrete nodes, whereas the penalty term usually is a kernel function of the differences between Y_i and Y_j. Such a pairwise energy function encourages nearby nodes to have similar labels conditioned on their similar observations. However, like the MN compatibility function, the CRF compatibility function is not limited to local smoothness. It captures the prior default relationships among the nearby labels, and its exact definition varies from task to task. Section 5.2.4 discusses specific CRF unary and pairwise energy functions for image segmentation. CRF models are often applied to image labeling, where the target variables $\mathbf{X}$ correspond to the labels. Image labeling is performed via MAP inference, that is,

$$\mathbf{x}^* = \arg\max_{\mathbf{x}} p(\mathbf{x}|\mathbf{y}).$$

Because label-observation MRFs capture the joint distribution of labels and observations, while CRFs capture the conditional distribution of labels given observations, they differ in the definition of both unary and pairwise potential functions. For CRF, the unary potential function captures the label posterior, whereas the unary potential function for the label-observation MRFs captures the label likelihood. Moreover, the pairwise potential function for the CRF model is conditioned on the label observations, whereas the pairwise potential function for the label-observation MRFs does not depend on the label observations. Besides the advantages mentioned before, CRF models require less training data since they only model the conditional distribution of $\mathbf{X}$. Furthermore, compared to the label-observation MN models, they do not assume that $\mathbf{Y}$ is independent given $\mathbf{X}$, and they allow arbitrary relationships among the observations, which is obviously more natural in reality. On the other hand, since CRF models $p(\mathbf{X}|\mathbf{Y})$, it cannot obtain the joint probability distribution $p(\mathbf{X}, \mathbf{Y})$ or the marginal probability distribution $p(\mathbf{X})$. Moreover, a CRF model cannot handle cases where $\mathbf{Y}$ is incomplete during inference. The generative models have no these limitations. Table 4.1 summarizes the similarities and differences between CRFs and MRFs.

4.4 High-order and long-range Markov networks

The standard pairwise MNs assume that the maximum clique size is two and that each node is only connected to its immediate neighbors. Although efficient in parameteriza-

Table 4.1 Similarities and differences between CRF and MRF.

Items	**MRF**	**CRF**
Models	1) model p(**X**,**Y**) 2) a generative model	1) model p(**X**|**Y**) 2) a discriminative model
Assumptions	1) **X** and **Y** follow the Markov condition 2) **Y** are independent given **X**	**X** follows Markov condition, given **Y** 2) **Y** are dependent
Parameterizations	1) The unary potential depends only on the local observation 2) the pairwise potential does NOT depend on the observations	Both the unary and pairwise potentials depend on all (or nearby) observations

tions, they can only capture relatively simple statistical properties of the data. In practice, to account for additional contextual information and to capture complex properties of data, the standard MNs are often extended. The extensions happen in two directions. First, the neighborhood is extended to include nodes that are not located in the immediate neighborhood, producing the so-called long-range MNs. For example, the neighborhood may be extended to a square patch of $K \times K$ ($K > 2$) centered on the current node so that every node in the patch is a neighbor of the center node. If we expand the square patch to include the entire MN field, this yields the so-called fully connected MRFs, where each node is the neighbor of every other node. Recent works [4–6] on image segmentation using fully connected MRFs/CRFs have shown their improved performance over standard pairwise MNs.

Another direction is including high-order terms in the energy functions to capture complex properties of data, producing high-order MRFs [7–10]. The additional n-wise potential functions involve n-order (> 2) cliques (instead of only pairwise cliques) such as the third-order terms $\psi_{ijk}(x_i, x_j, x_k)$ or even higher-order terms. Mathematically, a high-order potential function can be defined as follows:

$$\psi_c(\mathbf{x}_c) = \exp(-w_c E(\mathbf{x}_c)), \tag{4.26}$$

where $\mathbf{x}_c$ represent a set of nodes in clique c of size larger than 2. The high-order energy function $E_c(\mathbf{x}_c)$ can be defined differently. Instead of pairwise energy function, high order energy function can be used to define N-wise ($N>2$) energy functions. One form of high-order energy function is the $\mathcal{P}^n$ Potts model [7] defined as follows:

$$E_c(\mathbf{x}_c) = \begin{cases} 0 & \text{if } x_i = x_j \ \forall i, j \in c, \\ \alpha |c|^{\beta} & \text{otherwise,} \end{cases} \tag{4.27}$$

where $|c|$ represents the number of nodes in c, and α and β are parameters for the energy function.

High-order MRFs are increasingly employed for various CV tasks, including image segmentation [7] and image denoising [11], and they have demonstrated improved performance over the first-order MRFs. Although high-order and long-range MNs have a greater representation power, they also lead to complex learning and inference challenges. Recent efforts in CV focus on developing advanced methods to overcome these challenges. We discuss some of these methods in Section 5.2.3.5.

4.5 Markov network inferences

An MN can perform the same types of inference as BNs, including posterior probability inference, MAP inference, and likelihood inference. The most common inferences for an MN are the posterior and MAP inferences. Given observation vector $\mathbf{y}$, whereas the posterior inference computes $p(\mathbf{x}|\mathbf{y})$, the MAP inference is finding the best configuration for all nonevidence variables $\mathbf{x}$ that maximize $p(\mathbf{x}|\mathbf{y})$, that is,

$$\mathbf{x}^* = \arg\max_{\mathbf{x}} p(\mathbf{x}|\mathbf{y}). \tag{4.28}$$

For the log-linear potential function, maximizing the posterior probability is the same as minimizing the energy function, that is,

$$\mathbf{x}^* = \arg\min_{\mathbf{x}} E(\mathbf{x}, \mathbf{y}). \tag{4.29}$$

Like BNs, inference methods for MNs include exact methods and approximate methods. Because of the strong similarities between some of the BN and MN inference methods, we will briefly summarize them further.

4.5.1 Exact inference methods

For both posterior and MAP inferences, the same exact inference methods we discussed for BNs in Section 3.3.1 can be applied. Specifically, exact inference methods include variable elimination, belief propagation, and the junction tree method. In addition, a popular method for MAP inference in MNs is the graph cuts method. In the next sections, we briefly discuss the variable elimination, belief propagation, the junction tree, and the graph cuts methods.

4.5.1.1 Variable elimination method

For small and sparsely connected MNs, variable elimination methods can be applied to both MN posterior and MAP inferences. The variable elimination algorithms we introduced in Section 3.3.1.1 can be directly applied to MN inference. Specifically, Algorithm 3.1 may be applied to MN posterior inference, whereas Algorithm 3.2 may be applied to MN MAP inference. Section 9.3.1.2 and Section 13.2.2 of [2] provide additional details on the variable elimination methods.

4.5.1.2 Belief propagation method

For posterior inference, the belief propagation method for MNs is very similar to that for BNs. Each node first collects messages from its neighbors and then updates its belief. It then transmits messages to its neighbors. In fact, belief propagation for MNs is simpler than for BNs since we do not need to divide the neighboring nodes into child and parental nodes and use different equations to compute their messages to a node. Instead, all neighboring nodes are treated in the same way, and the same equations are applied to all nodes. Specifically, belief updating in an MN involves only two equations. First, each node needs to compute the messages it receives from its neighboring nodes. Let X_i be a node, and let X_j be a neighbor to X_i. The message that X_j sent to X_i can be computed as follows:

$$m_{ji}(x_i) = \sum_{x_j} \phi_j(x_j)\psi_{ij}(x_i, x_j) \prod_{k \in N_{x_j} \setminus x_i} m_{kj}(x_j), \tag{4.30}$$

where $\phi_j(x_j)$ is the unary potential for X_j, $\psi_{ij}(x_i, x_j)$ is the pairwise potential between X_i and X_j, and X_k is a different neighbor of X_i. Given the messages that X_i receives from its neighbors, it can then update its belief as follows:

$$Bel(x_i) = k\phi_i(x_i) \prod_{j \in N_{X_i}} m_{ji}(x_i). \tag{4.31}$$

After updating its belief, X_i can pass messages to its neighbors using Eq. (4.30). This process repeats for each node until convergence. Like belief propagation for a BN, the simple belief propagation method may not work well if the model contains loops. In this case the junction tree method can be employed.

For MAP inference, the sum operation in Eq. (4.30) is replaced by the maximum operation. After receiving messages from all its neighbors, each node still uses Eq. (4.31) to update its belief. This process repeats for each node until convergence. After convergence, each node contains the maximum marginal probability. We can then identify the MAP assignment for each node.

Different from BN inference, which is NP hard in general, inference for pairwise MRF with BP is linear to the number of edges and quadratic to the number of states for each variable. Inference for high order MRF, however, becomes NP hard. Examples of belief propagation for MRF can be found in Appendix 4.8.1.

4.5.1.3 Junction tree method

For posterior inference in a complex MN, in particular those with loops, the junction tree method can be applied. We can employ the same junction tree method discussed in Section 3.3.1.3.2 for BN inference. The only difference is that instead of following five steps for BN, the junction tree method for MN only involves four steps: triangulation, clique identification, junction tree construction, and junction tree parameterization. For junction tree parameterization, instead of using CPTs, we use the potential functions to parameterize the junction tree. Specifically, for a cluster C, its potential can be computed as the product of the potentials of its constituent variables minus the variables in its separator (S_c), that is,

$$\psi_c(c) = \prod_{x_i \in c \setminus S_c} \psi_i(x_i).$$

The joint probability can be computed as the product of the potentials of the cluster nodes,

$$p(\mathbf{x}) = \frac{1}{Z}\prod_{c\in\mathbf{C}} \psi_c(c), \tag{4.32}$$

where $\mathbf{C}$ represents the sets of cluster nodes, and Z is the normalization constant to ensure that $p(\mathbf{X})$ sums to one (note that this is not needed for a BN as its parameters are already probabilities). Given a parameterized junction tree, we can then follow Eqs. (3.29) and (3.30) for message calculation and belief updating. Additional details for the junction tree method can be found in the Shafer–Shenoy algorithm [12].

For MAP inference, we can follow the same belief updating procedure as for posterior probability inference. The only difference is in computing the message. Specifically, while calculating the message to pass, the sum operation in Eq. (3.29) is replaced with the max operation. Then message passing and belief updating can be performed the same way until convergence. At convergence, each cluster node contains its max marginal probability. We can then find the best assignment for the nodes in each cluster using it's maximum marginal probability. Further details on MAP inference in a junction tree can be found in Section 13.3 of [2]. In summary, the message passing equations for both posterior and MAP inferences in junction tree are essentially the same as those for MN inferences, with the only difference being that both messages passing and belief updating are with respect to the cluster nodes instead of individual variable nodes as for MNs.

Finally, like ancestral sampling for BN posterior inference, Monte Carlo (MC) sampling can also apply to MN to perform exact posterior inference. But MC sampling for MN is not as simple as BN since we do not have a topological order to follow during sampling. We can randomly select one node as the root node and sample it's unary. This is then followed by sampling other nodes that are adjacent to the node using their pairwise functions. This can continue until we have samples for all nodes.

4.5.1.4 Graph cuts method

As MAP inference for MNs can be formulated as an energy minimization problem, the graph cuts method can be used to solve for the MAP inference. First introduced to CV by Greig, Porteous, and Seheult [13], the graph cuts technique has been proven to guarantee producing the optimal solution in polynomial time for binary MRF if the energy function is submodular. Coupled with its simplicity, the graph cuts method is widely used among CV researchers for low-level CV tasks, including image segmentation, denoising, and point matching for stereo.

According to the graph cuts algorithm, energy minimization problems can be converted to the minimum cut/maximum flow problem in a graph. Find a set of X labels to swap using a min cut/max flow algorithm from network theory such that the flow from a source node s to a sink node t is maximized. Initial s and t are manually identified. In graph theory a cut divides the graph into two disjoint subsets $\mathbf{S}$ and $\mathbf{T}$. The set of edges that the cut goes through are referred to as the cut-through edges. Each cut-through edge has one end point in $\mathbf{S}$ and another in $\mathbf{T}$, as shown in Fig. 4.8.

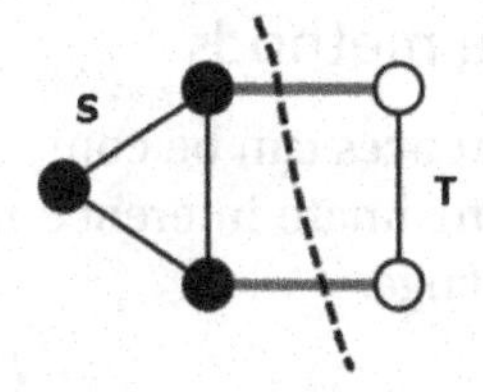

FIGURE 4.8 An illustration of the graph cuts algorithm, where the dotted curve represents the cut, and the two red (dark gray in print version) edges are the cut-through edges. Nodes to the left of the cut line have a label of 0, whereas nodes to the right have a label of 1. Figure courtesy of Wikipedia.

The total cost of the cut is the sum of the weights for all cut-through edges. For an undirected graph with equal weight for all edges, the total cost is the sum of the number of edges in the cut-through set. A minimum cut is the cut that divides the graph into two disjoint sets with minimum cut cost. For a binary MN, the MAP assignment for each node can be obtained by the minimum cut that divides the nodes into either **S**, where node labels are 0, or **T**, where node labels are 1. Given a defined energy function such as Eq. (4.15), the Edmonds–Karp algorithm [14] can be used to identify the minimum cut and hence the MAP label assignment for each node in polynomial time. Further information on the graph cuts algorithm for MAP inference with MN can be found in Section 13.6 of [2].

For nonbinary MNs, the graph cuts method can no longer can provide an optimal solution, and the time complexity, even with submodular energy functions, is NP-hard. In this case, approximate graph cuts methods such as the move-making algorithms may be employed. Starting from an initial label assignment, the move-making algorithms perform optimization by iteratively improving the label assignments using a greedy hill-climbing strategy. The two most popular such methods are the swap move algorithm and the expansion move algorithm [15]. For a pair of labels α, β, the swap move algorithm takes a subset of nodes currently given the label α and assigns them the label β, and vice versa. The algorithm continues the swap operation until no swap move produces a lower energy labeling. In contrast, the expansion move algorithm expands a label α to increase the set of nodes that are given this label. The algorithm stops when further expansion for any label does not yield a lower energy function.

4.5.1.5 Inference for continuous MNs

Like the inference for continuous Gaussian BNs, exact inference for Gaussian graphical models (GGMs) can be carried out directly from the model joint covariance matrix as discussed in Section 3.3.3. In fact, for Gaussian MNs, closed-form solutions exist for both posterior and MAP inference. Their inference complexity, however, is $\mathcal{O}(N^3)$, where N is the number of nodes. Furthermore, the methods for discrete MNs in general can be extended to GBNs. For exact inference, both variable elimination and belief propagation methods can be extended to GGMs [16–18]. For approximate inference, both sampling and variational methods can also be extended to GGMs [19].

4.5.2 Approximate inference methods

Both exact posterior and MAP inferences can be computationally expensive when $\mathbf{X}$ contains many variables. Various approximate inference methods have been introduced to overcome this computational challenge.

4.5.2.1 Iterated conditional modes

The iterated conditional modes (ICM) method [20] was proposed to obtain an approximate MAP estimate of $\mathbf{X}$ locally. Applying the pseudo-likelihood chain rule in Eq. (2.6) and the Markov condition, we have

$$\begin{aligned} p(\mathbf{x}|\mathbf{y}) &\approx \prod_{i=1}^{N} p(x_i|x_{-i},\mathbf{y}) \\ &= \prod_{i=1}^{N} p(x_i|N_{x_i},\mathbf{y}). \end{aligned} \tag{4.33}$$

Eq. (4.33) suggests that the conditional probability of $\mathbf{X}$ factorizes approximately over the conditional probability of X_i. As a result of this factorization, we can perform a MAP estimate individually for each X_i, that is,

$$x_i^* = \arg\max_{x_i} p(x_i|N_{x_i},\mathbf{y}), \tag{4.34}$$

where $p(x_i|N_{x_i},\mathbf{y})$ can be computed locally or globally via the joint probability $p(x_i, x_{-i}, \mathbf{y})$. Given the MAP estimate for X_i as x_i^*, the MAP estimate for $\mathbf{X}$ can be obtained as

$$\mathbf{x}^* = \{x_1^*, x_2^*, \ldots, x_N^*\}.$$

The ICM method works iteratively. Starting with an initialization of all nodes, the method then applies Eq. (4.34) to update the value for each node individually, based on the current values of other nodes. The process iterates until $\mathbf{x}^*$ converges. Algorithm 4.1 provides a pseudocode for the ICM method.

Algorithm 4.1 The ICM algorithm.

Input: $\mathbf{y}$ and $\mathbf{X} = \{X_1, X_2, \ldots, X_N\}$
Initialize $\mathbf{X}$ to $\mathbf{x}^0$
t=0
while not converging **do**
  for i=1 to N **do**
    $x_i^{t+1} = \arg\max_{x_i} p(x_i|\mathbf{y}, N_{x_i}^t)$
    $\mathbf{x}^{t+1} = \{x_1^t, x_2^t, \ldots, x_i^{t+1}, \ldots, x_N^t\}$
    t=t+1
  end for
end while
Output $\mathbf{x}^t$

The performance of the ICM method depends on the initialization $\mathbf{x}^0$.

4.5.2.2 Gibbs sampling

Gibbs sampling can be applied to both posterior and MAP inferences. Given an initial value of $\mathbf{x}^0$, Gibbs sampling works by sampling one node at a time given current values of other nodes, that is,

$$x_i^t \sim p(x_i | x_{-i}^{t-1}, \mathbf{y}), \tag{4.35}$$

where $p(x_i | x_1^{t-1}, x_2^{t-1}, \ldots, x_{i-1}^{t-1}, \ldots, x_N^{t-1}, \mathbf{y})$ can be estimated from the joint probability $p(x_i, x_1^{t-1}, x_2^{t-1}, \ldots, x_{i-1}^{t-1}, \ldots, x_N^{t-1}, \mathbf{y})$ with a normalization. Applying the Markov condition, we compute $p(x_i | x_1^{t-1}, x_2^{t-1}, \ldots, x_{i-1}^{t-1}, \ldots, x_N^{t-1}, \mathbf{y}) = p(x_i | N_{x_i})$ efficiently via $p(x_i | N_{x_i}) = \frac{\exp(-\alpha_i E_i(x_i) - \sum_{x_j \in N_{x_i}} w_{ij} E(x_i, x_j|))}{\sum_{x_i} \exp(-\alpha_i E_i(x_i) - \sum_{x_j \in N_{x_i}} w_{ij} E(x_i, x_j))}$. This suggests that $p(x_i | x_1^{t-1}, x_2^{t-1}, \ldots, x_{i-1}^{t-1}, \ldots, x_N^{t-1}, \mathbf{y})$ can be computed from neighbors of X_i as shown in Eq. 12.23 of [2]. This process repeats for each variable, and after some burn-in period t_0, we can collect samples $\mathbf{x}^t$ for $t = t_0 + 1, t_0 + 2, \ldots, T$.

Given the collected samples $\mathbf{x}^t$, posterior inference can be computed directly from the samples. MAP inference can also be approximately performed by finding the values of $\mathbf{x}^*$ corresponding to the mode of the sample distribution. Specifically, for continuous $\mathbf{X}$, $\mathbf{x}^*$ can be estimated as the sample mean of $\mathbf{x}^t$, $\mathbf{x}^* = \frac{1}{T} \sum_{t=t_0+1}^{T} \mathbf{x}^t$ or sample mode for a multimodal distribution. For discrete $\mathbf{X}$, $\mathbf{x}^*$ can be identified as the configuration with the most counts. Algorithm 4.2 provides a pseudocode for the Gibbs sampling method.

4.5.2.3 Loopy belief propagation

Like BN inference, loopy belief propagation can be applied to MN inference as well. For posterior inference, we can follow the same procedure as in Section 4.5.1.2. For MAP inference, the same belief propagation and updating procedure as discussed in Section 4.5.1.2 applies. As discussed in Section 3.3.2.1, for model with loops, LBP is not guaranteed to converge. But if it converges, then it provides sufficiently good solution.

4.5.2.4 Variational methods

Much like BN inference, variational methods can also be applied to both posterior and MAP inference for MN. The procedure remains the same, that is, finding a surrogate distribution $q(\mathbf{X}|\boldsymbol{\beta})$ to approximate $p(\mathbf{X}|\mathbf{y})$ by minimizing the KL divergence between q and p, that is,

$$q^*(\mathbf{x}|\boldsymbol{\beta}) = \arg\min_{\boldsymbol{\beta}} KL(q(\mathbf{x}|\boldsymbol{\beta}) || p(\mathbf{x}|\mathbf{y})). \tag{4.36}$$

Given q^*, the posterior inference can be easily done using $q^*(\mathbf{x}|\boldsymbol{\beta})$. A MAP estimate for $\mathbf{X}$ can also be obtained approximately using q, that is,

$$\mathbf{x}^* = \arg\max_{\mathbf{x}} q(\mathbf{x}|\boldsymbol{\beta}). \tag{4.37}$$

Algorithm 4.2 Gibbs sampling algorithm for posterior and MAP inference.

Input: $\mathbf{y}$ and $\mathbf{X} = \{X_1, X_2, \ldots, X_N\}$
Initialize $\mathbf{X}$ to $\mathbf{x}^0$
t=0
while $t < t_0$ **do** {t_0 is the burn-in period}
  for i=1 to N **do**
    $x_i^{t+1} \sim p(x_i|\mathbf{x}_{-i}^t, \mathbf{y})$ //obtain a sample
    $\mathbf{x}^{t+1} = \{x_1^t, x_2^t, \ldots, x_i^{t+1}, \ldots, x_N^t\}$
    t=t+1
  end for
end while
while $t < T$ **do** {T is the total number of samples to collect.}
  for i=1 to N **do**
    $x_i^{t+1} \sim p(x_i|\mathbf{x}_{-i}^t, \mathbf{y})$
    $\mathbf{x}^{t+1} = \{x_1^t, x_2^t, \ldots, x_i^{t+1}, \ldots, x_N^t\}$
    t=t+1
  end for
end while
For posterior inference, $p(\mathbf{x}|\mathbf{y})$ can be estimated from samples $\{\mathbf{x}^t\}_{t=t_0+1}^T$.
For MAP inference with continuous $\mathbf{X}$, $\mathbf{x}^* = \frac{1}{T-t_0}\sum_{t=t_0+1}^T \mathbf{x}^t$ or $\mathbf{x}^*$ = sample mode.
For MAP inference with discrete $\mathbf{X}$, $\mathbf{x}^*$ corresponds to the configuration of $\mathbf{x}$ with the most counts.

Since $q()$ is typically factorized, Eq. (4.37) can be solved individually for each element or a small subset of $\mathbf{X}$ independently. The simplest variational method is the mean field method, where we assume that all variables in $\mathbf{X}$ are independent, leading to a fully factorized function $q()$. See Section 3.3.2.3 for details on the mean field method. The only change is the way to compute $p(\mathbf{X}|\mathbf{Y})$. For an MN, we can use Eq. (4.15) to compute $p(\mathbf{X}|\mathbf{Y})$. For CRF, we can directly use Eq. (4.25).

4.5.3 Other MN inference methods

Besides the inference methods discussed, the MN inference, in particular, the MN-MAP inference is often formulated as discrete energy minimization problem in CV vision through well-established combinatorial optimization methods. These techniques include the globally optimal methods through integer programming and the approximate yet efficient linear programming relaxation algorithms methods, which can provide a lower bound for the optimum. Integer programming formulates the MAP inference problem as an integer linear program that optimizes a linear objective function over a set of integer variables subject to linear constraints. Although an integer program formulation does not solve the NP-hardness of the MAP inference, it allows leveraging the existing integer program solvers.

One of the often used methods to approximately solve linear integer program problems is the linear program relaxation, which converts a discrete linear optimization problem into a continuous linear program (LP) optimization for which efficient solutions exist. Further information on the LP methods for solving MN-MAP inference can be found in Section 13.5 of [2]. Andres et al. [21] provided a detailed review of 24 recent energy minimization techniques for different CV tasks. Their evaluation concluded that advanced linear programming and integer linear programming solvers are competitive in both inference accuracy and efficiency for both small and large numbers of labels for a wide range of CV tasks.

Another method of solving large-scale combinatorial optimization problems is the simulated annealing method [22]. Geman and Geman [23] proposed an algorithm based on simulated annealing to perform MRF-MAP inference. Algorithm 4.3 is the pseudocode (adapted from [24]) for simulated annealing.

Algorithm 4.3 The simulated annealing algorithm [24].

---

(1) Choose an initial temperature T
(2) Obtain an initial $\mathbf{x}^*$ by maximizing $p(\mathbf{y}|\mathbf{x})$
(3) Perturb $\mathbf{x}^*$ to generate $\mathbf{z}^*$
(4) Compute the potential difference $\Delta = \psi(\mathbf{z}^*|\mathbf{y}) - \psi(\mathbf{x}^*|\mathbf{y})$
if $\Delta > 0$ **then**
  replace $\mathbf{x}^*$ by $\mathbf{z}^*$
else
  replace $\mathbf{x}^*$ by $\mathbf{z}^*$ with probability $e^{\frac{\Delta}{T}}$
end if
(5) Repeat (3) N times
(6) Replace T by $\phi(T)$, where ϕ is a decreasing function
(7) Repeat (3)–(6) K times.

Finally, dynamic programming-based methods are also used to solve large combinatorial optimization problems. Dynamic programming (also called dynamic optimization) uses the divide-and-conquer strategy to recursively solve an optimization problem. It starts with solving a simper or smaller problem and then recursively solves the larger problem. Specifically, if the global energy (objective) function can be recursively decomposed into a sum of smaller energy functions, the minimization of the global energy function can be done exactly by recursively solving the smaller functions. One of the best examples of dynamic programming applied to MAP inference is the Viterbi algorithm introduced in Section 3.7.3.1.2. It recursively solves the decoding problem (max product inference), that is, finding the best hidden state configuration for T time slices by recursively identifying the best state at each time, starting from $t = 1$ until $t = T$.

4.6 Markov network learning

Like BN learning, MN learning also involves learning either the parameters of an MN or its structure or both. The parameters of an MN are those of the potential functions. Structure learning of an MN involves learning the links between nodes. In this section, we first focus on techniques for parameter learning and then discuss techniques for MN structure learning.

4.6.1 Parameter learning

MN parameter learning can be further divided into parameter learning under complete data and parameter learning under incomplete data.

4.6.1.1 Parameter learning under complete data

Like the parameter learning for BNs under complete data, we first study the case where complete data are given, that is, there are no missing values for each training sample. In this case, MN parameter learning can be stated as follows. Given a set of M i.i.d. training samples $\mathbf{D} = \{D_1, D_2, \ldots, D_M\}$, where $D_m = \{x_1^m, x_2^m, \ldots, x_N^m\}$ represents the mth training sample consisting of a vector of values for each node. The goal of parameter learning is to estimate the parameters $\boldsymbol{\Theta} = \{\boldsymbol{\Theta}_i\}$ such that the joint distribution represented by the MN with the estimated $\boldsymbol{\Theta}$ can best approximate the distribution of training data. For learning pairwise MN, $\boldsymbol{\Theta}_i$ is the parameters for the ith node, including parameters for both pairwise and unary potentials, that is, $\boldsymbol{\Theta}_i = \{w_{ij}, \alpha_i\}$. Note that additional parameters may exist within the unary and pairwise energy functions. These parameters are intrinsic to the unary and energy functions, and hence they are usually estimated separately before estimating the MN parameters $\boldsymbol{\Theta}$. However, it is also possible to learn them jointly with the MN parameters. Like BN parameter learning, parameter learning for MN can be accomplished by maximizing certain objective functions of the parameters given the data $\mathbf{D}$. The most commonly used objective functions are the likelihood and the posterior probability functions, which are discussed below.

4.6.1.2 Maximum likelihood estimation

Like BN parameter learning, MLE of MN parameters can be formulated as

$$\boldsymbol{\theta}^* = \arg\max_{\boldsymbol{\theta}} LL(\boldsymbol{\theta} : \mathbf{D}), \tag{4.38}$$

where $LL(\boldsymbol{\theta} : \mathbf{D})$ represents the joint log-likelihood of $\boldsymbol{\theta}$ given the data $\mathbf{D}$. Under complete data, the joint log-likelihood is concave with one unique global maximum. Given the i.i.d. samples D_m, the joint log-likelihood can be written as

$$LL(\boldsymbol{\theta} : \mathbf{D}) = \log \prod_{m=1}^{M} p(x_1^m, x_2^m, \ldots, x_N^m|\boldsymbol{\theta}). \tag{4.39}$$

For discrete pairwise MN with joint probability distribution specified by Eq. (4.8), Eq. (4.39) can be rewritten as

$$
\begin{aligned}
LL(\theta : \mathbf{D}) &= \log \prod_{m=1}^{M} p(x_1^m, x_2^m, \ldots, x_N^m | \theta) \\
&= \log \prod_{m=1}^{M} \frac{1}{Z} \exp\{-\sum_{(x_i^m, x_j^m) \in \mathcal{E}} w_{ij} E_{ij}(x_i^m, x_j^m) - \sum_{x_i^m \in \mathcal{V}} \alpha_i E_i(x_i^m)\} \\
&= \sum_{m=1}^{M} [-\sum_{(x_i^m, x_j^m) \in \mathcal{E}} w_{ij} E_{ij}(x_i^m, x_j^m) - \sum_{x_i^m \in \mathcal{V}} \alpha_i E_i(x_i^m)] - M \log Z,
\end{aligned} \tag{4.40}
$$

where Z is the partition function, it is a function of all parameters $\boldsymbol{\theta} = \{w_{ij}, \alpha_i\}$, and can be written as

$$
Z(\boldsymbol{\theta}) = \sum_{x_1, x_2, \ldots, x_N} \exp[-\sum_{(x_i, x_j) \in \mathcal{E}} w_{ij} E_{ij}(x_i, x_j) - \sum_{x_i \in \mathcal{V}} \alpha_i E_i(x_i)]. \tag{4.41}
$$

Given the objective function, we may use the gradient ascent method to iteratively estimate the parameters:

$$
\boldsymbol{\theta}^t = \boldsymbol{\theta}^{t-1} + \eta \frac{\partial LL(\boldsymbol{\theta} : \mathbf{D})}{\partial \boldsymbol{\theta}}, \tag{4.42}
$$

where

$$
\begin{aligned}
\frac{\partial LL(\boldsymbol{\theta} : \mathbf{D})}{\partial \boldsymbol{\theta}} &= \frac{\partial \sum_{m=1}^{M} [-\sum_{(x_i^m, x_j^m) \in \mathcal{E}} w_{ij} E_{ij}(x_i^m, x_j^m) - \sum_{x_i^m \in \mathcal{V}} \alpha_i E_i(x_i^m)]}{\partial \boldsymbol{\theta}} \\
&\quad - M \frac{\partial \log Z}{\partial \boldsymbol{\theta}}.
\end{aligned} \tag{4.43}
$$

Specifically, to update the parameter w_{ij} for each link, we have

$$
w_{ij}^t = w_{ij}^{t-1} + \eta \frac{\partial LL(\boldsymbol{\theta} : \mathbf{D})}{\partial w_{ij}}. \tag{4.44}
$$

Following Eq. (4.43), $\frac{\partial LL(\boldsymbol{\theta}:\mathbf{D})}{\partial w_{ij}}$ can be computed as

$$
\frac{\partial LL(\boldsymbol{\theta} : \mathbf{D})}{\partial w_{ij}} = -\sum_{m=1}^{M} E_{ij}(x_i^m, x_j^m) - M \frac{\partial \log Z}{\partial w_{ij}}. \tag{4.45}
$$

Plugging Z into Eq. (4.41), we can further derive the second term $\frac{\partial \log Z}{\partial w_{ij}}$:

$$
\frac{\partial \log Z}{\partial w_{ij}} = \sum_{x_1, x_2, \ldots, x_N} \frac{1}{Z} \exp[-\sum_{(x_i, x_j) \in \mathcal{E}} w_{ij} E_{ij}(x_i, x_j) - \sum_{x_i \in \mathcal{V}} \alpha_i E_i(x_i)][-E_{ij}(x_i, x_j)]
$$

$$
\begin{aligned}
&= \sum_{x_1,x_2,\ldots,x_N} p(x_1,x_2,\ldots,x_N)[-E_{ij}(x_i,x_j)] \\
&= \sum_{\mathbf{x}} p(\mathbf{x})[-E_{ij}(x_i,x_j)] \\
&= E_{\mathbf{x}\sim p(\mathbf{x})}[-E_{ij}(x_i,x_j)]. \qquad (4.46)
\end{aligned}
$$

Combining Eq. (4.45) with Eq. (4.46) yields:

$$\frac{\partial LL(\boldsymbol{\theta}:\mathbf{D})}{\partial w_{ij}} = -\sum_{m=1}^{M} E_{ij}(x_i^m, x_j^m) + M E_{\mathbf{x}\sim p(\mathbf{x})}[E_{ij}(x_i,x_j)]. \qquad (4.47)$$

Given its gradient, w_{ij} can be iteratively estimated via the gradient ascent method:

$$w_{ij}^{t+1} = w_{ij}^{t} + \eta \frac{\partial LL(\boldsymbol{\theta}:\mathbf{D})}{\partial w_{ij}}$$

We can apply similar derivations to iteratively estimate α_i.

It is clear from Eq. (4.47) that w_{ij} cannot be estimated independently since the second term (which is the derivative of the partition function Z) involves all parameters. This represents a major source of challenge with MN learning, in particular, when the number of nodes is large. Various approximate solutions have been proposed to address this computational difficulty. We further discuss three approximate solutions: the contrastive divergence method, the pseudolikelihood method, and the variational method.

4.6.1.2.1 Contrastive divergence method

According to the contrastive divergence (CD) method [25], instead of maximizing the log-likelihood to estimate $\boldsymbol{\theta}$, it maximizes the mean log-likelihood (MLL) over the training samples. The MLL can be written as

$$MLL(\boldsymbol{\theta}:\mathbf{D}) = \frac{LL(\boldsymbol{\theta}:\mathbf{D})}{M}.$$

As a result, following Eq. (4.47), we have

$$\frac{\partial MLL(\boldsymbol{\theta}:\mathbf{D})}{\partial w_{ij}} = -\frac{1}{M}\sum_{m=1}^{M} E_{ij}(x_i^m, x_j^m) + E_{\mathbf{x}\sim p(\mathbf{x})}[E_{ij}(x_i,x_j)], \qquad (4.48)$$

where the first term is called the sample mean of $E_{ij}(x_i,x_j)$, whereas the second term is called the current model mean of $E_{ij}(x_i,x_j)$. The sample mean can be easily computed from the training samples. The model mean in the second term is hard to compute due to its need to sum over all variables. Like Eq. (3.108) in Chapter 3, it can be approximated by sample average through Gibbs sampling. Gibbs samples can be obtained from $p(x_1,x_2,\ldots,x_N)$ based on the current model parameters $\boldsymbol{\theta}^t$. Given the samples, the model mean can be replaced by the average of the samples as shown in the following equation:

$$\frac{\partial MLL(\boldsymbol{\theta}:\mathbf{D})}{\partial w_{ij}} = -\frac{1}{M}\sum_{m=1}^{M} E_{ij}(x_i^m, x_j^m) + \frac{1}{|\mathbf{S}|}\sum_{x_i^s, x_j^s \in \mathbf{S}} E_{ij}(x_i^s, x_j^s), \tag{4.49}$$

where x_i^s is the sth sample obtained from sampling $p(\mathbf{X}|\boldsymbol{\theta}^t)$. In fact, it was shown in [26] that one sample ($s = 1$) is often sufficient to approximate the second term. Similarly, we can apply the same procedure to obtain the gradient of α_i as follows:

$$\frac{\partial MLL(\boldsymbol{\theta}:\mathbf{D})}{\partial \alpha_i} = -\frac{1}{M}\sum_{m=1}^{M} E_i(x_i^m) + \frac{1}{|\mathbf{S}|}\sum_{x_i^s \in \mathbf{S}} E_i(x_i^s). \tag{4.50}$$

Given the gradients in Eqs. (4.49) and (4.50), we can use gradient ascent to iteratively update w_{ij} and α_i. As the iteration continues, the sample mean and model mean become closer, and hence we have a smaller gradient until convergence. Based on this understanding, the CD algorithm is summarized in Algorithm 4.4

Algorithm 4.4 The CD algorithm.

Input: Training data $\mathbf{D} = \{D_m\}_{m=1}^M$ and initial parameters $\boldsymbol{\theta}^0 = \{w_{ij}^0, \alpha_i^0\}$
t=0
while not converging **do**
  for $\forall X_i \in \mathcal{V}$ **do**
    for $\forall (X_i, X_j) \in \mathcal{E}$ **do**
      Sample x_i and x_j from $p(\mathbf{X}|\boldsymbol{\theta}^t)$
      Compute $\frac{\partial LL(\boldsymbol{\theta}:\mathbf{D})}{\partial w_{ij}}$ and $\frac{\partial LL(\boldsymbol{\theta}:\mathbf{D})}{\partial \alpha_i}$ using Eq. (4.49) and Eq. (4.50)
      $w_{ij}^{t+1} = w_{ij}^t + \eta \frac{\partial LL(\boldsymbol{\theta}:\mathbf{D})}{\partial w_{ij}}$
      $\alpha_i^{t+1} = \alpha_i^t + \zeta \frac{\partial LL(\boldsymbol{\theta}:\mathbf{D})}{\partial \alpha_i}$
    end for
  end for
  t=t+1
end while

See Section 20.6.2 of [2] for further information on the contrastive divergence method.

4.6.1.2.2 Pseudolikelihood method

Another solution to the intractable partition function is the pseudolikelihood method [28], which assumes that the joint likelihood can be written as

$$p(x_1, x_2, \ldots, x_N|\boldsymbol{\theta}) \approx \prod_{i=1}^{N} p(x_i|\mathbf{x}_{-i}, \boldsymbol{\theta}), \tag{4.51}$$

where $\mathbf{x}_{-i}$ represent all nodes except for node X_i. While computing the parameters for node X_i, it assumes that parameters for all other nodes are given. Given the training data $\mathbf{D}$, the joint log pseudolikelihood can be written as

$$\begin{aligned} PLL(\mathbf{D}:\boldsymbol{\theta}) &= \log \prod_{m=1}^{M} \prod_{i=1}^{N} p(x_i^m | \mathbf{x}_{-i}^m, \boldsymbol{\theta}_i) \\ &= \sum_{m=1}^{M} \sum_{i=1}^{N} \log p(x_i^m | \mathbf{x}_{-i}^m, \boldsymbol{\theta}_i). \end{aligned} \tag{4.52}$$

For pairwise discrete MN, $p(x_i|\mathbf{x}_{-i}, \boldsymbol{\theta}_i)$ can be expressed as

$$\begin{aligned} & p(x_i|\mathbf{x}_{-i}, \boldsymbol{\theta}_i) = p(x_i|\mathbf{x}_{N_{x_i}}, \boldsymbol{\theta}_i) \\ = & \frac{p(x_i, \mathbf{x}_{N_{x_i}}, \boldsymbol{\theta}_i)}{p(\mathbf{x}_{N_{x_i}}, \boldsymbol{\theta}_i)} \\ = & \frac{p(x_i, \mathbf{x}_{N_{x_i}}, \boldsymbol{\theta}_i)}{\sum_{x_i} p(x_i, \mathbf{x}_{N_{x_i}}, \boldsymbol{\theta}_i)} \\ = & \frac{\frac{1}{Z}\exp(-\sum_{j \in N_{x_i}} w_{ij} E_{ij}(x_i, x_j) - \alpha_i E_i(x_i))}{\sum_{x_i} \frac{1}{Z}\exp(-\sum_{j \in N_{x_i}} w_{ij} E_{ij}(x_i, x_j) - \alpha_i E_i(x_i))} \\ = & \frac{\exp(-\sum_{j \in N_{x_i}} w_{ij} E_{ij}(x_i, x_j) - \alpha_i E_i(x_i))}{\sum_{x_i} \exp(-\sum_{j \in N_{x_i}} w_{ij} E_{ij}(x_i, x_j) - \alpha_i E_i(x_i))}. \end{aligned} \tag{4.53}$$

This computation is much easier to perform since the summation in the denominator involves only X_i. Given the pseudo-loglikelihood in Eq. (4.52), we can then compute its gradient with respect to w_{ij} and α_i as follows:

$$\begin{aligned} \frac{\partial PLL(\boldsymbol{\theta}:\mathbf{D})}{\partial w_{ij}} &= -\sum_{m=1}^{M} E_{ij}(x_i^m, x_j^m) + M \sum_{x_i} p(x_i|N_{x_i}, \boldsymbol{\theta}) E_{ij}(x_i, x_j), \\ \frac{\partial PLL(\boldsymbol{\theta}:\mathbf{D})}{\partial \alpha_i} &= -\sum_{m=1}^{M} E_i(x_i^m) + M \sum_{x_i} p(x_i|N_{x_i}, \theta) E_i(x_i), \end{aligned} \tag{4.54}$$

where $p(x_i|\mathbf{x}_{N_{x_i}}, \boldsymbol{\theta})$ can be computed using Eq. (4.53). Given the gradients, we can then update w_{ij} and α_i as follows:

$$\begin{aligned} w_{ij}^{t+1} &= w_{ij}^{t} + \eta \frac{\partial PLL(\boldsymbol{\theta}:\mathbf{D})}{\partial w_{ij}}, \\ \alpha_i^{t+1} &= \alpha_i^{t} + \zeta \frac{\partial PLL(\boldsymbol{\theta}:\mathbf{D})}{\partial \alpha_i}. \end{aligned} \tag{4.55}$$

The pseudolikelihood method is summarized in Algorithm 4.5.

Algorithm 4.5 The pseudolikelihood algorithm.

Input: Training data $\mathbf{D} = \{D_m\}_{m=1}^{M}$ and initial parameters $\boldsymbol{\theta}^0 = \{w_{ij}^0, \alpha_i^0\}$
t=0
while not converging **do**
  for $\forall X_i \in \mathcal{V}$ **do**
    for $\forall (X_i, X_j) \in \mathcal{E}$ **do**
      Compute $\frac{\partial PLL(\boldsymbol{\theta}^t:\mathbf{D})}{\partial w_{ij}}$ and $\frac{\partial PLL(\boldsymbol{\theta}^t:\mathbf{D})}{\partial \alpha_i}$ using Eq. (4.54)
      Update w_{ij}^t and α_i^t using Eq. (4.55)
    end for
  end for
  t=t+1
end while

One problem with the pseudolikelihood method is that the final joint probability may not sum to one.

4.6.1.2.3 Variational method

Besides the CD and pseudolikelihood methods, variational methods can also be employed to address the partition function challenge. This is accomplished by appropriating $p(\mathbf{x})$ in the second term of Eq. (4.47) by a factorable variational distribution $q(\mathbf{x}, \boldsymbol{\beta})$, where $\boldsymbol{\beta}$ are the variational parameters. By minimizing the KL divergence between $p(\mathbf{x})$ and $q(\mathbf{x})$, the variational parameters can be solved. One choice of $q(\mathbf{x}, \boldsymbol{\beta})$ is to make it fully factorable over $\mathbf{x}$, $q(\mathbf{x}, \boldsymbol{\beta}) = \prod_{n=1}^{N} p(x_i, \beta_i)$. This leads to the well-known mean field approximation. By replacing $p(\mathbf{x})$ by $q(\mathbf{x})$ in the second term of Eq. (4.47) the expectation can be easily solved.

4.6.1.3 Maximum A Posteriori (MAP) estimation of MN parameters

Like BN parameter learning, we can also perform MAP parameter learning by maximizing the posterior probability of the parameters:

$$\boldsymbol{\theta}^* = \arg\max_{\boldsymbol{\theta}} \log p(\boldsymbol{\theta}|\mathbf{D}), \tag{4.56}$$

where $\log p(\boldsymbol{\theta}|\mathbf{D})$ can be approximated by

$$\log p(\boldsymbol{\theta}|\mathbf{D}) = \log p(\mathbf{D}|\boldsymbol{\theta}) + \log p(\boldsymbol{\theta}), \tag{4.57}$$

where the first term is the log-likelihood term, which can be computed using Eq. (4.40), and the second term is the log prior probability of the parameters. Since the potential parameters $\boldsymbol{\theta}$ are not as meaningful as the BN parameters, which are the CPDs, it is hard to come up with a reasonable prior such as the conjugate prior for BN on the potential parameters. The most widely used prior for MN parameters is the zero-mean Laplacian distribution or zero-mean Gaussian distribution. For zero-mean Laplacian distribution,

$$p(\boldsymbol{\theta}|\beta) = \frac{1}{2\beta}\exp(-\frac{|\boldsymbol{\theta}|_1}{\beta}).$$

Substituting the prior into Eq. (4.57) and ignoring the constants yield

$$\log p(\boldsymbol{\theta}|\mathbf{D}) = \log p(\mathbf{D}|\boldsymbol{\theta}) - |\boldsymbol{\theta}|_1, \tag{4.58}$$

where the second term is a regularization term to the loglikelihood. It is implemented as the ℓ_1 norm. It imposes the sparsity constraint on $\boldsymbol{\theta}$. Similarly, if we assume that $p(\boldsymbol{\theta})$ follows zero-mean Gaussian distribution, it leads to ℓ_2-regularization. Whereas ℓ_2-regularization leads to small-value parameters, ℓ_1-regularization leads to sparse parameterization, that is, many parameters with zero values. Hence MN learning with ℓ_1-regularization may lead to a sparse MN structure.

With either ℓ_1- or ℓ_2-regularization, the objective function remains concave. The gradient (or subgradient) ascent method can be used to solve for the parameters. Specifically, with regularization, the parameter gradient $\nabla\boldsymbol{\theta}$ consists of two terms, the data term and regularization term. The data term gradient can be computed with either the CD method using Eq. (4.48) or the pseudolikelihood method using Eq. (4.54). For the ℓ_1-norm, the regularization gradient can be computed as $\text{sign}(\boldsymbol{\theta})$, where sign is the sign function, which equals to 1 if its argument is greater than 0 and -1 otherwise. When $\boldsymbol{\theta}$ is zero, its gradient can be approximated by a small positive constant or by the subgradient method. For the squared ℓ_2-norm, the regularization gradient is simply $2\boldsymbol{\theta}$. Given $\nabla\boldsymbol{\theta}$, the gradient ascent method can then be applied to iteratively update the parameters $\boldsymbol{\theta}$.

4.6.1.4 Discriminative learning

Both MLE and MAP estimation are generative in that they try to find the parameters maximizing the joint probability distribution. As we discussed in Section 3.4.1.3, for classification task, discriminative learning may produce better performance. Following the convention we defined for BN discriminative learning in Section 3.4.1.3, we can divide $\mathbf{X}$ into $\mathbf{X}_t$ and $\mathbf{X}_F$, that is, $\mathbf{X} = \{\mathbf{X}_t, \mathbf{X}_F\}$, where $\mathbf{X}_t$ represents a set of nodes in the MN that we want to estimate, and $\mathbf{X}_F$ are the remaining nodes of the MN that represent the features. MN discriminative learning can be formulated as finding the parameters $\boldsymbol{\theta}$ by maximizing the log conditional likelihood:

$$\boldsymbol{\theta}^* = \arg\max_{\boldsymbol{\theta}} \sum_{m=1}^{M} \log p(\mathbf{x}_t^m|\mathbf{x}_F^m, \boldsymbol{\theta}). \tag{4.59}$$

We can follow the derivations in Section 3.4.1.3 of Chapter 3 and the CRF learning in Section 4.6.1.5 to derive the equations for discriminative MN learning.

Another method for discriminative learning is the margin-based approach, where the goal is to find the parameters by maximizing the probability margin δ defined as

$$\delta(\boldsymbol{\theta}, m) = \log p(\mathbf{x}_t^m | \mathbf{x}_F^m, \boldsymbol{\theta}) - \max_{\mathbf{x}_{t'}, \mathbf{x}_{t'} \neq \mathbf{x}_t^m} \log p(\mathbf{x}_{t'} | \mathbf{x}_F^m, \boldsymbol{\theta}). \tag{4.60}$$

Given the margin definition, the margin-based method finds $\boldsymbol{\theta}$ by maximizing the overall margins for all samples, that is,

$$\boldsymbol{\theta}^* = \arg\max_{\boldsymbol{\theta}} \sum_{m=1}^{M} \delta(\boldsymbol{\theta}, m). \tag{4.61}$$

One major advantage of the margin-based method is its elimination of the need to deal with the partition function. Through the subtraction in the margin definition, the partition function is effectively subtracted out. Further information on the margin-based method can be found in Section 20.6.2.2 of [2].

4.6.1.5 Parameter learning for CRF

Whereas MN parameter learning is generative by maximizing the loglikelihood, CRF parameter learning is discriminative by maximizing the conditional loglikelihood. Let $\mathbf{X}$ be the unknown label nodes, and let $\mathbf{y}$ represent the observation nodes. Given the training data $\mathbf{D} = \{\mathbf{x}^m, \mathbf{y}^m\}_{m=1}^M$, CRF learning can be formulated as finding its parameters $\boldsymbol{\theta}$ by maximizing the log conditional likelihood:

$$\boldsymbol{\theta}^* = \arg\max_{\boldsymbol{\theta}} \sum_{m=1}^{M} \log p(\mathbf{x}^m | \mathbf{y}^m, \boldsymbol{\theta}). \tag{4.62}$$

For pairwise CRF with loglinear potential function, $\log p(\mathbf{x}^m | \mathbf{y}^m, \boldsymbol{\theta})$ in Eq. (4.62) can be rewritten as

$$\begin{aligned} \log p(\mathbf{x}^m | \mathbf{y}^m, \boldsymbol{\theta}) &= -\sum_{m=1}^{M} \sum_{(x_i, x_j) \in \mathcal{E}} w_{ij} E_{ij}(x_i^m, x_j^m | \mathbf{y}^m) \\ &\quad - \sum_{m=1}^{M} \sum_{x_i \in \mathcal{V}} \alpha_i E_i(x_i^m | \mathbf{y}^m) - M \log Z. \end{aligned} \tag{4.63}$$

The log conditional likelihood remains concave. It therefore admits one unique optimal solution for $\boldsymbol{\theta}$. We can use the gradient ascent method to iteratively estimate $\boldsymbol{\theta}$. The remaining challenge is computing the gradient of the partition function. We can use the CD or the pseudolikelihood method to solve this problem. One key difference between CD for MRF and CD for CRF is that the sample average of $E_{ij}(x_i, x_j | \mathbf{y})$ varies with the values of $\mathbf{y}$ for CRF.

4.6.1.6 MN parameter learning under incomplete data

In this section, we briefly discuss MN parameter learning when the training data are incomplete. Like the BN parameter learning under incomplete data, this can be done in two ways, the direct method that directly maximizes the log marginal likelihood or the EM method that maximizes the lower bound of the log marginal likelihood through the EM approach. For the direct approach, the challenge remains with the partition function Z. We can use the CD or the pseudolikelihood method to address this challenge. For the EM approach, the formulation is the same as that of BNs. For MNs, however, the solution to the M-step does not admit a closed-form solution like for BN. It needs to be solved iteratively through a separate gradient ascent. As a result, the advantages of EM over the gradient ascent approach for MN learning are not as apparent as for BN learning. Finally, since the log marginal likelihood is no longer concave, there exist local maxima. The performance of both approaches depends on initialization. Details about MN parameter learning under incomplete data can be found in Section 20.3.3 of [2].

4.6.2 Structure learning

Like BN structure learning, MN structure learning is learning the links among nodes. Given random variables $\mathbf{X} = (X_1, X_2, \ldots, X_N)$ associated with N nodes in an MN, MN structure learning is to estimate the underlying graph $\mathcal{G}$ (i.e., the links) from M i.i.d. samples $\mathbf{D} = (x_1^m, x_2^m, \ldots, x_N^m)$, $m = 1, 2, \ldots, M$.

Like BN structure learning, MN structure learning can be done by either the constraint-based approach or the score-based approach. The constraint-based approach determines the absence or presence of a link via a conditional independency test. A link characterizes the dependency between two variables. The score-based approach defines a score function and finds the structure $\mathcal{G}$ that maximizes the score function. We will focus on the score-based approach.

4.6.2.1 Score-based approach

The score-based approach defines a score function. One such score is the marginal likelihood function like Eq. (3.92),

$$\mathcal{G}^* = \arg\max_{\mathcal{G}} p(\mathbf{D}|\mathcal{G}). \tag{4.64}$$

Following the derivation in Section 3.4.2.1.1, the marginal likelihood score can be written as the BIC score

$$S_{BIC}(\mathcal{G}) = \log p(\mathbf{D}|\mathcal{G}, \hat{\boldsymbol{\theta}}) - \frac{\log M}{2} Dim(\mathcal{G}), \tag{4.65}$$

where $\hat{\theta}$ is the ML estimate of $\boldsymbol{\theta}$ for $\mathcal{G}$, and $Dim(\mathcal{G})$ is the degree of freedom of $\mathcal{G}$, that is, the number of independent parameters of $\mathcal{G}$. Whereas the first term is the joint likelihood, which ensures that $\mathcal{G}$ fits the data, the second term is a penalty term that favors simple structures. Besides the BIC score, alternative score functions have been proposed (see Section 3.4.2.1.1 for details). Unlike the BIC score for BN, the BIC score for MN is not decomposable due to the partition function. Hence, the structure for each node cannot be learned separately.

Given the definition of the score function, MN structure learning can be carried out by combinatorially searching the MN structural space for the structure that maximizes the score function. Although brute-force search is possible for special MN structures such as trees, it becomes computationally intractable for general MNs since the structural space for a general MN is exponentially large. For general structures, a heuristic greedy search may be employed, which, given the current structure, varies the local structure for each node to identify the change that leads to the maximum score improvement. This process repeats until convergence. Here a similar search method, such as the hill-climbing method for BN structure learning, can be employed. Different from BN structure learning, computation of the score for each node for MN structure learning can be much more computationally expensive again due to the partition function Z. The likelihood term in the BIC score can be approximated by the pseudo-likelihood, yielding the pseudo-BIC score, which can be computed more efficiently. As a result, much work in MN structure learning focuses on reducing the computational cost of evaluating the score function.

The global score computation is intractable. But because of the non-decomposable score function, MN structure can also be approximately learned locally by learning the local structure for each node independently. The local structure for each node consists of its neighbors, that is, its MB. We use the same score function to learn the local structure. Although the local approach is computationally less complex, it cannot yield optimal structure since it assumes that the structure learning is decomposable, which as we know is not. Hence, a better approach may be hybrid, whereby a local method is used to learn an initial structure and a global method can be used to refine the initial structure.

4.6.2.2 Structure learning via parameter regularization

Another common approach for MN structure learning is through parameter regularization. The basic idea is to assume the structure is fully connected initially and then begin to prune the links through learning. It formulates structure learning as parameter learning with ℓ_1-regularization on the parameters as shown in the equation

$$\boldsymbol{\theta}^* = \arg\max_{\boldsymbol{\theta}} \log p(\mathbf{D}|\boldsymbol{\theta}) - \eta|\boldsymbol{\theta}|_1, \tag{4.66}$$

where the first term is the loglikelihood, and the second term is the ℓ_1-regularization term (also called a graphical lasso). By adding an ℓ_1-regularization the parameter learning is subject to a sparsity constraint on the pairwise parameters. The ℓ_1-regularization forces many pairwise parameters $w_{ij} = 0$, which indirectly means there is no link between nodes X_i and X_j. This assumption holds if we assume the MN follows Gibbs distribution, with a log-linear pairwise potential function. Note that Eq. (4.66) is the same as Eq. (4.58), and it remains concave. We can solve it iteratively via subgradient ascent.

For continuous MNs such as a Gaussian BN, we can use the precision matrix **W** (or the weight matrix) to determine the structure. The zero entries of **W** correspond to the absences of edges in $\mathcal{G}$.

Finally, we can also employ the constraint based approach by performing independence tests. Following the pairwise independence property, the necessary and sufficient condition for the absence of a link between two nodes is that they are independent, given all other nodes. Alternatively, we can use the complement of the pairwise independence property to ascertain the presence of a link between the two nodes, i.e., there exists a link between two nodes if and only if they are dependent, given any subset (including empty set) of the remaining nodes.

4.7 Markov networks versus Bayesian networks

Both BNs and MNs are used to graphically capture the joint probability distribution among a set of random variables. Because of their built-in conditional independencies, they can compactly represent the joint probability distribution. They share similar local and global independence properties. In particular, the Markov condition is the key to both models. As a result of the Markov condition, the joint probability distribution of all nodes can be factorized as products of local functions (CPDs for BN and potential functions for MN). Both are generative models.

Despite their similarities, they also significantly differ. First, BNs are mostly used to model the causal dependencies among random variables, whereas MNs are mainly used to model the mutual dependencies among random variables. MNs can therefore capture a wider range of relationships. MNs have been widely used for many CV tasks. With complex global and nonmodular coupling, MNs, on the other hand, are hard to generate data from. In contrast, with clear causal and modular semantics, BNs are easy to generate data from. BNs are suitable for domains where the interactions among variables have a natural direc-

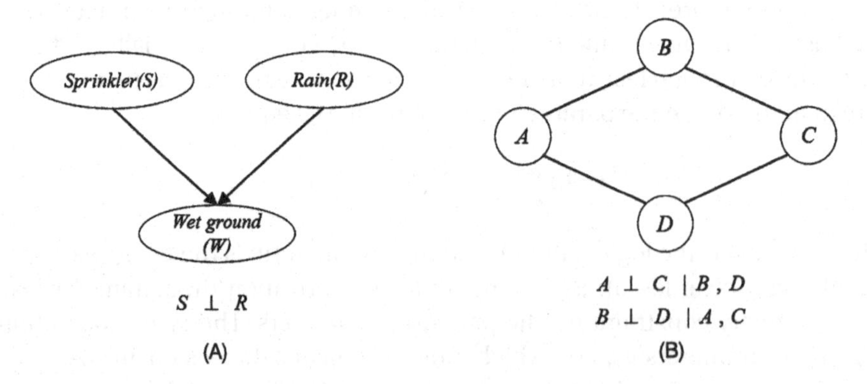

FIGURE 4.9 Illustration of the modeling limitations of BNs and MNs: (A) a classical BN example that captures the "explaining-away" relationships. MNs cannot model such relationships. (B) A simple MN example whose cyclic dependencies cannot be fully represented by BNs.

Table 4.2 Comparison of BNs and MNs.

Items	BNs	MNs
Graph	DAG	Undirected graph
Allowing Loops	No directed loops	Allow undirected loops
Relationships	Causal/one-way	Mutual/correlation
Local regions	Markov blanket	Neighborhood
Local functions	CPDs	Potential functions
Parameter learning	MLE, MAP, and EM	Pseudolikelihood, CD, MAP, EM
Inference	VE, BP, Junction tree, Gibbs, Variational	ICM, Graph Cuts, Gibbs, Variational, etc.
Computational cost	High	High
Joint probability	$p(\mathbf{X}) = \prod_{i=1}^{N} p(X_i \mid \pi(X_i))$	$p(\mathbf{X}) = \frac{1}{Z} \prod_{c \in C} \psi_c(X_c)$

tionality such as time series data, where temporal causality exists naturally. Moreover, the explaining-away relationship (i.e., the V-structure) is unique in BNs, and they allow capturing dependencies among variables that MNs cannot capture. On the other hand, BNs cannot capture cyclic dependencies represented by MNs. Fig. 4.9 illustrates their differences in representation.

Second, BNs are locally specified by conditional probabilities, whereas MNs are locally specified by potential functions. Compared to conditional probabilities, potential functions are more flexible. But, a normalization is required to ensure that the joint distribution will be a probability. The normalization can cause significant problems during both inference and learning since it is a function of all potential function parameters. As a result, MNs have different learning and inference methods to specifically handle the partition function. Third, learning and inference are relatively easier for BNs than for MNs due to the strong independencies embedded in BNs. In fact, unlike the BN parameter learning, which is decomposable and has closed form solutions, there is no closed form solution to MN parameter learning and learning is not decomposable. Table 4.2 compares BNs with MNs.

BNs and MNs are also convertible. We can convert a BN to its corresponding MN and vice versa. One way of converting a BN to an MN is producing the moral graph of the BN. A moral graph of a BN is an undirected graph obtained by changing all links in the BN to undirected links and adding undirected links to spouse nodes, that is, nodes that share the same child as discussed in Section 3.3.1.3.2.

4.8 Appendix

4.8.1 Examples for belief propagation in MRF

Given the binary MRF in Fig. 4.10,

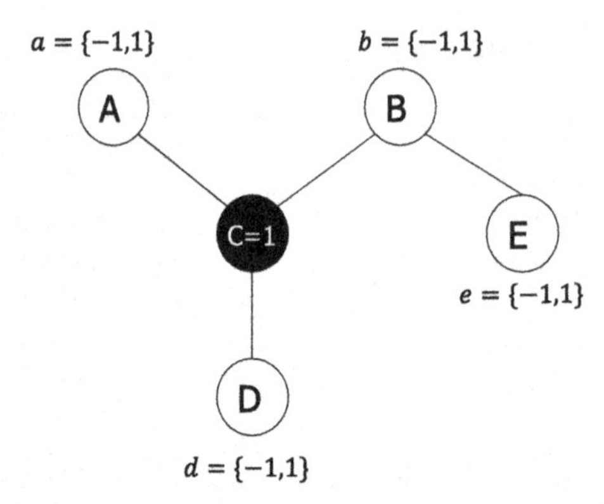

FIGURE 4.10 Structure of a binary MRF with evidence

it's unary potentials,

$$
\begin{aligned}
\phi_A(a) &= \begin{bmatrix} \exp(-1.2) \\ \exp(2) \end{bmatrix} \\
\phi_B(b) &= \begin{bmatrix} \exp(0.8) \\ \exp(-0.2) \end{bmatrix} \\
\phi_C(c) &= \begin{bmatrix} \exp(-1.3) \\ \exp(-0.2) \end{bmatrix} \\
\phi_D(d) &= \begin{bmatrix} \exp(0.2) \\ \exp(-0.2) \end{bmatrix} \\
\phi_E(e) &= \begin{bmatrix} \exp(-0.5) \\ \exp(0.5) \end{bmatrix}
\end{aligned}
\tag{4.67}
$$

and pairwise potentials,

$$
\begin{aligned}
\psi_{AC}(a,c) &= \begin{bmatrix} \exp(2) & \exp(-1) \\ \exp(-1) & \exp(2) \end{bmatrix} \\
\psi_{BC}(b,c) &= \begin{bmatrix} \exp(-0.3) & \exp(1.2) \\ \exp(1.2) & \exp(-0.3) \end{bmatrix} \\
\psi_{BE}(b,e) &= \begin{bmatrix} \exp(0.5) & \exp(1) \\ \exp(1) & \exp(0.5) \end{bmatrix} \\
\psi_{DC}(d,c) &= \begin{bmatrix} \exp(0.9) & \exp(-0.2) \\ \exp(-0.2) & \exp(0.9) \end{bmatrix}
\end{aligned}
\tag{4.68}
$$

we show below the process for performing belief propagation for both sum-product and max-product inference for this MRF.

4.8.1.1 Sum-product inference with evidence

We now perform sum-product inference, given $c = 1$. During initialization, the entries of the unary and pairwise potential functions involving the evidence node C corresponding to $c = 0$ are set to zero and remain unchanged otherwise, i.e.

$$\begin{aligned}
\phi_C(c) &= \begin{bmatrix} 0 \\ \exp(-0.2) \end{bmatrix} \\
\psi_{AC}(a,c) &= \begin{bmatrix} 0 & \exp(-1) \\ 0 & \exp(2) \end{bmatrix} \\
\psi_{BC}(b,c) &= \begin{bmatrix} 0 & \exp(1.2) \\ 0 & \exp(-0.3) \end{bmatrix} \\
\psi_{DC}(d,c) &= \begin{bmatrix} 0 & \exp(-0.2) \\ 0 & \exp(0.9) \end{bmatrix}
\end{aligned} \tag{4.69}$$

Messages for all nodes are initialized to ones. For each node, we update the messages it receives from its neighbors based on their current messages. We calculate messages and belief of each node using Eq. (4.30) and (4.31) as shown below.

Node A: The message node A receives from its neighbor node C is calculated as

$$m_{CA}(a) = \sum_c \phi_C(c)\psi_{AC}(a,c)m_{BC}(c)m_{DC}(c) = \begin{bmatrix} 0.3012 \\ 6.0496 \end{bmatrix} \tag{4.70}$$

where m_{BC} and m_{DC} take on their current values (i.e., initial values). The belief of node A given current message is then

$$Bel(a) = \alpha\phi_A(a)m_{CA}(a) = \begin{bmatrix} 0.0020 \\ 0.9980 \end{bmatrix} \tag{4.71}$$

where α is the normalization constant.

Node B: The messages node B receives from its neighbor nodes C and E are calculated as

$$m_{CB}(b) = \sum_c \phi_C(c)\psi_{BC}(b,c)m_{AC}(c)m_{DC}(c) = \begin{bmatrix} 2.7183 \\ 0.6065 \end{bmatrix} \tag{4.72}$$

$$m_{EB}(b) = \sum_e \phi_E(e)\psi_{BE}(b,e) = \begin{bmatrix} 5.4817 \\ 4.3670 \end{bmatrix} \tag{4.73}$$

where m_{AC} and m_{DC} assume their current values. The belief of node B given current messages is then

$$Bel(b) = \alpha\phi_B(b)m_{CB}(b)m_{EB}(b) = \begin{bmatrix} 0.9386 \\ 0.0614 \end{bmatrix} \tag{4.74}$$

Node C: The messages node C receives from its neighbor nodes A, B and, D are calculated as

$$m_{AC}(c) = \sum_a \phi_A(a)\psi_{AC}(a,c) = \begin{bmatrix} 0 \\ 54.7090 \end{bmatrix} \tag{4.75}$$

$$m_{BC}(c) = \sum_b \phi_B(b)\psi_{BC}(b,c)m_{EB}(b) = \begin{bmatrix} 0 \\ 43.1532 \end{bmatrix} \tag{4.76}$$

$$m_{DC}(c) = \sum_d \phi_D(d)\psi_{DC}(d,c) = \begin{bmatrix} 0 \\ 3.0138 \end{bmatrix} \tag{4.77}$$

where $m_{EB} = [5.4817; 4.3670]$ is calculated in Eq. (4.73). The belief of node C given current messages is then

$$Bel(c) = \alpha\phi_C(c)m_{AC}(c)m_{BC}(c)m_{DC}(c) = \begin{bmatrix} 0 \\ 1 \end{bmatrix} \tag{4.78}$$

Since node C is the evidence node with $c = 1$, its belief doesn't change over iterations.

Node D: The message node D receives from its neighbor node C is calculated as

$$m_{CD}(d) = \sum_c \phi_C(c)\psi_{CD}(c,d)m_{AC}(c)m_{BC}(c) = \begin{bmatrix} 1582.5 \\ 4754.2 \end{bmatrix} \tag{4.79}$$

where $m_{AC} = [0; 54.7090]$ is calculated in Eq. (4.75) and $m_{BC} = [0; 43.1532]$ is calculated in Eq. (4.76). The belief of node D given current message is then

$$Bel(d) = \alpha\phi_D(d)m_{CD}(d) = \begin{bmatrix} 0.3318 \\ 0.6682 \end{bmatrix} \tag{4.80}$$

Node E: The message node E receives from its neighbor node B is calculated as

$$m_{BE}(e) = \sum_b \phi_B(b)\psi_{BE}(b,e)m_{CB}(b) = \begin{bmatrix} 11.3240 \\ 17.2634 \end{bmatrix} \tag{4.81}$$

where $m_{CB} = [2.7183; 0.6065]$ is calculated in Eq. (4.72). The belief of node E given current message is then

$$Bel(e) = \alpha\phi_E(e)m_{BE}(e) = \begin{bmatrix} 0.1944 \\ 0.8056 \end{bmatrix} \tag{4.82}$$

We now finish the first iteration. We repeat the process for several times and observe that the messages do not change in the third iteration. Thus, the belief propagation converges after the second iteration. In the end, we obtain the belief of each node given the evidence as

$$\begin{aligned} Bel(a) &= p(a|c=1) = [0.0020; 0.9980] \\ Bel(b) &= p(b|c=1) = [0.9386; 0.0614] \\ Bel(d) &= p(d|c=1) = [0.3318; 0.6682] \\ Bel(e) &= p(e|c=1) = [0.1944; 0.8056] \end{aligned} \tag{4.83}$$

4.8.1.2 Max-product inference with evidence

To perform max-product inference, we only need to replace the summation operation in the sum-product inference with the maximization operation, and the remaining calculations remain the same. We follow the same procedure to initialize messages for all nodes to ones. Unary and pairwise potential functions corresponding to the unobserved state of the evidence node are set to 0, and remain unchanged otherwise. Max-product propagation can then start.

Node A: The message node A receives from its neighbor node C is calculated as

$$m_{CA}(a) = \max_c \phi_C(c)\psi_{AC}(a,c)m_{BC}(c)m_{DC}(c) = \begin{bmatrix} 0.3012 \\ 6.0496 \end{bmatrix} \tag{4.84}$$

where m_{BC} and m_{DC} are all ones as initialized. The belief of node A given current message is then

$$Bel(a) = \alpha\phi_A(a)m_{CA}(a) = \begin{bmatrix} 0.0020 \\ 0.9980 \end{bmatrix} \tag{4.85}$$

Node B: The messages node B receives from its neighbor nodes C and E are calculated as

$$m_{CB}(b) = \max_c \phi_C(c)\psi_{BC}(b,c)m_{AC}(c)m_{DC}(c) = \begin{bmatrix} 2.7183 \\ 0.6065 \end{bmatrix} \tag{4.86}$$

$$m_{EB}(b) = \max_e \phi_E(e)\psi_{BE}(b,e) = \begin{bmatrix} 4.4817 \\ 2.7183 \end{bmatrix} \tag{4.87}$$

where m_{AC} and m_{DC} are all ones as initialized. The belief of node B given current messages is then

$$Bel(b) = \alpha\phi_B(b)m_{CB}(b)m_{EB}(b) = \begin{bmatrix} 0.9526 \\ 0.0474 \end{bmatrix} \tag{4.88}$$

Node C: The messages node C receives from its neighbor nodes A, B, and D are calculated as

$$m_{AC}(c) = \max_a \phi_A(a)\psi_{AC}(a,c) = \begin{bmatrix} 0 \\ 54.5982 \end{bmatrix} \tag{4.89}$$

$$m_{BC}(c) = \max_b \phi_B(b)\psi_{BC}(b,c)m_{EB}(b) = \begin{bmatrix} 0 \\ 33.1155 \end{bmatrix} \tag{4.90}$$

$$m_{DC}(c) = \max_{d} \phi_D(d)\psi_{DC}(d,c) = \begin{bmatrix} 0 \\ 2.0138 \end{bmatrix} \tag{4.91}$$

where $m_{EB} = [4.4817; 2.7183]$ is calculated in Eq. (4.87). The belief of node C given current messages is then

$$Bel(c) = \alpha\phi_C(c)m_{AC}(c)m_{BC}(c)m_{DC}(c) = \begin{bmatrix} 0 \\ 1 \end{bmatrix} \tag{4.92}$$

Since node C is the evidence node with $c = 1$, its belief doesn't change over iterations.

Node D: The message node D receives from its neighbor node C is calculated as

$$m_{CD}(d) = \max_{c} \phi_C(c)\psi_{CD}(c,d)m_{AC}(c)m_{BC}(c) = \begin{bmatrix} 1212.0 \\ 3641.0 \end{bmatrix} \tag{4.93}$$

where $m_{AC} = [0; 54.5982]$ and $m_{BC} = [0; 33.1155]$ are calculated in Eq. (4.89) and Eq. (4.90) respectively. The belief of node D given current message is then

$$Bel(d) = \alpha\phi_D(d)m_{CD}(d) = \begin{bmatrix} 0.3318 \\ 0.6682 \end{bmatrix} \tag{4.94}$$

Node E: The message node E receives from its neighbor node B is calculated as

$$m_{BE}(e) = \max_{b} \phi_B(b)\psi_{BE}(b,e)m_{CB}(b) = \begin{bmatrix} 9.9742 \\ 16.4446 \end{bmatrix} \tag{4.95}$$

where $m_{CB} = [2.7183; 0.6065]$ is calculated in Eq. (4.86). The belief of node E given current message is then

$$Bel(e) = \alpha\phi_E(e)m_{BE}(e) = \begin{bmatrix} 0.1824 \\ 0.8176 \end{bmatrix} \tag{4.96}$$

We now finish the first iteration. We repeat the process for several iterations and observe that the messages do not change in the third iteration. Thus, the belief propagation converges after the second iteration. Given the beliefs, we have the MAP configuration as

$$[1, -1, 1, 1] = \arg\max_{a,b,d,e} p(a,b,d,e|c=1) \tag{4.97}$$

where $x^* = \arg\max_x Bel(x)$ for $x \in \{A, B, D, E\}$.

References

[1] J. Besag, Spatial interaction and the statistical analysis of lattice systems, Journal of the Royal Statistical Society. Series B (Methodological) (1974) 192–236.

[2] D. Koller, N. Friedman, Probabilistic Graphical Models: Principles and Techniques, MIT Press, 2009.

[3] R. Salakhutdinov, G. Hinton, Deep Boltzmann machines, in: Artificial Intelligence and Statistics, 2009, pp. 448–455.

[4] P. Krähenbühl, V. Koltun, Efficient inference in fully connected CRFs with Gaussian edge potentials, in: Advances in Neural Information Processing Systems, 2011, pp. 109–117.
[5] P. Krähenbühl, V. Koltun, Parameter learning and convergent inference for dense random fields, in: International Conference on Machine Learning, 2013, pp. 513–521.
[6] V. Vineet, J. Warrell, P. Sturgess, P.H. Torr, Improved initialization and Gaussian mixture pairwise terms for dense random fields with mean-field inference, in: BMVC, 2012, pp. 1–11.
[7] P. Kohli, M.P. Kumar, P.H. Torr, P3 & beyond: solving energies with higher order cliques, in: IEEE Conference on Computer Vision and Pattern Recognition, 2007, pp. 1–8.
[8] A. Fix, A. Gruber, E. Boros, R. Zabih, A graph cut algorithm for higher-order Markov random fields, in: IEEE International Conference on Computer Vision, ICCV, 2011, pp. 1020–1027.
[9] V. Vineet, J. Warrell, P.H. Torr, Filter-based mean-field inference for random fields with higher-order terms and product label-spaces, International Journal of Computer Vision 110 (3) (2014) 290–307.
[10] Z. Liu, X. Li, P. Luo, C.-C. Loy, X. Tang, Semantic image segmentation via deep parsing network, in: Proceedings of the IEEE International Conference on Computer Vision, 2015, pp. 1377–1385.
[11] B. Potetz, T.S. Lee, Efficient belief propagation for higher-order cliques using linear constraint nodes, Computer Vision and Image Understanding 112 (1) (2008) 39–54.
[12] P.P. Shenoy, G. Shafer, Axioms for probability and belief-function propagation, in: Uncertainty in Artificial Intelligence, 1990.
[13] D.M. Greig, B.T. Porteous, A.H. Seheult, Exact maximum a posteriori estimation for binary images, Journal of the Royal Statistical Society. Series B (Methodological) (1989) 271–279.
[14] T.H. Cormen, Introduction to Algorithms, MIT Press, 2009.
[15] Y. Boykov, O. Veksler, R. Zabih, Fast approximate energy minimization via graph cuts, in: Proceedings of the Seventh IEEE International Conference on Computer Vision, vol. 1, IEEE, 1999, pp. 377–384.
[16] Y. Weiss, W.T. Freeman, Correctness of belief propagation in Gaussian graphical models of arbitrary topology, Neural Computation 13 (10) (2001) 2173–2200.
[17] D.M. Malioutov, J.K. Johnson, A.S. Willsky, Walk-sums and belief propagation in Gaussian graphical models, Journal of Machine Learning Research 7 (Oct 2006) 2031–2064.
[18] Gaussian belief propagation resources [online], available: http://www.cs.cmu.edu/~bickson/gabp/index.html.
[19] M.J. Wainwright, M.I. Jordan, Graphical models, exponential families, and variational inference, Foundations and Trends® in Machine Learning 1 (1–2) (2008) 1–305.
[20] J. Besag, On the statistical analysis of dirty pictures, Journal of the Royal Statistical Society. Series B (Methodological) (1986) 259–302.
[21] J. Kappes, B. Andres, F. Hamprecht, C. Schnorr, S. Nowozin, D. Batra, S. Kim, B. Kausler, J. Lellmann, N. Komodakis, et al., A comparative study of modern inference techniques for discrete energy minimization problems, in: Proceedings of the IEEE Conference on Computer Vision and Pattern Recognition, 2013, pp. 1328–1335.
[22] P.J. Van Laarhoven, E.H. Aarts, Simulated annealing, in: Simulated Annealing: Theory and Applications, Springer, 1987, pp. 7–15.
[23] S. Geman, D. Geman, Stochastic relaxation, Gibbs distributions, and the Bayesian restoration of images, IEEE Transactions on Pattern Analysis and Machine Intelligence 6 (1984) 721–741.
[24] R. Dubes, A. Jain, S. Nadabar, C. Chen, MRF model-based algorithms for image segmentation, in: 10th International Conference on Pattern Recognition, vol. 1, 1990, pp. 808–814.
[25] M.A. Carreira-Perpinan, G. Hinton, On contrastive divergence learning, in: AISTATS, vol. 10, Citeseer, 2005, pp. 33–40.
[26] G.E. Hinton, Training products of experts by minimizing contrastive divergence, Neural Computation 14 (8) (2002) 1771–1800.
[27] J. Lafferty, A. McCallum, F. Pereira, Conditional random fields: probabilistic models for segmenting and labeling sequence data, in: International Conference on Machine Learning, 2001, pp. 282–289.
[28] J. Besag, Efficiency of pseudolikelihood estimation for simple Gaussian fields, Biometrika (1977) 616–618.

Computer vision applications

5.1 Introduction

Computer vision (CV) involves interpreting and understanding the world from their images or videos. CV tasks can be hierarchically organized into low, middle, and high levels. Whereas the low-level CV tasks focus on image processing for image enhancement and information extraction, the middle-level tasks involve estimating the properties of the objects in the images, including geometric properties, motion, and object categories. High-level CV tasks focus on interpretation and understanding events or activities in the images/videos. PGMs have been employed to address a wide range of CV tasks at each level. These tasks range from low-level CV tasks, such as feature extraction and image segmentation, and middle-level CV tasks including object detection, tracking, and recognition to high-level CV tasks including human activity recognition.

Rather than providing an exhaustive list of PGM applications in CV, in this chapter, we will summarize the major representative work. To better organize this discussion, we will group CV tasks into three levels: low, middle, and high. Within each level, we identify typical tasks and then discuss applications of the PGMs to each task. For each task, we will start with a generic definition of the problem as well as its input and output variables. We will then show a basic PGM model, which can be used to solve the task, including the model architecture and associated learning and inference methods. Finally, we discuss the related representative work, which improves upon the basic model from different aspects.

5.2 PGM for low-level CV tasks

Low-level CV tasks typically analyze the images and videos to enhance the images or extract basic information to represent the raw images. Typical low-level tasks include image segmentation, image denoising, and feature extraction.

5.2.1 Image segmentation

Image segmentation involves dividing images into homogeneous regions that correspond to distinct parts of the scene. Each region is homogeneous with respect to certain image property. The segmented image can be used to concisely represent the original raw images. Image segmentation can be further divided into binary and multiclass segmentations. Binary segmentation, also called figure-ground segmentation, divides an image into foreground and background. Multiclass image segmentation, on the other hand, assigns each pixel to an object label representing different types of objects (such as sky, road, tree, etc.)

Probabilistic Graphical Models for Computer Vision. https://doi.org/10.1016/B978-0-12-803467-5.00010-1

in the scene. Such image segmentation is also called semantic segmentation or semantic labeling. Fig. 5.1 shows an example of image segmentation, including figure-ground image segmentation (A) and (B) and multiclass semantic image segmentation (C) and (D).

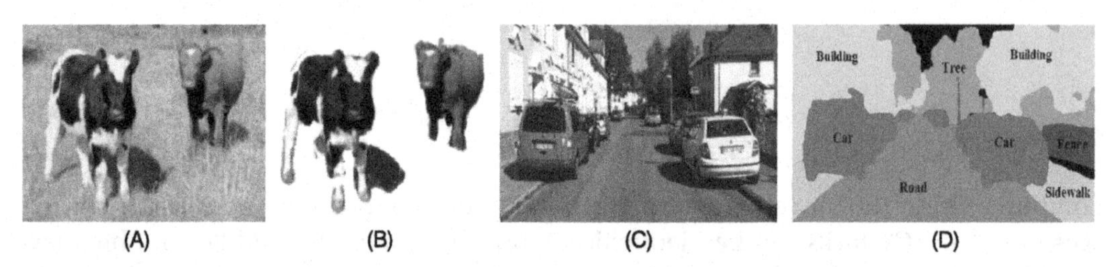

FIGURE 5.1 Example of image segmentation: (A) the original image; (B) figure-ground segmentation; (C) the original image; and (D) multi-class segmentation. Figures (A) and (B) adapted from [1], and (C) and (D) adapted from [2].

Image segmentation is among the most successful and one of the earliest CV tasks to which PGMs have been applied. Image segmentation with PGMs has some natural advantages over other approaches, as PGMs can naturally represent 2D layout of an image as well as the spatial relationships among image pixels. Among different PGM models, the Markov random fields (MRFs) (i.e., Markov networks (MNs)) have long been employed for image segmentation [81,5,82–84]. PGM applications to image segmentation have evolved over the years. MRF was first introduced in 1984 by Geman and Geman [85], when they laid the foundation for future research in image analysis with MRF. The earlier inference methods for MRF were limited to local methods, including the ICM method, Gibbs sampling method, and the simulated annealing method. In the early 2,000s, graph-based methods, such as graph-cuts and normalized cuts, were introduced for MRF inference, which can achieve global optimality for certain types of MRFs. Further developments in the late 2,000s focused on developing high-order/long-range complex MRF models to capture more contextual information. Recently, efforts have been focused on developing inference methods, including deep learning-based methods, for efficient inference in complex MRF models.

5.2.2 Image denoising

Image denoising focuses on restoring a noisy image by reducing its noise. Specifically, for image denoising, we observe a corrupted image in which the intensities of a certain percentage of image pixels are perturbed by random noise. The goal of image denoising is recovering the original clean image. Fig. 5.2 shows an example of image denoising.

5.2.3 Image labeling with MRFs

Both image segmentation and image denoising can be formulated as a multilabel image labeling problem. Mathematically, an image labeling problem can be stated as follows. Given an input vector $\mathbf{X} = \{X_n\}_{n=1}^{N}$, $n \in \mathcal{V} = \{1, 2, \ldots, N\}$, where X_n represent the intensity (or color) or features of an image pixel or a group of image pixels, $\mathcal{V}$ represents the im-

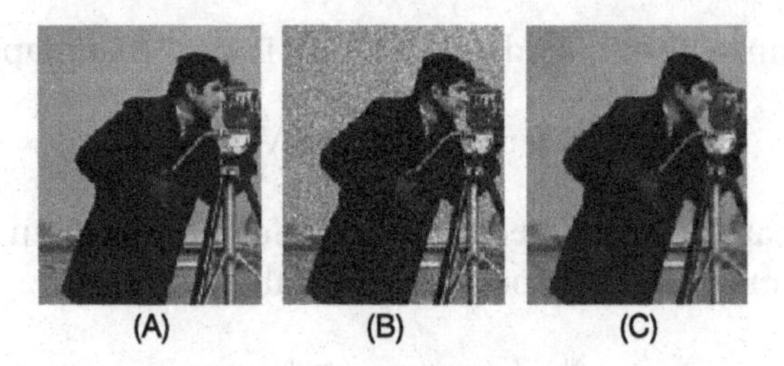

FIGURE 5.2 An example of image denoising: (A) original image; (B) noisy image; and (C) restored image. Figure adapted from [3].

age domain, that is, the 2D image grid, and an output vector $\mathbf{Y} = \{Y_n\}_{n=1}^N$, where Y_n is the label for X_n, and it assumes one of K possible values, that is, $Y_n \in \mathcal{L} = \{l_1, l_2, \ldots, l_K\}$. For image segmentation, $\mathcal{L}$ represents a set of object labels that each image pixel can take. For image denoising, $\mathcal{L}$ represents the set of intensity values that an image pixel can take. Furthermore, edge detection can be formulated as a particular case of image labeling with labels taking two values: edge or not edge. The goal of image labeling is to infer the most likely values for **Y** given **X**. Specifically, for image segmentation, the input **X** is an image, and output **Y** is a segmented image, where each pixel takes its label value. For image denoising, given an observed image **X**, the goal is recovering the original uncorrupted image $\mathbf{Y} = \{Y_1, Y_2, \ldots, Y_N\}$. For both image segmentation and image denoising, the underlying assumption is that the output labels are locally smooth.

For image labeling with PGMs, in the simplest case, we can employ a two-layer 2D lattice graph, as shown in Fig. 5.3. The top layer is the label layer, where each node Y_n rep-

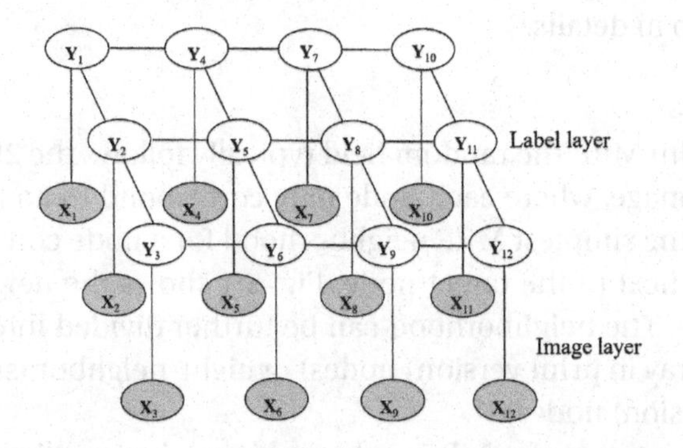

FIGURE 5.3 Two-layer graphical model for image segmentation. Figure adapted from Fig. 11.2 of [4].

resents the output label for each pixel or a group of pixels (super-pixels). The bottom layer is the image layer where each node X_n represents the intensity or features of an image pixel or a group of image pixels. The links capture the dependencies among nodes. The graph captures the joint probability distribution of **X** and **Y**. Using the graph, the goal of image labeling with PGMs can be formulated as learning a PGM $\mathcal{G}$ that maps $\mathbf{X}$ into $\mathbf{Y}$, that is,

$$\mathcal{G} : \mathbf{X} \mapsto \mathbf{Y}, \ \forall n \in \mathcal{N}.$$

Mathematically, this can be formulated as a MAP inference problem by finding the values of the label layer that maximizes the posterior probability, that is,

$$\mathbf{y}^* = \arg\max_{\mathbf{y}} p_G(\mathbf{y}|\mathbf{x}), \tag{5.1}$$

where p_G represents the joint probability of $\mathbf{X}$ and $\mathbf{Y}$ determined by $\mathcal{G}$.

The most commonly used PGMs for image labeling are pairwise label-observation MRFs introduced in Section 4.2.2. MRFs possess several characteristics, which are useful in image labeling. Through their pairwise potential function, MRFs can capture the interactions among neighboring pixels and encode their expected dependencies. Properties such as local smoothness and continuity of regions over an entire image can be systematically enforced by dependencies among local neighbors. Through its unary potential function, an MRF can capture the dependencies between a pixel label and its image features. A detailed discussion of MRF for image analysis can be found in [86].

Image labeling with pairwise label-observation MRFs is formulated as a MAP-MRF inference problem. It typically includes several steps. First, a local neighborhood is defined. The neighborhood defines connections among the label nodes. Given the neighborhood, the next step is defining the architecture of the MRF model. The third step is learning the parameters of the MRF model, which consist of the parameters for the unary and pairwise potential functions. Finally, an optimization method is employed to perform the MAP inference for image labeling. MRF methods for image labeling differ in these steps. We further discuss each step in details.

5.2.3.1 Neighborhood

For image labeling with MRF, the random field typically follows the 2D lattice graph corresponding to the 2D image, where each node may correspond to an image pixel. With the lattice random field, the simplest MRF neighborhood for a node consists of nodes that are located immediately next to the target node. Fig. 5.4 shows the neighborhood nodes for the white target node. The neighborhood can be further divided into four-neighbor (first order, the red (dark gray in print version) nodes) or eight-neighbor (second order, the blue (mid gray in print version) nodes).

Traditional MRFs are pairwise MRFs, where the maximum clique size is 2, and each pixel is only directly connected to its immediate neighbors. These MRFs can only capture relatively simple first-order statistical properties of the image, because they are based on pairwise relations between pixels. In practice, to account for additional contextual information, a simple neighborhood is often spatially extended to include pixels that are far away, yielding the so-called long-range MRFs [87]. For example, we may extend the neighbor to a square patch of $K \times K$ centered on the current pixel such that every pixel in the

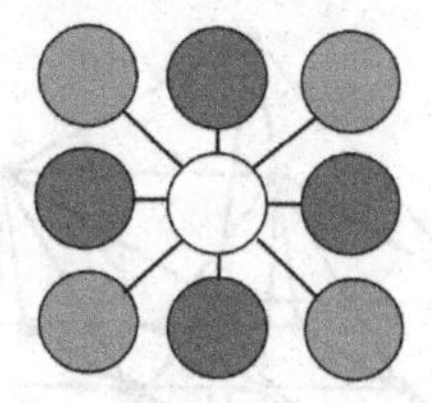

FIGURE 5.4 First-order neighborhood (red (dark gray in print version) nodes) and second-order neighborhood (blue (mid gray in print version) nodes) for target node (white). Figure adapted from Wikipedia.

patch is a neighbor of the target pixel. Expanding the square patch to include the entire image yields the ultimate long-range MRFs, that is, the fully connected MRFs, where a pixel is the neighbor of all other pixels.

5.2.3.2 MRF architecture

The simplest MRF architecture for image labeling is the two-layer MRF as shown in Fig. 5.5, where the top layer **Y** represents the label layer, and the bottom layer **X** represents the image measurements. For image denoising, X_n represents noisy pixel intensity (or features),

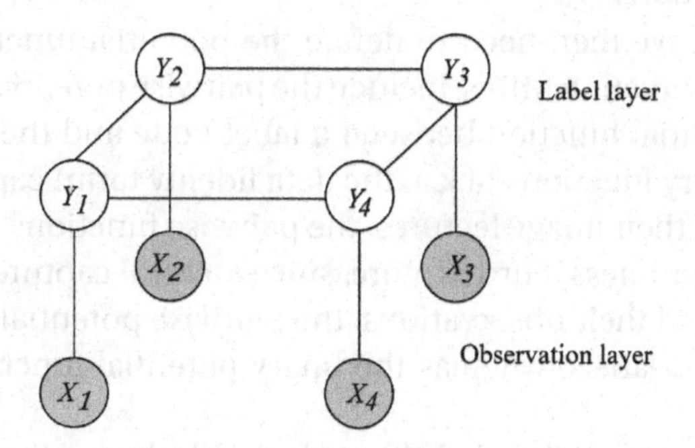

FIGURE 5.5 The basic two-layer MRF for image labeling.

and Y_n represents the corresponding unperturbed intensity. For image segmentation, to reduce computational complexity, X_n often represents the features for a group of pixels, and Y_n represents the labels for all pixels in the group. Pixel grouping is achieved by first performing an oversegmentation of the image. Each segment of the oversegmented image is called a superpixel. Each X node represents the features for each superpixel, and each Y node represents its label assuming that the labels for all pixels in each superpixel are the same.

The basic two-layer MRF model has been extended to more complex structures. In [5, 88] the propose a multiscale hierarchical model, where multiple label layers at different scales are stacked on top of one another in a coarse-to-fine manner to form a quadtree structure (pyramid) as shown in Fig. 5.6. The random variables at two adjacent layers form a Markov chain. This multiscale hierarchical model can enforce interactions of labels at different scales, often yielding improved performance over the basic two-layer MRF model.

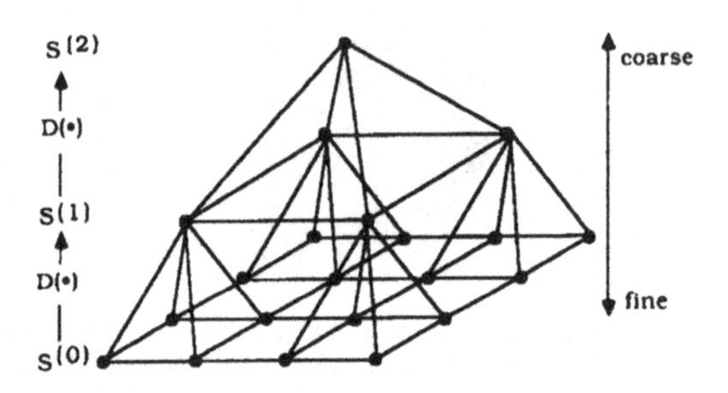

FIGURE 5.6 A multiscale quadtree MRF for image segmentation [5].

The fixed structure of a quadtree multiscale random field can lead to blocky segmentation [89,90]. Irving et al. [91] alleviated this drawback. Instead of assuming the nodes at one layer to be nonoverlapping, they propose an overlapping tree model, where the sites at each layer correspond to overlapping parts in the image. This approach can make the spatially adjacent sites more likely to have common parents at the next coarser layer.

5.2.3.3 Potential functions

Given the architecture, we then need to define the potential functions, which, for standard pairwise label-observation MRFs, include the pairwise potential function for the label nodes and unary potential function between a label node and the corresponding image node. Whereas the unary functions (a.k.a. the data fidelity term) capture the relationships between the labels and their image features, the pairwise functions usually capture the local coherence and smoothness. Furthermore, since an MRF captures the joint probability distribution of labels and their observations, the pairwise potential function captures the prior distribution of the labels, whereas the unary potential function captures the label likelihood.

Following the potential function definitions for label-observation MRF in Section 4.2.2, the pairwise potential function for the label layer can be specified by a loglinear model as

$$\psi_{ij}(y_i, y_j) = \exp(-w_{i,j} E_{ij}(y_i, y_j)),$$

where the pairwise energy function $E_{ij}(y_i, y_j)$ can be specified by the Ising model for binary labels or the Potts model for multivalue labels. Other nondecreasing functions [92] of the label difference such as $E_{ij}(y_i, y_j) = \min(|y_i - y_j|^k, \alpha)$ have also been used to specify the pairwise energy function to penalize the local smoothness violation, where α is a positive constant, and $k = 1, 2$. In fact, the pairwise energy function is often chosen carefully to have certain properties such as submodularity so that an exact inference solution can be obtained with the graph cuts method.

For image labeling task, the unary energy function typically equals the negative log-likelihood of the label given its observation, that is, $E_i(x_i|y_i) = -\log p(x_i|y_i)$. Hence, for a loglinear model, the unary potential function equals the class conditional probability $p(x_i|y_i)$. Depending on the type of image features, the unary potential function can be a unary Gaussian function if X_i is a continuous feature or a multivariate Gaussian if X_i represents a vector of features. The image features typically characterize the object shape, color, and location. For a unary continuous image feature X_i, the unary potential function for label l can be written as

$$\phi_i(x_i|y_i) = p(x_i|y_i = l) = \frac{1}{\sqrt{2\pi}\sigma_l}\exp(-\frac{(x_i - \mu_l)^2}{2\sigma_l^2}),$$

where the unary energy function $E_i(x_i|y_i = l)$ can be represented by the quadratic term

$$E_i(x_i|y_i = l) = \frac{(x_i - \mu_l)^2}{2\sigma_l^2} + \frac{1}{2}\log\sigma_l^2.$$

Similarly, the unary potential can be expressed as a multivariate Gaussian for a vector of image features $\mathbf{X}_i$. If X_i is discrete, then the unary potential function $\phi_i(x_i|Y_i = l)$ for label l can be specified by a binomial distribution (α_l) for unary X_i or multinominal distribution ($\boldsymbol{\alpha}_l$) for multivariate X_i. Given these definitions, following Eq. (4.15) for label-observation pairwise MRF, the joint probability distribution of **X** and **Y** can be written as

$$p(\mathbf{x}, \mathbf{y}) = \frac{1}{Z}\exp(-\sum_{i \in \mathcal{V}_y} \alpha_i E_i(x_i|y_i) - \sum_{i,j \in \mathcal{E}_y} w_{i,j} E_{ij}(y_i, y_j)), \tag{5.2}$$

where Z is the partition function, and $\mathcal{V}_y$ and $\mathcal{E}_y$ represent all nodes of Y and links among Y nodes, respectively. Note that the bias terms in Eq. (4.15) are often ignored for image labeling tasks.

Aside from standard unary and pairwise energy functions, high-order energy functions are being increasingly employed to capture important properties of natural images, producing high-order MRFs [93]. As discussed in Section 4.4, high-order MRFs involve higher-order terms such as the third-order term $\psi_{ijk}(y_i, y_j, y_k)$ or even higher-order terms. Recent research has demonstrated that high-order MRFs improved performance compared to first-order MRFs.

5.2.3.4 MRF learning methods

Given the specification of the unary and pairwise potential functions, the MRF learning methods discussed in Section 4.6, including the contrastive divergence method and the pseudolikelihood method introduced in Section 4.6.1.2, can be used to learn the parameters of the potential functions. The parameters include the pairwise parameters $w_{i,j}$ and unary parameters α_i. Before we learn the parameters $w_{i,j}$ and α_i, the parameters for the unary energy parameters such as μ_l, σ_l, and α_{kl} can be learned separately for each node X_i and Y_i. They are intrinsic to the energy functions and hence are the same for all nodes. The maximum likelihood method can be employed to learn these parameters.

5.2.3.5 Inference methods

Given the learned MRF model, image labeling for a query image can be accomplished through the MAP inference. In Section 4.5, we introduced different MRF-MAP inference methods. They include the exact methods such as variable elimination, belief propagation, the junction tree method, and the graph-cuts methods. For approximate MAP inference, we may employ loopy belief propagation, the ICM method, the Gibbs sampling method, and the variational methods. In addition, MRF-MAP inference can also be formulated as a combinatorial optimization problem, for which we may employ the existing integer programming, linear programming, and dynamic programming methods. Dubes et al. [94] reviewed different MRF inference techniques for MRF-based image segmentation and empirically evaluated their performance on simple noisy images.

In [95,92] the authors evaluated different MRF inference methods for different CV tasks. Szeliski et al. [92] evaluated four classical MRF inference methods, namely, graph cuts, loopy belief propagation, message passing, and the ICM algorithm for stereo, image stitching, and interactive segmentation. Their evaluation shows that three energy minimization methods, graph cuts, loopy belief propagation, and message passing, significantly outperform the ICM method. More recently, Kappes et al. [95] performed a more thorough comparison of different energy minimization techniques. They performed an empirical comparison of 24 state-of-the-art MRF inference techniques, including polyhedral, message passing, and max-flow methods, for 20 diverse CV applications in terms of both accuracy and efficiency. Their evaluation showed that polyhedral methods and integer programming solvers are competitive in terms of runtime and solution quality over a large range of model types. The polyhedral methods approximately formulate a discrete energy minimization problem by a set of linear inequalities and then efficiently solve the problem as a linear programming problem.

With the increasing use of high-order and long-range MRFs, recent efforts [93] have focused on developing efficient methods for inference in high-order and/or long-range MRFs, as the standard inference methods discussed before cannot be directly applied due to computational complexity. In [93] the authors introduced a graph cuts method for inference in high-order binary MRFs. By jointly transforming the high-order energy terms into first-order terms, their method improves the previous methods for inference in high-order MRFs and produced better results for stereo and segmentation tasks.

5.2.4 Image segmentation with CRFs

The MRF models assume label similarity for pixels nearby, regardless of their values. This assumption works well within a segmented region, but does not work well for pixels located near boundaries of different regions. Moreover, MRF models assume image observations are independent, given their labels. The CRF model was introduced to overcome these limitations so that the label smoothness assumption is conditioned on their image measurements, that is, image measurements are no longer variables but rather observations. As discussed in Section 4.3, CRF models the conditional probability distribution $p(\mathbf{Y}|\mathbf{X})$ of the image labels, where $\mathbf{X}$ represents the entire image. Hence, CRF can capture

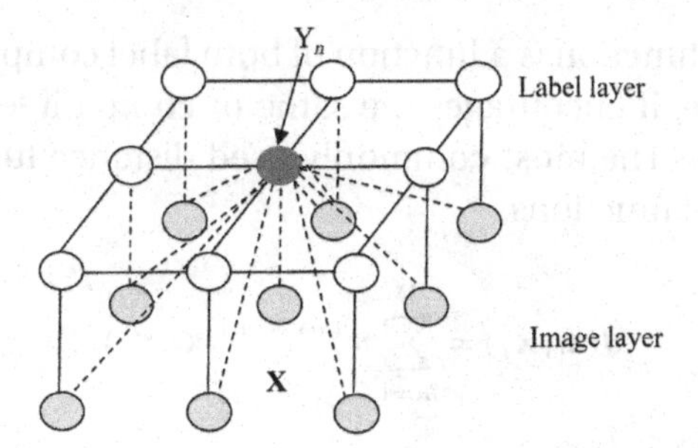

FIGURE 5.7 A simple CRF model for image segmentation, where, for a label node Y_n, its image observations **X** consist of all shaded nodes in the image layer.

the dependencies among the local image observations instead of assuming their independencies as with MRF. Furthermore, compared to the MRF model, CRF models require fewer labeled images during training, as they only model conditional label distribution. Finally, CRF models, in general, produce more accurate results than the MRF models, because their learning criterion is more relevant to the task of inferring labels. As a result, the CRF model has achieved wide application in CV. Fig. 5.7 shows a simple two-layer CRF model for image segmentation.

Following the discussion in Section 4.3, a CRF model over random variables $\mathbf{Y} = \{Y_1, \ldots, Y_N\}$ conditioned on an image $\mathbf{X}$[1] can be defined as follows. We assume that there is a random variable associated with each pixel in the image $\mathcal{V} = \{1, \ldots, N\}$, and each random variable in $\mathbf{Y}$ takes values from a label set $\mathcal{L} = \{l_1, \ldots, l_L\}$. Then we can express a pairwise CRF as

$$\begin{aligned} p(\mathbf{y}|\mathbf{x}) &= \frac{1}{Z(\mathbf{x})} \exp(-E(\mathbf{y}|\mathbf{x})), \\ E(\mathbf{y}|\mathbf{x}) &= \sum_{i \in \mathcal{V}} w_i E_i(y_i|\mathbf{x}) + \sum_{(i,j) \in \mathcal{E}} w_{ij} E_{ij}(y_i, y_j|\mathbf{x}), \end{aligned} \quad (5.3)$$

where $E(\mathbf{y}|\mathbf{x})$ is the energy associated with a configuration $\mathbf{Y}$ conditioned on $\mathbf{X}$, and $E_i(y_i|\mathbf{x})$ and $E_{ij}(y_i, y_j|\mathbf{x})$ are respectively the conditional unary and pairwise energy functions. The pairwise energy encourages local smoothness for pixels with similar intensity values, and it typically takes the form of the product of a label compatibility function and a distance function of the observation differences as follows

$$E_{ij}(y_i, y_j|\mathbf{X}) = \mu(y_i, y_j) d(\mathbf{x}_i, \mathbf{x}_j), \quad (5.4)$$

where μ is an arbitrary label compatibility function (such as the Potts model). Other possible label compatibility functions can be also used or even be learned from the data. $d(\mathbf{x}_i, \mathbf{x}_j)$ is a distance function that measures the differences between nearby observations. Unlike the pairwise function for MRF, which consists of only a label compatibility function,

[1] Note that the roles of **X** and **Y** are swapped, compared to Section 4.3.

the CRF pairwise energy function is a function of both label compatibility and label observation differences. Hence, it encourages the same or compatible labels for nearby pixels with similar appearances. The most commonly used distance function is the sum of the weighted Gaussian kernel functions,

$$d(\mathbf{x}_i, \mathbf{x}_j) = \sum_{m=1}^{M} w^{(m)} G^{(m)}(\mathbf{x}_i, \mathbf{x}_j),$$

where $G^{(m)}$ is the mth Gaussian kernel (also called an RBF kernel) for the mth-type observation, which can be defined as

$$G^{(m)}(\mathbf{x}_i, \mathbf{x}_j) = \exp(-\frac{1}{2}(\mathbf{x}_i - \mathbf{x}_j)^\top \Lambda^{(m)}(\mathbf{x}_i - \mathbf{x}_j)),$$

where $\Lambda^{(m)}$ is the precision matrix (i.e., the inverse of the covariance matrix), which defines the kernel shape. Specifically, if $\mathbf{x}_i$ includes appearance features $\mathbf{I}_i$ and position features $\mathbf{p}_i$, then the Gaussian kernel distance function can be written as

$$\begin{aligned} d(\mathbf{x}_i, \mathbf{x}_j) &= w^{(I)} \exp(-\frac{1}{2}(\mathbf{I}_i - \mathbf{I}_j)^\top \Lambda^{(I)}(\mathbf{I}_i - \mathbf{I}_j)) \\ &\quad + w^{(p)} \exp(-\frac{1}{2}(\mathbf{p}_i - \mathbf{p}_j)^\top \Lambda^{(p)}(\mathbf{p}_i - \mathbf{p}_j)). \end{aligned}$$

Besides the Gaussian kernel distance function, other distance functions, such as the simple squared difference $||\mathbf{x}_i - \mathbf{x}_j||_2^2$, can be also used.

Like label-observation MRF, the unary energy functions are used to relate pixel labels to image observations (shape, texture, location). However, unlike the unary function for the label-observation MRF model, which typically captures the distribution of image observations given the image labels (i.e., label likelihood), the unary function for CRF usually captures the distributions of the labels, given the image observations, that is, label posterior. The unary function is typically computed independently for each pixel by a classifier that produces a distribution over the label assignment Y_i given image features X_i. Hence one commonly used CRF unary function is the negative log label posterior probability given the observation, that is, $E_i(y_i|\mathbf{x}) = -\log p(y_i|\mathbf{x})$, which can be generated by different types of discriminative classifiers such as logistic regression [96], SVM, and boosting methods. For example, a TextonBoost classifier was used in [97] to generate the unary energy function. Recently, researchers leverage the latest developments in deep learning to generate stronger unary energy functions. For instance, Chen et al. [98] use a deep CNN model to serve as a unary potential function.

Given a CRF model that defines $p(\mathbf{Y}|\mathbf{X})$, image segmentation can be performed via a MAP inference with the CRF model,

$$\mathbf{y}^* = \arg\max_{\mathbf{y}} p(\mathbf{y}|\mathbf{x}).$$

Much work has demonstrated the success of CRF models and their advantages over the MRF models in image segmentation. They differ in the types of image features used, in the definition of unary and pairwise energy functions, and in learning and inference methods. Kumar and Hebert [96] proposed the applications of a CRF model to natural image modeling and binary classification. A generalized linear logistic regression is used to generate the unary energy function, whereas the Ising model multiplied by an image feature vector is used to capture the pairwise interactions. The parameters of the CRF model are learned using the maximum pseudo-likelihood method. The graph cuts method is used to perform the MAP inference. Experiments on both synthetic and real-world images show the improved performance of their CRF model compared to the corresponding MRF model. Shotton et al. [99] proposed the use of a CRF model to combine appearance, shape, and context information for joint image segmentation and object recognition. A boosted classifier is used for feature selection and for generating the unary potential functions, based on combining the shape, location, and color information. A pairwise potential function is implemented with edge compatibility of neighboring pixels using the Potts model, multiplied by a Gaussian kernel of pixel color features. To overcome the computational challenges with training the CRF model on large datasets, Shotton et al. [100] propose following the piecewise training method to train each CRF component (kernel parameter, unary function parameters, and pairwise function parameters) independently. The alpha-expansion graph-cuts method [101] is used to perform MAP inference. To capture additional contextual information, Ren et al. [102] employed the CRF model for figure/ground labeling by integrating edge and region features at different levels. Their CRF model consists of edge and region labels. The unary energy functions, based on image cues at different levels, including texture, brightness, and edge energy, are generated to capture object similarity, continuity, and familiarity. The Potts model multiplied by a region image difference function is used to construct the pairwise energy function for the region labels. A simple indicator function is used to capture the pairwise compatibility between region labels and the corresponding edge label. Maximum likelihood with gradient ascent is used to learn the model parameters. Loopy belief propagation is used to approximately perform the MAP inference.

Complex CRF models have also been introduced for image labeling and segmentation. He et al. [105,103] introduced a multiscale CRF model for segmenting static images. They introduced an additional hidden layer through an RBM model to capture both regional and global contextual knowledge (e.g., the scene context). They have demonstrated that their CRF model generally outperforms the corresponding MRF models in their experiments. Awasthi et al. [6,104] introduced the hierarchical tree-structured CRF model that contains additional layers of hidden variables to capture label relationships and to impose consistency constraints for image labels at different scales. As shown in Fig. 5.8, their CRF model contains three hidden layers H_3, H_2, and H_1 that capture the label relationships at pixel, region, and global levels.

Wang and Ji [7] presented a dynamic conditional random field (DCRF) model for object segmentation in image sequences. Spatial and temporal dependencies within the segmen-

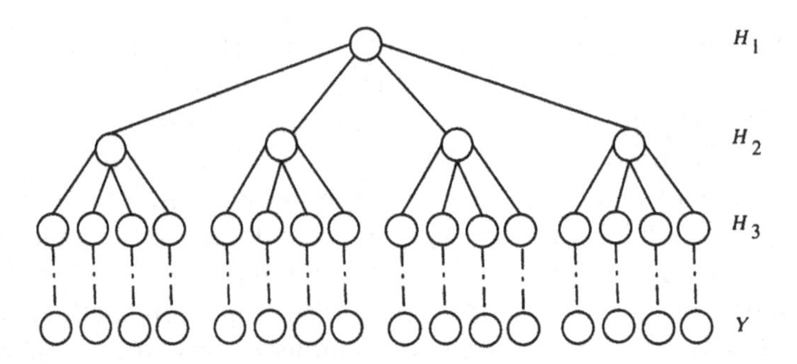

FIGURE 5.8 The tree CRF model [6], where Y is the label layer, and H_3, H_2, and H_1 are the hidden layers at different levels. They capture the label relationships at pixel, region, and global levels.

tation process are unified by a dynamic probabilistic framework based on the conditional random field. Specifically, let $\mathbf{Y}^t$ and $\mathbf{X}^t$ represent the image labels and image observations at the tth image frame, respectively . The temporal image segmentation can then be formulated as the filtering problem

$$\mathbf{y}^{*t} = \arg\max_{\mathbf{y}^t} p(\mathbf{y}^t|\mathbf{x}^{1:t}). \tag{5.5}$$

Following the Bayesian filtering derivations, $p(\mathbf{y}^t|\mathbf{x}^{1:t})$ can be computed recursively as follows:

$$p(\mathbf{y}^t|\mathbf{x}^{1:t}) = p(\mathbf{x}^t|\mathbf{y}^t)\int p(\mathbf{y}^t|\mathbf{y}^{t-1})p(\mathbf{y}^{t-1}|\mathbf{x}^{1:t-1})d\mathbf{y}^{t-1},$$

where $p(\mathbf{x}^t|\mathbf{y}^t)$ is the label observation or likelihood model, and $p(\mathbf{y}^t|\mathbf{y}^{t-1})$ is the label transition model. Using the loglinear model, the transition model $p(\mathbf{y}^t|\mathbf{y}^{t-1})$ is parameterized as follows:

$$p(\mathbf{y}^t|\mathbf{y}^{t-1}) \propto \exp(-\sum_i(\sum_{j\in N_i} E_{ij}(y_i^t, y_j^t) + \sum_{k\in M_i} E_i(y_i^t|y_k^{t-1}))),$$

where i represents the ith pixel in the image, N_i is the spatial neighbor for the ith pixel at time t, and M_i defines temporal neighbor at time $t-1$ for the ith pixel at time t. Fig. 5.9 shows the 5-pixel temporal and 4-pixel spatial neighborhoods Wang and Ji defined. The terms $E_{ij}(Y_i^t, Y_j^t)$ and $E_i(Y_i^t|Y_k^{t-1})$ are the pairwise label energy functions that capture the spatial and temporal dependencies among image labels.

Similarly, the observation model $p(\mathbf{x}^t|\mathbf{y}^t)$ can be parameterized as follows:

$$p(\mathbf{x}^t|\mathbf{y}^t) \propto \exp(-\sum_i(w_i E_i(x_i^t|y_i^t) + \sum_{j\in N_i} w_{ij} E_{ij}(x_i^t, x_j^t|y_i^t, y_j^t))),$$

where $E_i(x_i^t|y_i^t)$ is the unary energy function that measures the likelihood of label y_i given its image observation x_i, and $E_{ij}(x_i^t, x_j^t|y_i^t, y_j^t)$ are the pairwise energy functions that mea-

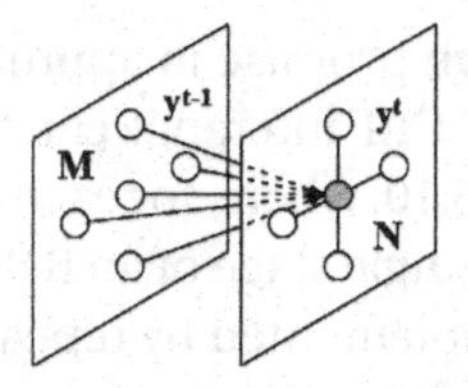

FIGURE 5.9 The 5-pixel temporal neighbor (M) and the 4-pixel spatial neighbor (N) for the DCRF model [7].

sure the interactions between two neighboring labels, given their image observations. Through Eq. (5.5), their model employs both intensity and motion cues and it combines dynamic information and spatial interaction of the observed data. Given this formulation, they introduced a filtering algorithm based on the mean field approximation to efficiently solve Eq. (5.5) for temporal image segmentation. Experimental results show that their model effectively fuses contextual constraints in video sequences and improves the accuracy of object segmentation. Strictly speaking, their DCRF model is not a dynamic CRF model but rather a dynamic MRF model as the label dependencies model $p(\mathbf{y}^t|\mathbf{y}^{t-1})$ is independent of the image observation $\mathbf{x}$, and the observation model $p(\mathbf{x}^t|\mathbf{y}^t)$ captures label likelihood instead of label posterior.

Like MRFs, to improve labeling accuracy, the first-order pairwise CRF models have also been extended to high-order CRF models to capture additional image information [105–107,97]. High-order CRF models can be constructed by including not only pairwise but also high-order (e.g., third-order) terms in energy functions in a similar fashion as in Eq. (4.27). The introduction of higher-order terms can significantly expand the expressive power of CRF models and improve their performance, as demonstrated by recent research [108,109]. Specifically, the importance of higher-order information has been demonstrated for image segmentation in [110] and for image denoising in [111]. Besides high-order CRFs, long-range CRFs (i.e., fully connected CRFs) [112,113,97] and a combination of high-order and long-range CRF models have also been introduced.

High-order and long-range CRFs also significantly increase inference complexity and their applications are hence are limited to hundreds of image regions or fewer. To overcome this challenge, recent efforts [97,114,115,108,116] have focused on developing efficient inference methods for high-order and long-range CRF models. Variants of mean field methods, in particular, filter-based mean-field inference, have shown great promise. Krähenbühl and Koltun [97] introduced an approximate inference method based on mean-field approximation for inference in a fully connected CRF model. They show that a mean field update of all variables in a fully connected CRF can be performed using Gaussian filtering in the feature space through a series of message updating steps, each of which updates a single variable by aggregating information from all other variables, leading to a sublinear inference algorithm. A number of cross bilateral Gaussian filter-based methods have been proposed for problems such as object class segmentation [97], denoising [117], stereo, and optical flow [118], which permit substantially faster inference in these problems and offer performance gains over competing methods. In addition, deep learning-based methods have shown promise in approximating the CRF inference. Zheng et al. [8] showed that the mean-field CRF inference can be reformulated as a recurrent neural network (RNN) as shown in Fig. 5.10, where, for each iteration of the mean field function Q, estimation is implemented as a single stage of an RNN with multiple CNN layers. Multiple mean-field iterations can be implemented by repeating each RNN stage in such a way that each iteration takes the value of Q estimated from the previous iteration.

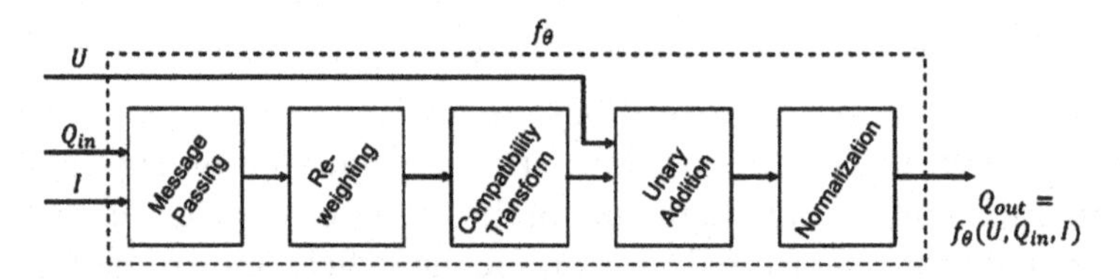

FIGURE 5.10 A mean-field iteration as a stage of an RNN [8]. A single iteration f_θ of the mean-field algorithm can be modeled as a stack of common CNN layers, where I is the input image, U is the unary function, and Q is the mean-field function.

5.2.5 Image segmentation with Bayesian networks

Whereas MRF and CRF models are widely used for region-based image segmentation, BNs can also be used to perform image segmentation as natural causal relationships exist among image entities. Fig. 5.11 shows the two-layer BN equivalent to the two-layer MRF in Fig. 5.3. Like the MRF model, the BN model captures the joint probability $p(\mathbf{X}, \mathbf{Y})$. Different from the MRF model, the links in the BN model are replaced by directed links, and the link potential parameters are replaced by conditional probabilities. Given the BN, we can use the methods discussed in Section 3.4 to learn its parameters from the data. Given the learned BN, we can apply the BN inference methods discussed in Section 3.3 to perform MAP inference of the most likely values for **Y** given **X**.

Although the BN in Fig. 5.11 represents a direct adaptation of the MRF model in Fig. 5.3, its directed links are somewhat arbitrary and do not carry any causal meaning. To fully tap the power of the BN for image segmentation, we need to use a BN to explicitly capture the natural causal relationships among elements of an image. To this end, Zhang and Ji [1] introduced a multilayer BN for edge-based image segmentation. The BN is constructed to represent the image entities in an oversegmented edge map. Specifically, an oversegmented image consists of superpixel regions $\{y_i\}$, the edge segments $\{e_j\}$ that represent the boundaries of superpixels, and the vertices $\{v_k\}$ that represent the intersections of edge segments. Natural causal relationships exist among these image entities: the intersections of image regions produce (cause) edge boundaries, and the intersections of the edge boundaries, in turn, produce (cause) vertices. A BN can, therefore, be used to capture such causal relationships for image segmentation.

To illustrate this application of the BN, we use Fig. 5.12 as an example, where Fig. 5.12A represents an oversegmented image composed of superpixels. Fig. 5.12B is a synthetic edge map corresponding to the annotated region in Fig. 5.12A. The edge map consists

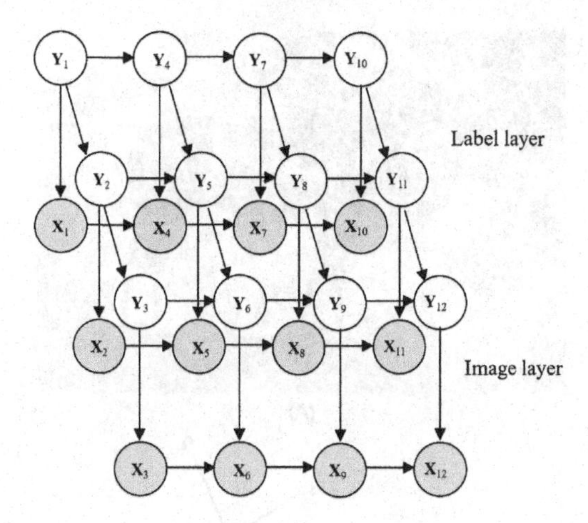

FIGURE 5.11 Two-layer grid pairwise BN for image segmentation, where the top layer represents the pixels labels, and the bottom layer represents image observations/features. Figure adapted from [4].

of six superpixel regions $\{\mathbf{y}_i\}_{i=1}^{6}$, seven edge segments $\{\mathbf{e}_j\}_{j=1}^{7}$, and two vertices $\{\mathbf{v}_k\}_{k=1}^{2}$. Fig. 5.12C shows the BN model that captures the relationships among different image entities in Fig. 5.12B. The BN model contains three layers: the region, edge, and vertex layers.

The region layer contains all the superpixel region nodes **y**. The edge layer contains all the edge segments **e**. The vertex layer contains all vertices **v**. All nodes are discrete. A region node y_i takes one of different values to represent an object label. Both edge and vertex nodes take binary values to represent their presence or absence. The links capture the causal relationships among image entities. For example, the link between an edge node (e.g., e_1) and two neighboring region nodes (e.g., y_1 and y_2) captures their causal relationships, that is, the interaction of two neighboring region nodes y_1 and y_2 produces the edge e_1. Similarly, the edge and vertex nodes are also causally linked. The parents of a vertex node are the edges whose intersection forms the vertex. For example, the intersection of edge nodes e_1 and e_2 produces the vertex node v_1.

All nodes have corresponding image measurements. We denote the measurement of the region node y_i as M_{y_i}, the measurement of edge node e_j as M_{e_j}, and the measurement of the vertex node v_t as M_{v_t}. The region/edge/vertex nodes are causally connected to their measurement nodes through directed links. For region nodes, their measurements can be pixel intensities or extracted image features. For edge measurements, we can use complex edge features (e.g., edgelets), the edge probability or simply the average gradient. For vertex measurements, we can use the output of the Harris corner detector [119] to measure the likelihood of a vertex being a corner.

In general, given an oversegmented edge image, let **e** represent all the edge nodes $\{e_j\}_{j=1}^{m}$, **y** represent all the region nodes $\{y_i\}_{i=1}^{n}$, and **v** represent all the vertex nodes $\{v_k\}_{k=1}^{l}$. Let $\mathbf{M}_y$ represent the measurements for all region nodes, $\mathbf{M}_e$ represent all the measurements for the edge nodes, and $\mathbf{M}_v$ represent all the measurements for the vertex nodes.

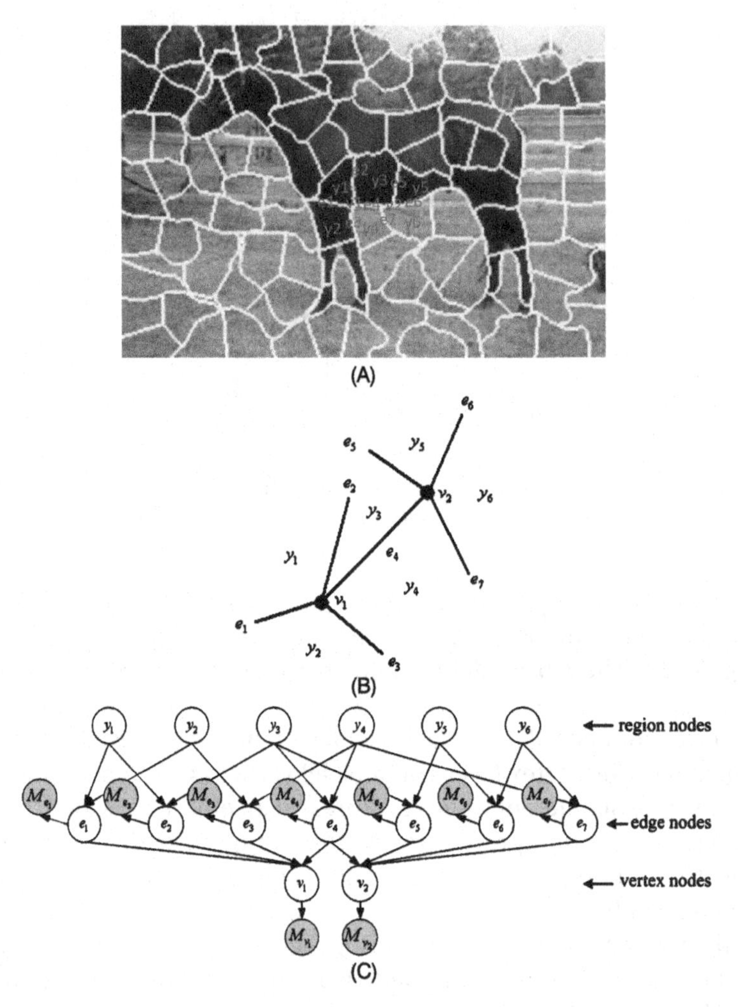

FIGURE 5.12 (A) An oversegmented image with a local marked region. (B) The edge map corresponding to the marked region in (A). (C) An BN that models the statistical relationships among superpixel regions (**y**), edge segments (**e**), vertices (**v**), and their measurements (Ms) in (B). Figure from [1].

Using the oversegmented edge map, a BN can be manually constructed to represent the causal relationships among regions, edges, vertices, and their measurements. Given the structure of the BN, its parameters can be either manually specified or learned from data. For example, the prior probability of each region node y_i is set to be uniform, that is, $p(y_i = k) = \frac{1}{K}$, where K is the number of possible segment labels. For an edge node e_j and its parent region nodes, its CPT $p(e_j|\pi(e_j))$ can be specified as

$$P(e_j = 1|\pi(e_j)) = \begin{cases} 0.8 & \text{if the parent region labels are different, and} \\ 0.2 & \text{otherwise.} \end{cases} \tag{5.6}$$

For each vertex node v_t, we can specify its $p(v_t|\pi(v_t))$ similarly. For the measurement nodes, their conditional probabilities can be specified using Gaussian distributions. For example, for the measurement nodes M_{e_j}, its conditional probability $P(M_{e_j}|e_j)$ is parameterized using a multivariate Gaussian distribution with mean $\boldsymbol{\mu}_e$ and covariance matrix Σ_e, which can be learned from the training data. Similarly, for the region measurement node M_{y_i}, its conditional probability $p(M_{y_i}|y_i)$ can be specified following a Gaussian distribution with mean $\boldsymbol{\mu}_y$ and covariance matrix Σ_y. They can also be learned from data. Finally, for the vertex measurement M_{v_k}, we can first derive the Harris corner matrix A (i.e., the structure tensor), and then measure the strength of a corner by the corner response function $R = det(A) - k \cdot trace(A)^2$. Given the corner response, we can specify the conditional probabilities of vertex nodes using a sigmoid function: $p(M_{v_t}|v_t = 1) = \sigma(R)$, where σ is the sigmoid function.

Given a fully specified BN and the measurement vectors $\mathbf{M}_y$, $\mathbf{M}_e$, and $\mathbf{M}_v$, image segmentation can be performed by a MAP inference to find the most probable values for the region, edge, and vertex nodes:

$$(\mathbf{e}^*, \mathbf{y}^*, \mathbf{v}^*) = \arg\max_{\mathbf{e,y,v}} P(\mathbf{e,y,v}|\mathbf{M}_y, \mathbf{M}_e, \mathbf{M}_v). \tag{5.7}$$

Details about the BN model and experimental results can be found in [1].

Other similar BN models have been proposed to perform edge-based image segmentation. Mortensen et al. [9] propose a semiautomatic segmentation technique based on a two-layer BN as shown in Fig. 5.13, where the nodes in the top layer represent the edge segments, and the nodes in the bottom layer represent the corresponding vertex nodes. The BN hence captures the relationships between edge segments and vertices. The BN parameters include the prior probabilities for the edge nodes and conditional probabilities for the vertex nodes. The conditional probabilities for the vertex nodes are manually specified to enforce the simplicity and closure constraint on the boundary, whereas the prior probabilities for the edge nodes are heuristically specified based on the image measurements of their neighborhood. Given the BN and the image measurements (gradient and curvature) of the edge segments, Mortensen et al. employed the MAP inference to find a sequence of most likely edge segments that form the boundary of the objects in the image. The MAP inference is implemented approximately by the minimum-path spanning tree graph search method.

Alvarado et al. [120] use a BN model to capture all available knowledge about the real composition of a scene for segmenting a hand-held object. Their BN model combines high-level cues such as the possible locations of the hand in the image with low-level image measurements to infer the probability of a region belonging to the hand. Feng et al. [10] proposed employing a fixed multiscale BN-quadtree model as shown in Fig. 5.14 to capture the label dependencies at different scales similar to the multiscale MRF model in Fig. 5.6. A neural network was used to generate image measurements of the labels. The EM

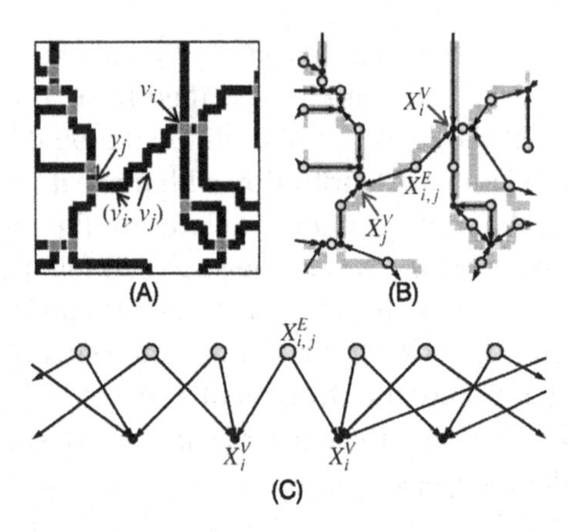

FIGURE 5.13 The two-layer BN in [9] for edge-based image segmentation. (A) A section of an edge map; (B) the corresponding directed graph; and (C) the corresponding two-layer BN, where $X_{i,j}^E$ represents the edge segment between vertices X_i^V and X_j^V.

method was employed to learn the model parameters. Given the image measurements, pixel labels are inferred using MAP inference with the BN model.

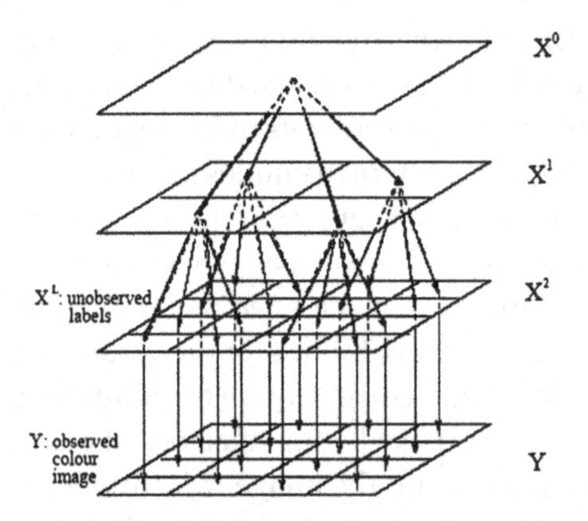

FIGURE 5.14 A multiscale quadtree structured BN model for image segmentation [10], where Y represents image observations, and Xs represent image labels at different scales.

To overcome the fixed structure with the quadtree BN model as in Fig. 5.14, Todorovic et al. [90,121] developed a dynamic multiscale tree BN model that simultaneously infers the optimal structure (i.e., the links between nodes at adjacent layers) and random variable states. Their experiments demonstrated the promise of such an approach. However, their model is complex and requires inference of a large number of random variables.

All these prior studies have demonstrated the capability of BN models to integrate multiple sources of information and constraints to disambiguate the image labeling problem.

5.3 PGM for middle-level CV tasks

In this section, we will focus on the applications of PGMs to middle-level CV tasks. Typical middle-level CV tasks include object detection and tracking, object recognition, 3D reconstruction, and object pose estimation. For example, for human faces, face detection localizes the face in the image, whereas face recognition determines the identity of the detected face (e.g., George W. Bush). 3D face reconstruction involves reconstructing the 3D face models from their 2D images. Face pose estimation is concerned with determining the 3D position and orientation of the face relative to the camera coordinate frame.

5.3.1 Object detection and recognition

Much like the low-level segmentation task, PGMs have been successfully applied to object detection and recognition. Object detection and recognition represent some the most important tasks in CV. The goal of object detection is detection of an instance of a known object in an image and determines its location and spatial scope. Object recognition, on the other hand, involves classifying a detected object into a predefined object class. Object detection typically precedes object recognition. Whereas object detection can be treated as a binary classification problem (object versus nonobject), object recognition is often formulated as a multiclass classification problem. Despite their differences, object detection and recognition are often interchangeable. When the goal of object detection is to detect a specific class of object, it becomes binary object recognition. As a result, methods for object detection and recognition can be similar. In this section, we will focus on object detection techniques and will discuss object recognition techniques where necessary.

Over the years, much progress has been made in object detection and recognition. In fact, as a result of the latest developments in deep learning, automatic object recognition techniques are approaching or have even surpassed human ability on some benchmark datasets. For a review of recent developments in object recognition, we refer the readers to [122]. Object detection starts with constructing an object model from a set of training examples. The object model should capture both the object appearance and the shape variations under different image acquisition conditions. Object models can be appearance-based, shape-based, or hybrid. The appearance-based models capture the photometric properties of an object, whereas the shape-based models capture its geometric properties. The hybrid object model simultaneously captures the appearance and shape properties of an object. In addition, the object model can be holistic or local. A holistic approach uses the entire image to construct an object model, whereas a local (part-based) approach represents an object by its primitive parts or local key points (features). Given the object model, object detection can be accomplished globally or locally. Global object detection involves either manual enumerating all possible object positions (e.g., sliding a window across the image) or employing a discrete optimization to identify the image locations that best match the object model. In contrast, local object detection starts from an initial object position and then searches the nearby region to find the image position that best matches the object model. The initial object positions are often generated by an object proposer that generates image region candidates based on their probability to contain the target objects.

Object detection methods can be classified into deterministic and probabilistic approaches. The deterministic approach uses a set of fixed image features to represent the object and formulate object detection as a deterministic classification problem. The local feature-based methods are currently the dominant approach. Different types of local features are extracted to represent the object. The local features are robust under various image acquisition conditions, under occlusion, and they can be made invariant under viewpoint changes and affine transformations. The most commonly used local features include multiscale Harris features, the HoG features, LBP features, and Lowe's SIFT features. As an extension to local features, bag-of-features-based methods [123] are a dominant local approach for object detection and recognition, wherein the object models are learned by building a codebook of local features. Each codebook consists of a vocabulary of words (feature clusters). They have achieved state-of-the-art performance on many benchmark datasets. Despite their success, bag-of-features methods ignore the spatial information among the image features. As a result, features that do not belong to the object can be incorporated into the object model, leading to overfitting. They are, therefore, less robust in handling noisy images and images with occlusions. Using the extracted local features, deterministic methods typically formulate object detection as a binary classification problem, whereby a binary classifier is constructed to discriminate between subimages containing the object instance and those not containing the object instance. Most existing object detection methods belong to this category. One of the widely used discriminative methods is based on AdaBoost [124], which selects a small number of discriminative visual features and yields efficient classifiers. It achieves excellent performance for face detection. Although deterministic methods achieve state-of-the-art performance, they require large training sets and cannot easily incorporate prior or contextual information into the learning process.

Probabilistic approaches, on the other hand, represent objects by the probabilistic distributions of image features. We will focus on probabilistic part-based object models, which represent an object by its parts and their probabilistic relationships. Specifically, the probabilistic part-based methods capture the probabilistic distributions of object part appearance variation and the variation in the spatial configurations among the object parts. Given the probabilistic object models, object detection is formulated as a probabilistic inference problem at each image location, whereby the likelihood of the object model is computed and compared against that of the background, based on which a decision is made as to whether the image patch contains the object instance or not. Compared to deterministic models, probabilistic models can better capture a large range of object shape and appearance variations. These variations result from object pose change, illumination variation, camera parameter change, and nonrigid object shape deformation. Compared to the holistic representation, the part-based object representation can more efficiently encode object appearance and capture the spatial relationships among object parts. The spatial relationships not only aid in locating object parts efficiently, but also allow object detection even with occlusion of some parts. In fact, there is a long history of part-based object representation in CV, including pictorial structure models, constellation models, and the recent attribute-based object representation.

In the following sections, we will focus on a part-based probabilistic object model. Assuming that an object consists of K parts (or key points), the object parts can be manually specificized or automatically learned from data. Specifically, let $\mathbf{X} = \{X_1, X_2, \ldots, X_K\}$ be a shape vector that represents the relative locations of K object parts. $\mathbf{X}$ may represent the silhouette, contour, or median axis (skeleton) of an object. Depending on how each part is represented, X_k can simply specify the image coordinates of the center of the kth part. X_k can also be a vector that represents the geometric properties of each part, including its position, size, orientation, and scope (bounding box). To be translation and rotation invariant, $\mathbf{X}$ may be normalized by subtracting it from the centroid of all points and by rotating around the major axis of the points. Let $\mathbf{I} = \{\mathbf{I}_1, \mathbf{I}_2, \ldots, \mathbf{I}_K\}$ be the corresponding appearance vector that represents part appearance, where $\mathbf{I}_k$ represents the corresponding image features to capture the appearance of the kth part X_k. Note that instead of using the raw pixel intensities, $\mathbf{I}$ often represents image features extracted from the image patch, including handcrafted features like SIFT and LBP as well as features learned by deep models. Then we can construct a hybrid probabilistic object model $p(\mathbf{X}, \mathbf{I}|\mathbf{\Theta})$ that captures both the object shape and appearance variation, where $\mathbf{\Theta}$ are the model parameters, and they can be learned from training data. By combining the appearance models for the parts and structural relations between parts the part-based object model is powerful and flexible to represent different types of objects, including both rigid and nonrigid objects.

Given the object model, object detection can be performed via a MAP inference to determine the location of each body part to maximize $p(\mathbf{x}|\mathbf{i}, \mathbf{\Theta})$:

$$\mathbf{x}^* = \arg\max_{\mathbf{x}} p(\mathbf{x}|\mathbf{i}, \mathbf{\Theta}). \tag{5.8}$$

For object recognition, we can develop one object model for each class and perform object detection using each object model. We further discuss two dominant local probabilistic object models: the pictorial structure model and the constellation model.

5.3.1.1 Pictorial structure model

In [125], Fischler and Elschlager introduced the pictorial structure (PS) model for object detection and recognition. A PS model for an object is given by a collection of object parts arranged spatially in a deformable configuration to capture the global object shape distribution in the image, for example, to consider human faces consisting of two eyes, a nose, and a mouth. These parts appear in a deformable arrangement, which depends on facial geometry, pose, expression, and the viewpoint of the user. The PS model represents an object by its parts. It captures the conditional probability distribution $p(\mathbf{X}|\mathbf{I}, \mathbf{\Theta})$ of the shape of an object given its image, where $\mathbf{X} = \{X_1, X_2, \ldots, X_K\}$ represents the image locations of the K object parts, and $\mathbf{I} = \{\mathbf{I}_1, \mathbf{I}_2, \ldots, \mathbf{I}_K\}$ represents the corresponding image features for K parts. Mathematically, $p(\mathbf{X}|\mathbf{I}, \mathbf{\Theta})$ can be implemented by a two-layer pairwise label-observation MN as shown in Fig. 5.15, where the nodes in the top layer represent the locations for each part, and the nodes in the bottom layer represent the appearance of each corresponding object part.

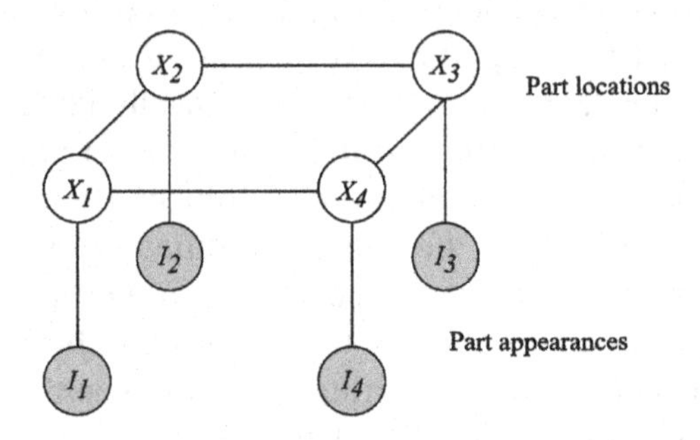

FIGURE 5.15 A two-layer label-observation MRF model for part-based object representation.

Given such a model, the joint probability for all nodes can be represented as

$$p(\mathbf{x}|\mathbf{i}, \Theta) = \frac{1}{Z}\exp\left(\sum_{k=1}^{K} -\alpha_k E_k(\mathbf{i}_k|x_k) - \sum_{(k,j)\in\mathcal{E}} w_{k,j} E_{kj}(x_k, x_j)\right), \tag{5.9}$$

where $\mathcal{E}$ represents all edges in the MRF model. The unary energy function $E_k(\mathbf{i}_k|x_k)$ encodes the negative loglikelihood of appearance of the kth part at location x_k, whereas the pairwise energy function $E_{kj}(x_k, x_j)$ captures the negative log joint probability that the jth and kth parts appear at positions x_j and x_k, respectively. These energy functions can be specified as

$$\begin{aligned} E_k(\mathbf{i}_k|x_k) &= -\log p(\mathbf{i}_k(x_k)), \\ E_{kj}(x_k, x_j) &= -\log p(x_k, x_j), \end{aligned} \tag{5.10}$$

where the likelihood and joint probabilities can be parameterized with Gaussian distributions:

$$\begin{aligned} p(\mathbf{i}_k) &= \mathcal{N}(\mu_k, \Sigma_k), \\ p(x_k, x_j) &= \mathcal{N}(\mu_{kj}, \Sigma_{kj}), \end{aligned} \tag{5.11}$$

where $p(\mathbf{i}_k)$ captures the appearance distribution of the kth body part, and $p(x_k, x_j)$ captures the relative spatial distribution for parts k and j, that is, the distribution of $x_k - x_j$. The part appearance parameters (μ_k, Σ_k) and the part relationship parameters (μ_{kj}, Σ_{kj}) can be learned separately for each object part through MLE. The model parameters $\Theta = \{\alpha_k, w_{k,j}\}$ for the unary and pairwise energy functions can be learned from training data using the methods introduced in Section 4.5. Given the model and a testing image $\mathbf{I}$, a MAP inference can be performed to perform object detection in the image by identifying the locations of the object parts (i.e., the object configuration) that maximize $p(\mathbf{x}, \mathbf{i}|\Theta)$:

$$\mathbf{x}^* = \arg\max_{\mathbf{x}} p(\mathbf{x}|\mathbf{i}, \Theta). \tag{5.12}$$

This can be accomplished equivalently by minimizing the energy function:

$$\mathbf{x}^* = \arg\min_{\mathbf{x}}[-\sum_{k=1}^{K} \alpha_k E_k(\mathbf{i}_k|x_k) - \sum_{(j,k)\in\mathcal{E}} w_{k,j} E_{kj}(x_k, x_j)].$$

The challenge with the MAP inference is the large search space of **X**, that is, the possible location of each object part is large, ultimately taking each pixel in the image. This computational complexity can be alleviated by restricting the possible positions for each object part (e.g., initializing part positions near the optimal positions) and by restricting the MRF model to certain special structures (e.g., tree structure).

In [11], Felzenszwalb and Huttenlocher adopted the PS model for face detection. They introduced a pairwise MRF consisting of nine facial landmark points, including four eye corners, two eye centers, two mouth corners, and one nose point as shown in Fig. 5.16A. To reduce the computational complexity for MAP inference, a tree MRF model is learned to capture the relationships among the landmark points as shown in Fig. 5.16B, where the nodes represent the landmark points, and the links between the nodes capture the relative spatial relationships between the two neighboring facial landmark points. Both unary and pairwise energy functions are parameterized as Gaussian models as in Eqs. (5.10) and (5.11). Training data were collected to learn the parameters for both unary and pairwise

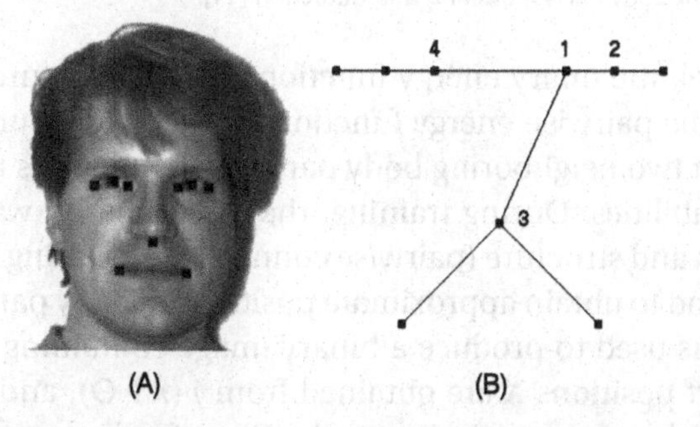

FIGURE 5.16 The pictorial face model [11]. (A) Face image and nine landmark points, where the numbers represent the index to the landmark points; and (B) The tree-structured graphical model that captures the spatial relationships among nine landmark points. Note that, for clarity, the image measurements for each landmark point are omitted.

energy function parameters as well as the model structure using the MLE method. During the detection, instead of performing a local search to solve for the MAP inference in Eq. (5.12) near an initial object position, the authors introduced a Viterbi-like dynamic programming algorithm to perform an efficient global search over the entire image to decide the optimal positions for each object part. By requiring the pairwise energy function to represent the Mahalanobis distance between two parts their search algorithm can achieve linear time with respect to the number of possible part positions.

To further demonstrate their model, Felzenszwalb and Huttenlocher extended their face PS model to human body detection. A body model is represented by ten body parts, corresponding to the torso, head, two parts per arm, and two parts per leg, as shown in Fig. 5.17. Each body part X_k is represented by a rectangular box characterized by its position, size, and orientation, $\mathbf{X}_k = (x_k, y_k, s_k, \theta_k)$. Neighbor body parts are connected through joints.

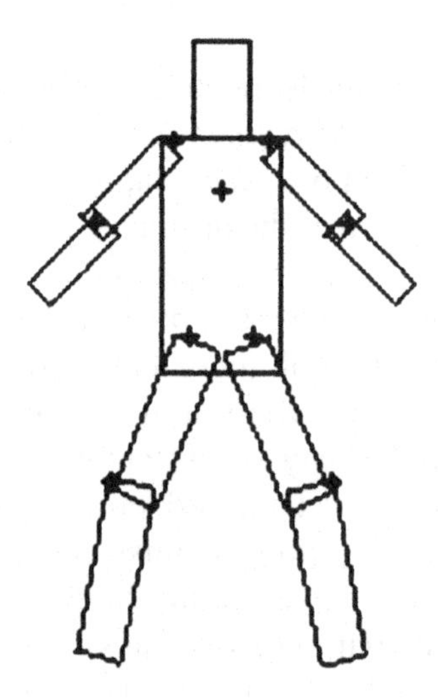

FIGURE 5.17 The PS model for body parts modeling and detection [11].

Like the face model, the unary energy function $E_k(\mathbf{i}_k|x_k)$ captures the appearance for each body part, and the pairwise energy function $E_{jk}(x_j, x_k)$ captures the relative spatial relationships between two neighboring body parts. Both functions are specified similarly using Gaussian probabilities. During training, the MLE method was used to learn both the model parameters and structure (pairwise connections). During detection, to improve detection efficiency and to obtain approximate positions of body parts, a background subtraction technique was used to produce a binary image containing a human body. Then samples of object part positions $\mathbf{X}$ are obtained from $p(\mathbf{x}|\mathbf{i}, \mathbf{\Theta})$, and the sample that best matches the binary body image, according to the Chamfer distance[2], is selected as the final position of the body.

In [126], Kumar and Torr extended the PS model in [125] by including both object appearance and object shape in the unary functions and extending the model from tree structure to fully connected graph. Furthermore, they relax the Mahalanobis distance requirement on the pairwise energy function by using the Potts model. During detection, they first perform an object part detection, which can significantly reduce the possible

[2] Chamfer distance is a robust distance metric that measures the average mismatch to a target.

part positions. Then they used max-product loopy belief propagation to search near the detected part positions to find the final part positions. Experiments show that this method improved the PS model in terms of both object detection accuracy and efficiency. The PS model typically assumes that the parts do not (partially) occlude each other, which makes the PS model unsuitable for images with occluded object parts. To overcome this limitation, Kumar et al. [127] extended the traditional PS model by introducing the layered pictorial structure (LPS) model to address the part occlusion and appearance changes. Besides the shape and appearance, each part is also assigned a layer number to represent its relative depth.

5.3.1.2 Constellation models

Besides the PS model, the constellation model [128] has also been applied to object detection/recognition. Like the PS model, the constellation model assumes that an object consists of rigid parts, that are spatially organized in a deformable configuration. The deformation of the part spatial configuration in the image may be due to the object itself (e.g., a deformable object) or due to variations in pose, view angles, and camera parameters. Objects belonging to the same class share a similar object part appearance and spatial configuration. Different from the PS model, the constellation model captures the object appearance distribution in terms of object part appearances. Moreover, for constellation models, object parts are often treated as latent variables, and they are learned automatically from unsegmented training images instead of manually specified as in a PS model. The only supervised information provided is that the image contains an instance of the object class. As a result, EM learning methods are often employed for constellation models. Furthermore, constellation models explicitly account for shape variations and for the randomness in the presence/absence of features due to occlusion and detector errors.

Burl et al. [128] introduced a probabilistic approach to object detection/recognition. They assume that an object consists of parts with a distinctive appearance, arranged in some deformable configuration. They proposed employing a PGM to model the appearance distribution $p(\mathbf{i}|\mathbf{\Phi})$ of an object, which can be further decomposed into an object part appearance distribution $p(\mathbf{i}|\mathbf{x}, \Phi)$ and the spatial distribution $p(\mathbf{x}|\mathbf{\Theta})$ among object parts:

$$p(\mathbf{i}|\mathbf{\Theta}, \mathbf{\Phi}) = \sum_{\mathbf{x}} p(\mathbf{i}|\mathbf{x}, \mathbf{\Phi})p(\mathbf{x}|\mathbf{\Theta}).$$

Given an image patch **I**, object detection can be performed by identifying the image patch that maximizes $p(\mathbf{I}|\mathbf{\Theta}, \mathbf{\Phi})$. To avoid summation over all possible configurations of **X**, Burl et al. employ the winner-take-all strategy by replacing the summation operator with the maximum operator assuming that the summation is dominated by one term corresponding to a specific configuration $\mathbf{x}^*$. As a result, $p(\mathbf{I}|\mathbf{\Theta})$ can be approximated as follows:

$$p(\mathbf{i}|\mathbf{\Theta}, \mathbf{\Phi}) \approx p(\mathbf{i}|\mathbf{x}^*, \mathbf{\Phi})p(\mathbf{x}^*|\mathbf{\Theta}). \tag{5.13}$$

They further assume that part appearances are independent, leading to

$$p(\mathbf{i}|\mathbf{x}^*, \boldsymbol{\Phi}) = \prod_{k=1}^{K} p(\mathbf{i}_k|\mathbf{x}_k^*, \boldsymbol{\Phi}),$$

where $\mathbf{I}_k$ is the image feature vector that models the appearance of the kth part, and $p(\mathbf{i}_k|\mathbf{x}_k^*, \boldsymbol{\Phi})$ is modeled as the ratio of the likelihood of the kth object part to that of the background. To be invariant to translation, rotation, and scale, they proposed to geometrically normalize $\mathbf{X}$ with respect to a baseline part pair. The normalized $\mathbf{X}$ is assumed to follow a multivariate Gaussian distribution. Likewise, the appearance for each part $p(\mathbf{i}_k|\mathbf{x}_k)$ is also assumed to follow a multivariate Gaussian distribution. Given such parameterizations, the joint object appearance distribution $p(\mathbf{i}|\boldsymbol{\Theta})$ in Eq. (5.13) can be implemented by a two-layer pairwise Gaussian MN, where the top layer consists of nodes corresponding to body part $\mathbf{X}_k$ and is fully connected. The bottom layer consists of nodes corresponding to part appearance $\mathbf{I}_k$, and they are connected to corresponding $\mathbf{X}_k$. Given the initial detection of $\mathbf{x}^0$ through an independent part detector, $\mathbf{x}^*$ is found via a gradient ascent method to maximize $p(\mathbf{i}|\Theta, \boldsymbol{\Phi})$. Burl et al. demonstrated their method for facial landmark detection. However, the method requires the object parts be manually identified and labeled in training data.

Weber et al. [129] extended the work in [128] to object classification. They represent an object as a flexible constellation of rigid parts, where an object class is defined as a collection of parts sharing features or parts that are visually similar and occur in a similar spatial configuration. Instead of manually identifying and labeling object parts like [128], they propose an unsupervised learning method to automatically learn object parts via clustering from an unsegmented image. Specifically, part learning starts with an interest point detector, which is then followed by vector quantization to identify a small number of discriminative image patches as potential object parts. EM learning and model validation are then performed over the candidate object parts to greedily select a small number (<10) of the most discriminative parts as final part representations of the object class and to learn the object shape distribution $p(\mathbf{X})$ as the object class model. Then given the learned object class models, they employ the same likelihood ratio criterion as in [128] to perform object class detection. Weber et al.'s method shows the feasibility of automatically learning an object part model from unsegmented images and applies the learned model to object detection. Fergus et al. [130] further extended the work in [129] to learn the object class model that simultaneously captures object shape, appearance, and scale variations. During training, feature detection is first performed for each training image to detect regions of interests (ROIs), their scales, and their appearance features. Then EM learning is first performed to associate detected ROIs with object parts and then to learn the joint probability distribution of the object class shape, appearance, and scale. Given the object model, object detection is performed by calculating the likelihood ratio to determine the presence or absence of the object within the image. Their method assumes that the feature detector outputs a sparse set of ROIs and each ROI is only associated with one object part. In addition, model validation may be needed to determine the number of parts for each object class. Li et al. [12] proposed a Bayesian approach for learning constellation models for object category classification from just a few training images, where they treated object shape and appearance parameters as random variables and modeled their distributions via hyperparameters. Their object category model consists of P latent parts, and each latent part is characterized by its appearance and shape as shown in Fig. 5.18.

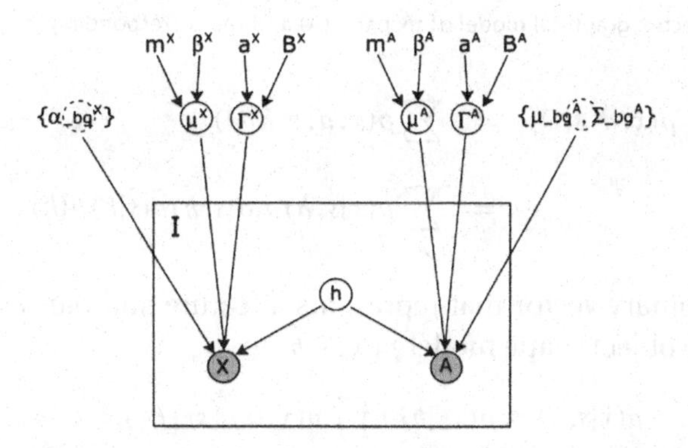

FIGURE 5.18 A graphical model representation of the constellation model for one object part, where A represents part appearance, X object shape, and h latent object parts [12]. They follow a Gaussian distribution with mean μ and covariance matrix Γ and hyper-parameters (m, α, β, B).

They introduced both a batch and an incremental method, based on a variational Bayesian method, to learn object model hyperparameters based on which the same likelihood ratio criterion is applied during object detection. The classical constellation models do not explicitly model the part structural relationships; they typically assume that parts are fully connected to model $p(\mathbf{X})$. Furthermore, a fully connected graph can lead to expensive training and inference as the number of parts becomes large. As a result, the number of parts for many existing constellation models is limited to a small number (< 10), and this limits their representation power. To overcome this issue, various heuristics have been proposed to decrease the connectivity of the graphical models or to simplify other aspects of the problem. Fergus et al. [13] present a "parts and structure" model for object category recognition. Instead of assuming a fully connected graph, they use a sparse graphical representation of the object. Specifically, they replace the fully connected graphical model in Fig. 5.19A with a star model as in Fig. 5.19B, which is a tree of depth one. The root of the star model is the landmark part, and other nodes represent the non-landmark parts. There are no direct connections among the nodes for the nonlandmark parts. As a result, the nonlandmark parts are independent of each other, given the landmark part, which allows the joint object shape model to be factorized as the product of each part.

Specifically, let X, D, and S represent the object shape, appearance, and scale, respectively. The object model can be written as

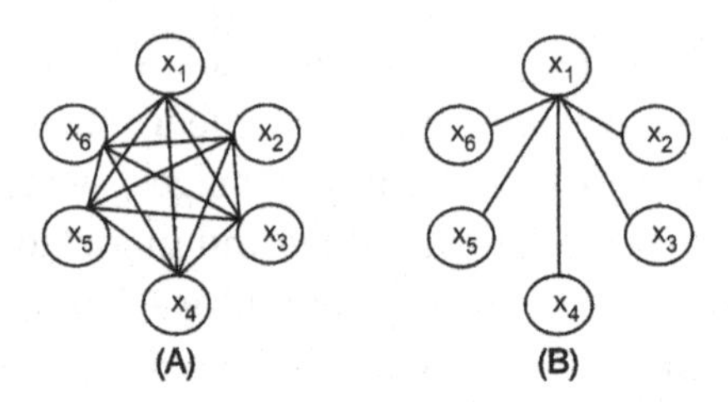

FIGURE 5.19 A fully connected graphical model of six parts (A) and the corresponding star model (B) [13].

$$\begin{aligned} p(x,d,s|\Theta) &= \sum_h p(x,d,s,h|\Theta) \\ &= \sum_h p(x|s,h)p(d|s,h)p(s|h)p(h), \end{aligned} \tag{5.14}$$

where h is a latent binary vector that represents a feature and part assignment. The star model simplifies the object shape model $p(X|S,h)$ to

$$p(x|s,h) = p(x_L|h_L) \prod_{j \neq L} p(x_j|x_L, s_L, h_j), \tag{5.15}$$

where X_L and s_L represent the position and scale of the landmark part, and x_j represents the shape of the jth nonlandmark part. The model can be learned efficiently in a semisupervised manner from example images containing category instances without requiring segmentation from background clutter. Moreover, as a result of the model simplification, the inference complexity is reduced to $O(N^2P)$ from $O(NP)$, where P is the number of parts. Given the star model, recognition proceeds in a similar fashion using the likelihood ratio criterion.

Loeff et al. [131] present a similar idea to improve the constellation model learning efficiency and to increase the number of features for each part. Specifically, for each object, they assume that there exists an unobserved center. Given the center, the detected object feature points are independent of each other. They further introduced a 2D part-feature assignment variable that allows one to assign multiple features to an object part. To significantly speed up the model learning and inference, they employed a variational EM method to learn the model parameters. The same likelihood ratio criterion is applied to the query image for object detection. Their model is able to deal with occlusion and errors in feature extraction, and it makes the inference linear in the number of parts of the model and the number of features in the image.

5.3.1.3 Other graphical models

Aside from PS and constellation models, standard PGM models such as MRF and CRF have also been applied to object detection. Using MRF, object detection and recognition can be formulated as an image labeling problem [86]. Gupta et al. [132] present an MRF-based method for simultaneous image restoration and recognition of digits from blurred images. The MRF models the joint distribution of a blurred image and the corresponding unblurred image. The unary potential function is parameterized as a Gaussian mixture, whose parameters are learned from training data. Nonparametric belief propagation based on Gibbs sampling was employed to infer the most likely unblurred values for each pixel from their blurred image. Digit recognition is subsequently performed on the unblurred image using the k-nearest neighbor method. One novelty of this work is the proposed nonparametric belief propagation based on Gibbs sampling and stochastic integration. Another contribution is the iterative procedure that alternates between image restoration and recognition so that the confidence score produced by image recognition can be used to help select the relevant training images to learn the unary potential parameters. Hence, this iterative process improves both image restoration and recognition. The primary use of MRF is for image restoration (a kind of image labeling), and image recognition is based on the entire restored image. The model lacks explicit object shape models and is, therefore, susceptible to image restoration errors.

In their work on Obj cut, Kumar et al. [14] present a Bayesian approach for simultaneous image labeling (figure-ground segmentation) and object category recognition. Specifically, as shown in Fig. 5.20, their model consists of an MRF model on the top for image labeling and a global shape prior at the bottom based on the layer pictorial structure (LPS) model. The model captures the joint conditional probability distribution of the image labels and the object shape given an image:

$$E(\mathbf{m}, \Theta) = \sum_{x} [\phi(\mathbf{D}|m_x) + \phi(m_x|\Theta) + \sum_{y} (\phi(\mathbf{D}|m_x, m_y) + \phi(m_x, m_y))]. \tag{5.16}$$

Specifically, Eq. (5.16) is the model energy function, where the first term is an intensity constrained unary energy term, the second term is a shape-constrained unary energy term, and the third and fourth terms are the intensity dependent and independent pairwise energy terms, respectively. Through the second unary energy term, the LPS model captures the object global shape through object parts and their spatial configurations. It relates to the MRF model through an additional unary potential function that measures the likelihood of the global shape model given the pixel labels. The MRF and LPS models are separately learned. While learning the MRF model, the LPS model parameters Θ in Eq. (5.16) are treated as latent variables and marginalized out through an EM framework. The method discussed in [127] was used to learn the LPS model. During segmentation and recognition given an image, for each object category, samples of its LPS model parameters Θ are obtained from its posterior distribution. Figure-ground segmentation can then be performed using the minimal cut algorithm [133] by minimizing the sum of the weighted energy function (Eq. (5.16)), weighted by the likelihood of each LPS model parameter sample. By biasing the segmentation toward a particular object category their model thus achieves simultaneous object segmentation and object category recognition. Whereas the shape prior can help reduce segmentation errors, rigid shape models can also introduce errors when the object deviates from standard shape models.

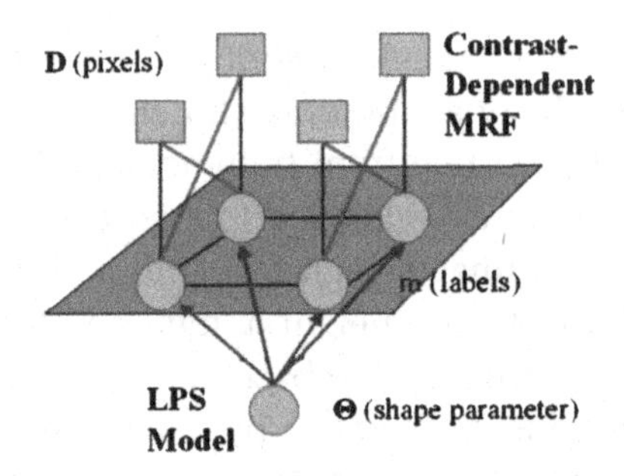

FIGURE 5.20 Graphical model representation of the object category specific MRF [14], where the top represents a pairwise MRF, and the bottom represents a global shape prior parameterized with an LPS model.

To capture the hierarchical contexts of objects and parts, the hierarchical Bayes model introduced in Section 3.8.1 has also been employed for object detection and recognition. Sudderth et al. [15] propose a hierarchical Bayes model with both latent variables and hyperparameters as shown in Fig. 5.21 for object detection and recognition in a cluttered scene. This model is essentially a Bayesian latent topic model that incorporates the hierarchy of objects and parts. As shown in Fig. 5.21, the root node O represents the object category, and its value represents one of the K object categories. The root node's value is given during training. It has two latent child nodes, r and z. The latent node r stands for the image location of the object. The latent node z stands for the object parts represented as clusters of features that appear in similar locations and have a similar appearance. The cardinality of z (i.e., the number of parts) is fixed to P. Both r and z are unknown during training and testing. Node z has two observed child nodes w and x that represent the appearance and position of the part relative to the object location r, respectively. They are assumed to be conditionally independent given z. A set of SIFT descriptors are used to generate appearance measurements w and position measurements x. In the graph of this model, nodes O, z, and w are discrete. They are parameterized by θ and ϕ, which in turn are parameterized by hyperparameters α and β. Nodes r and x are continuous and follow Gaussian distributions parameterized by $\{\varsigma, \Phi\}$ and $\{\mu, \Lambda\}$, respectively. They are in turn parameterized by hyperparameters ν_o, Δ_o, ν_p, and Δ_p. With the introduction of the object category node O, the model allows objects of different categories to share body parts and image features. During training, all hyperparameters are fixed. For each training image, given current model parameters, the object category labels o, and their image measurements x and w, Gibbs sampling is used to generate samples for each possible assignment of latent variable z by sampling the likelihood $p(z|w, x, o)$. Gibbs sampling is repeated for each candidate assignment P of z. The generated z samples are then combined with training data x and w, and they are used to learn the model parameters Θ and Φ through Bayesian estimation. Then Sudderth et al. extend their Gibbs sampling method to infer object position latent variable r. For this, they assume that r follows a Gaussian distribution and then apply the EM method to obtain its distribution parameters ξ and Φ. Given current distribution parameters for r, they can incorporate it into the likelihood model $p(z|w, x, o)$. Given the updated likelihood model, Gibbs sampling can proceed as before to generate samples for z. During the detection and recognition given an image and its observations w and x, the likelihood for each object category, that is, $p(w, x|o)$, is evaluated, and the object category corresponding to the highest likelihood is the recognized object label.

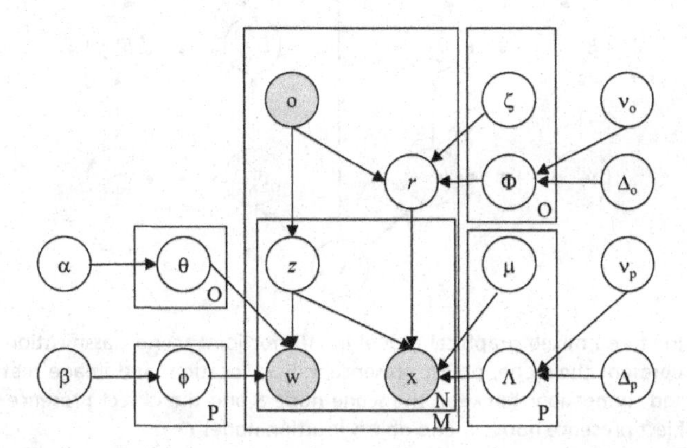

FIGURE 5.21 Hierarchical probabilistic model proposed by Sudderth et al. [15] for describing the hierarchy of objects and parts for object detection and recognition.

Besides the generative models, discriminative PGM models such as CRFs have also been successfully applied to object detection and recognition. Kumar and Hebert [134] introduced discriminative random fields (DRFs; same as the CRF model) to model the high-level context between the labels for detecting man-made structures in an image. They propose using a logistic regression function to represent the unary potential function. To explicitly account for observation and interaction between the neighboring sites, the pairwise potential function is the weighted sum of two terms, a data-independent pairwise potential function (Ising model) and a data-dependent pairwise potential function expressed as a logistic function of neighboring site labels and their image observations. The data-dependent term acts as a discontinuity adaptive model that moderates smoothing when the data from two sites is "different". This form of interaction favors piecewise constant smoothing of the labels while simultaneously considering the discontinuities in the observed data explicitly. Parameter learning is performed through the pseudo-likelihood method. Given the model, object detection and recognition is performed with the local MAP inference using the ICM method.

Murphy et al. [16] use a conditional tree-structured model for simultaneously detecting and classifying objects. Specifically, they introduce a tree-structured mixed graphical model as shown in Fig. 5.22 for joint object detection and scene classification. Their model captures the joint distribution of a scene, object class presence, and object class locations given image features. The joint conditional probability distribution is parameterized by the conditional prior probability of the scene generated by a boosting classifier, the conditional probability of the object location produced by a sigmoid function, and a pairwise

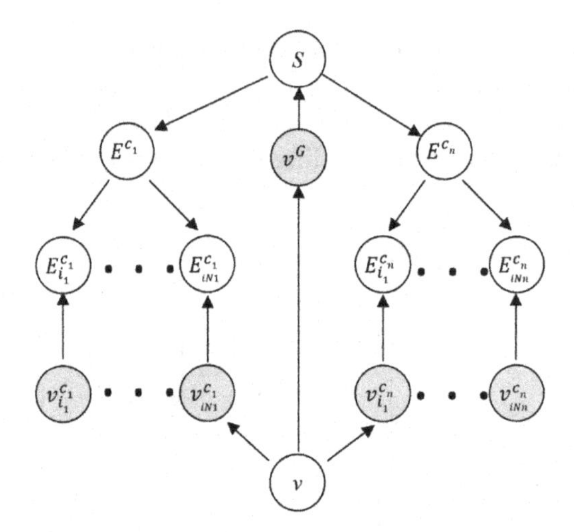

FIGURE 5.22 The tree-structured mixed graphical model in [16] for joint scene classification and object detection, where S, E, O, and v represent the scene, object presence, object location, and image features, respectively. The model employs undirected connections between the scene node S and the object presence nodes E and directed connections between object presence nodes E and object location nodes O.

potential between the scene label and object class presence label, specified by a table of counts. The MLE method was used to learn the model parameters. Given the model and a query image, Murphy et al.'s model can perform simultaneous object detection and scene classification. Their model shows how the global scene information can be used to improve object class detection and vice versa.

Variants of hidden CRF models have been introduced for part-based object recognition. These models typically include a hidden layer to represent object parts. Instead of manually specifying the object parts, they are learned automatically from data. These hidden CRF models therefore represent the discriminative version of the constellation model. Szummer et al. [17] present an extended CRF model that learns object part labeling in an unsupervised manner for classifying hand-drawn diagrams. Different from the two-layer CRF model shown in Fig. 5.7, they introduced a CRF model with a hidden layer **h**, that associates each pixel with the corresponding object part. Specifically, as shown in Fig. 5.23, the nodes in the hidden layer **h** represent part labels for all pixels; **x** represents the input image; and **y** represents object labels for all pixels. The model captures the joint conditional dis-

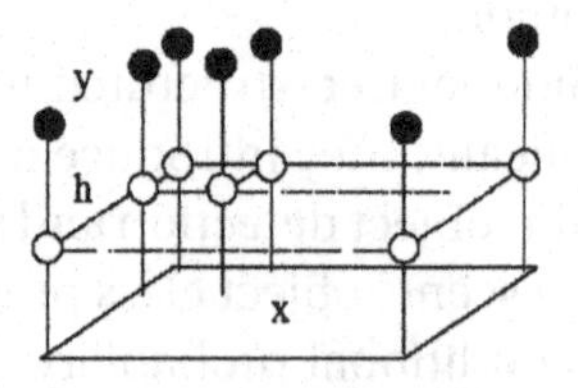

FIGURE 5.23 A hidden random field (HRF) [17] for part-based object recognition, where **x** represents the image, **h** represents the part labels, and **y** represents the object labels.

tribution of object labels and object part labels, given the input image. It is parameterized by a unary potential function $\phi_i(h_i, \mathbf{x})$ that captures the part label posterior, given the input image; a pairwise potential function $\psi_{ij}(h_i, h_j, \mathbf{x})$ that captures the interaction among the labels of neighboring object parts, given the image; and another pairwise potential function $\psi_{hy}(h_i, y_i, \mathbf{x})$ that captures the interactions between the object label and object part labels. Instead of using the EM method, they learn the model parameters by directly maximizing the logarithm of the conditional marginal likelihood of the object labels using the BFGS quasi-Newton method. Using the junction tree algorithm, a marginal MAP inference is performed independently for each pixel to estimate their object labels.

Quattoni et al. [135] introduced a similar CRF model with hidden layer for part-based object recognition. In addition to the object label and input image, the model also includes a hidden layer $\mathbf{h}$ that captures the object part labels for each pixel. For exact inference and efficient learning, they assume that edges of $\mathbf{h}$ form a tree structure. Their CRF model captures the joint conditional distribution of object label and hidden object parts, conditioned on the image. The model total energy function $\Psi(\mathbf{y}, \mathbf{h}, \mathbf{x}; \theta)$ captures the joint distribution of object labels, part labels, and the image. It consists of a unary energy function $\theta(x_j, h_j)$ for each object part and its image observation, a pairwise energy function $\theta(y, h_j)$ that captures the compatibility between the object label and part label, and a triple-wise energy function $\theta(y, h_j, h_k)$ that captures the compatibility between the object label and two neighboring part labels. Different from [17], they introduce a triple-wise energy term to capture the interactions between the pairs of part labels and their object labels. Conjugate gradient ascent was used to learn the model parameters by maximizing the log-likelihood of the parameters subject to ℓ_2 regularization. Given an image, MAP inference with belief propagation is then performed to infer posterior probability of the object label (foreground object, background), which is obtained by marginalizing out the hidden part labels from the joint posterior of object label and object parts. Because of the tree structure, both learning and inference can be performed efficiently.

To provide further contextual information on the object parts, Kapoor and Winn [18] extended the hidden CRF model in [17] by introducing the located hidden random field (LHRF), as shown in Fig. 5.24 for simultaneous part-based labeling and object detection.

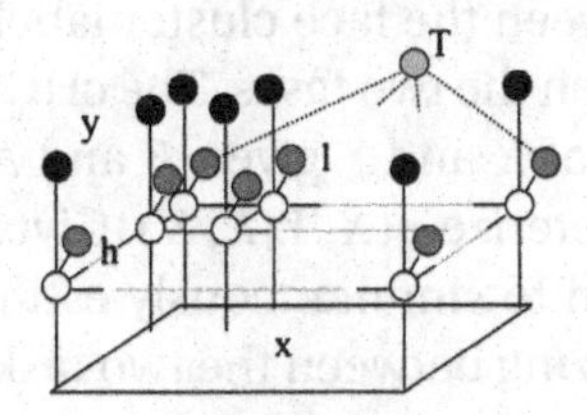

FIGURE 5.24 The located hidden random field in [18], where image **x** and the shaded vertices are observed during training time. The part labels **h**, denoted by unfilled white circles, are not observed and are learned during the training. They are connected to their locations **l** and the object labels **y**. The object location variable, *T*, is connected to all the part locations **l**.

LHRF is a three-layer hierarchical CRF model, including a hidden layer **h** to represent object part labels. However, different from [17], the hidden object part layer is connected not only to object labels **y** but also to part positions **l**. In addition, an object location variable **T** is introduced to constrain the object part locations. LHRF captures a joint conditional probability distribution $p(\mathbf{h}, \mathbf{y}, \mathbf{l}, T|\mathbf{x})$ of object part labels **h**, object labels **y**, part locations **l**, and object location **T** given image **x**. Different from [17], two additional potential functions $\psi_{hl}(h_i, l_i)$ and $\delta(l_i, T)$ are introduced to capture the interactions between the part label and their locations and to capture the interactions between part locations and object location, respectively. By introducing the global position of the object as a latent variable, the LHRF models the long-range spatial configuration of these parts and their local interactions. Given a training set of images with segmentation masks (object level labels) for the object of interest, LHRF automatically learns a set of parts that are both discriminative in terms of appearance and informative about the location of the object. Given training images including object labels and part locations, model parameters are learned by maximizing the posterior probability of the model parameters (with Gaussian parameter prior) through a gradient ascent method. Given an image, a MAP inference of object label for each pixel is then performed through loopy belief propagation. Experiments on benchmark datasets show that the use of discriminative parts leads to improved detection and segmentation performance, with the additional benefit of obtaining the labels of the object parts. Based on a model similar to that in [18], Winn et al. [136,137] further presented the layout consistency relationships among parts for recognizing and segmenting partially occluded objects in both 2D and 3D cases.

Finally, Zhang et al. [19] introduced a coupled hidden CRF (cHCRF) for face clustering and naming in videos. Given a set of face tracks F and a set of names N extracted from an input video, they need to perform two tasks, face clustering and face naming. Face clustering groups the face tracks into clusters X so that each cluster corresponds to one person's face, whereas face naming associates each face track with a specific name Y. Zhang et al. propose using a hidden conditional random field model to solve the problem. An HCRF model consists of a latent label layer and an observation layer. Each task can be formulated with one HCRF, where the hidden layer corresponds to face cluster labels for face track clustering and face track naming tasks. To capture the correlations between the two tasks, they connect two HCRF models through their label layers to form the cHCRF shown in Fig. 5.25, where the links between the face cluster labels X and the face name labels Y capture the dependencies between the two tasks. The cHCRF hence models the joint probability distribution $p(X, Y|F, N)$ of X and Y given F and N. Unary and pairwise potential functions are specified to parameterize $p(X, Y|F, N)$. Given a video sequence, Zhang et al. propose an optimization method to simultaneously estimate the model parameters and their labels by alternately optimizing between the two tasks.

5.3.1.4 Multiobject detection

So far, our discussions have focused on detecting and recognizing a single object. We can extend the techniques to perform joint detection of multiple objects. PGMs can be em-

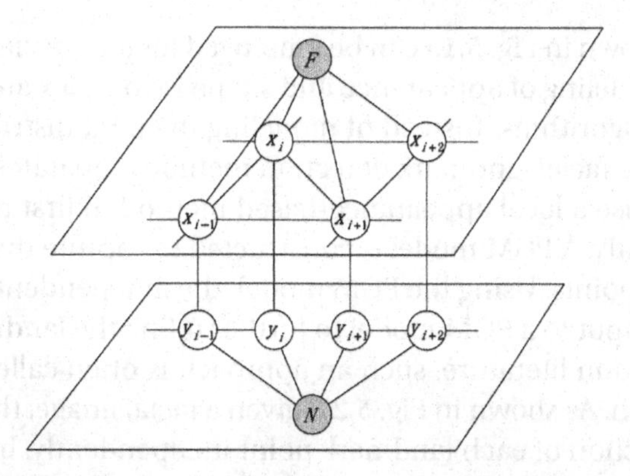

FIGURE 5.25 The coupled hidden conditional random field model for simultaneous face clustering and naming [19]. The gray nodes denote observable variables, whereas the white nodes indicate hidden variables. The top layer represents face clustering; the bottom layer represents name assignment. The hidden nodes are fully connected. Some links are omitted for clarity.

ployed to capture the spatial and temporal relationships among different objects. One such example is facial landmark point detection. As shown in Fig. 5.26A, facial landmark points represent dominant image points around facial components, and they can be used to concisely represent the facial shape. Facial landmark detection involves automatically locating each facial landmark point in the image, as shown in Fig. 5.16B. Facial landmark detection is a prerequisite for many facial analysis tasks, including facial recognition, facial expression recognition, and face pose estimation.

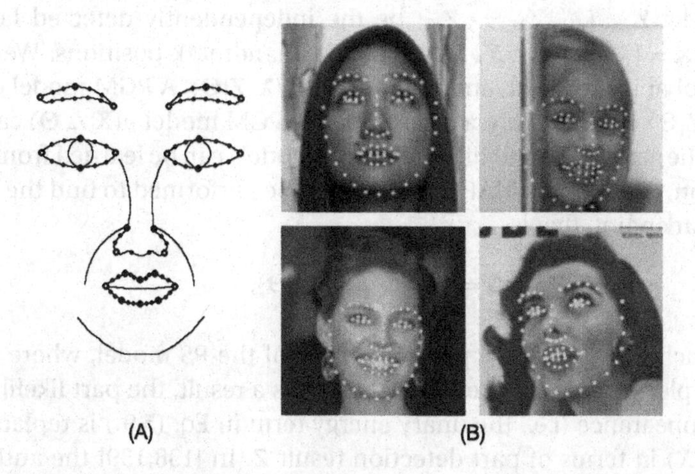

FIGURE 5.26 Facial landmark points (A) and their detection results (B). Figure adapted from [20].

Facial landmark points can be accomplished independently or collectively. PGMs are widely used to perform joint facial landmark detection by capturing the spatial relationships among the landmark points. In fact, a similar probabilistic object shape and appearance model to that shown in Fig. 5.15 can be employed for joint facial landmark detection. However, the joint modeling of appearance and shape is complex and can lead to computationally expensive algorithms. Instead of modeling the joint distribution of appearance and shape, most of the facial landmark detection methods separately perform their modeling. Typically, they use a local appearance-based method to first perform the landmark detection independently. A PGM model is constructed to capture the spatial relationships among the landmark points. Using the PGM model, the independently detected landmark points then serve as input to a PGM model to further refine the landmark positions. In the facial landmark detection literature, such an approach is often called a constrained local model (CLM) approach. As shown in Fig. 5.27, given a facial image, the CLM approach first performs a local detection of each landmark point independently, based on the independent local appearance information around each landmark. The detected landmark points are then refined by a global facial shape model to yield the final landmark positions.

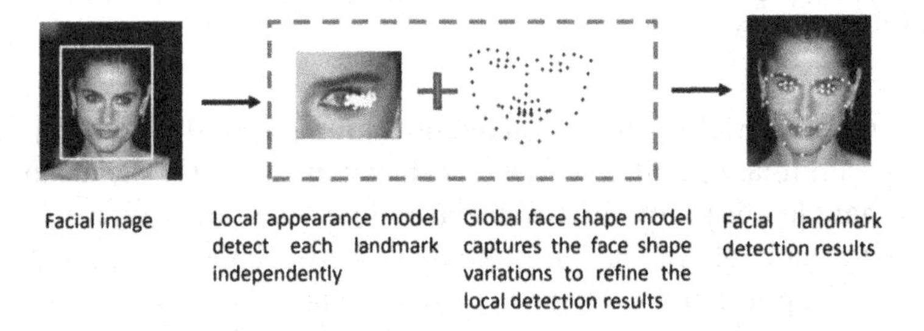
Facial image
Local appearance model detect each landmark independently
Global face shape model captures the face shape variations to refine the local detection results
Facial landmark detection results

FIGURE 5.27 Local constrained model for facial landmark detection. Figure adapted from [20].

Specifically, let $\mathbf{Z} = \{Z_1, Z_2, \ldots, Z_K\}$ be the independently detected facial landmark points, and let $\mathbf{X} = \{X_1, X_2, \ldots, X_K\}$ be the actual landmark positions. We can construct a generative probabilistic facial landmark model $p(\mathbf{X}, \mathbf{Z}|\mathbf{\Theta})$. A PGM model can be used to represent $p(\mathbf{X}, \mathbf{Z}|\mathbf{\Theta})$. Alternatively, a discriminative PGM model $p(\mathbf{X}|\mathbf{Z}, \mathbf{\Theta})$ can also be constructed. Both the generative and discriminative models can be learned from training data. During detection, given $\mathbf{Z}^*$, an MAP inference can be performed to find the final positions for each landmark point, that is,

$$\mathbf{x}^* = \arg\max_{\mathbf{x}} p(\mathbf{x}|\mathbf{z}^*, \mathbf{\Theta}). \tag{5.17}$$

Such an approach is essentially a particular case of the PS model, where the image observation $\mathbf{I}$ is replaced by the detection results $\mathbf{Z}$. As a result, the part likelihood $p(I|X)$ in terms of part appearance (i.e., the unary energy term in Eq. (5.9)) is replaced by the part likelihood $p(Z|X)$ in terms of part detection result Z. In [138,139] the authors proposed using a pairwise MRF combined with a support vector regression (SVR) for detecting 22 facial landmark points. Based on the local appearance for each facial landmark point, the SVR provides initial landmark locations. The MRF model captures the relative spatial relationships among the landmark points to ensure feasible spatial configurations. During detection, given the detected initial facial landmark points by the SVR, the MRF model then refines the initial landmark positions to produce the final landmark position through an MAP inference with belief propagation.

To address the challenges of detecting facial landmark points under large head pose and facial expression variations, more complex PGM models have been proposed. Wu and Ji [21,22] proposed a discriminative face shape model based on the three-way RBM. The model explicitly handles face pose and expression variations by decoupling the face shapes into a head pose related part and an expression related part. Compared to the other probabilistic face shape models, the three-way RBM can better handle large facial expressions and pose variations within a unified model. Specifically, as shown in Fig. 5.28, the

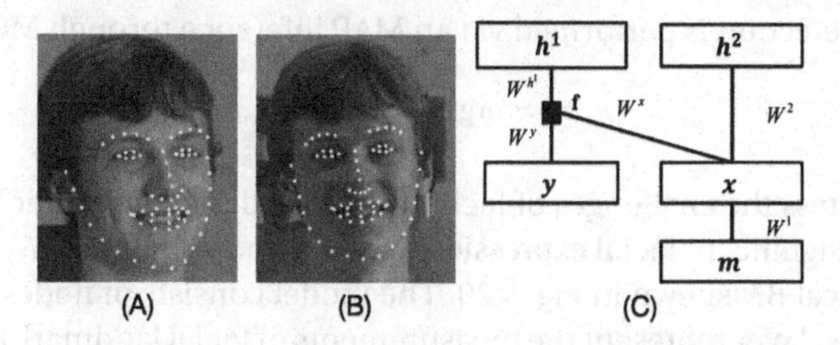

FIGURE 5.28 The three-way RBM model for landmark detection [21,22]. (A) Nonfrontal images with both pose and expression represented by **x** and its measurements **m**; (B) Corresponding frontal image with the same expression, represented by **y**; and (C) the three-way RBM face shape model efficiently encodes the interactions between **x**, **y**, and $\mathbf{h}^1$ through the factor node **f**. Images from Multi-PIE dataset [23].

face shape model consists of three layers of nodes, where **x** denotes the ground truth facial point locations that we want to infer, and **m** is their measurements from the local point detectors. In the middle layer, node **y** represents the corresponding frontal face shapes of **x** with the same facial expression. In the top layer, there are two sets of binary hidden nodes, including $\mathbf{h}^1$ and $\mathbf{h}^2$. The interactions between **x** and **y** are captured efficiently through the latent factor node **f** (the black square node), which allows the three-way pairwise interactions between **x**, **y**, and $\mathbf{h}^1$ to be factorized as the sum of three two-way interactions. The model captures two levels of information. The first level of information refers to the face shape patterns captured with the nodes in the two top layers, including **x**, **y**, $\mathbf{h}^1$, and $\mathbf{h}^2$. The second level of information is the input from the measurements **m**. In the model, it jointly combines the top-down information from face shape patterns and the bottom-up information from the measurements. Quantitatively, the model captures the joint conditional distribution $p(\mathbf{x}, \mathbf{y}, \mathbf{h}^1, \mathbf{h}^2|\mathbf{m}, \boldsymbol{\theta})$. Following the parameterizations for RBM in Eq. (4.23), the joint conditional distribution can be specified by the unary energy functions for **x**, **y**, $\mathbf{h}^1$, and $\mathbf{h}^2$ by the pairwise energy functions between **x** and $\mathbf{h}^2$ and by the three-way pairwise energy functions between **x**, **y**, and $\mathbf{h}^1$:

$$
\begin{aligned}
p(\mathbf{x}, \mathbf{y}, \mathbf{h}^1, \mathbf{h}^2|\mathbf{m}, \boldsymbol{\theta}) &= \frac{1}{Z(\boldsymbol{\theta})} \exp\{\boldsymbol{\alpha}_x^\top E_x(\mathbf{x}, \mathbf{m}) + \boldsymbol{\alpha}_y^\top E_y(\mathbf{y}, \mathbf{m}) + \boldsymbol{\alpha}_{h^1}^\top E_{h^1}(\mathbf{h}^1, \mathbf{m}) \\
&+ \boldsymbol{\alpha}_{h^2}^\top E_{h^2}(\mathbf{h}^2, \mathbf{m}) + E_x^\top(\mathbf{x}, \mathbf{m}) \mathbf{W}_{xh^2} E_{h^2}(\mathbf{h}^2, \mathbf{m})
\end{aligned}
$$

$$+ \sum_{\mathbf{f}} E_x^\top(\mathbf{x}, \mathbf{m})\mathbf{W}_{xf} E_f(\mathbf{f}, \mathbf{m}) + \mathbf{y}^\top \mathbf{W}_{yf} E_f(\mathbf{f}, \mathbf{m}))$$

$$+ E_{h^1}^\top(\mathbf{h}^1, \mathbf{m})\mathbf{W}_{h^1 f} E_f(\mathbf{f}, \mathbf{m})\}, \quad (5.18)$$

where the energy function $E_*(*, \mathbf{m})$ for each term is needed to make the variable $\mathbf{z}$ a function of $\mathbf{m}$. In addition, as $\mathbf{x}$ and $\mathbf{y}$ are both continuous, Gaussian functions may be used to specify their unary and pairwise energy functions. Using the training data, the model parameters $\boldsymbol{\theta}$ are learned by maximizing the joint log conditional likelihood of $\mathbf{x}$ and $\mathbf{y}$ (by marginalizing out the hidden variables $\mathbf{h}^1$ and $\mathbf{h}^2$) via the contrastive divergence method in Section 4.6.1.2.1. Given the model and given landmark measurements $\mathbf{m}$ for an image, landmark point detection is performed via an MAP inference through MCMC sampling:

$$\mathbf{x}^* = \arg\max_{\mathbf{x}} p(\mathbf{x}|\mathbf{m}).$$

To further address the challenges of facial landmark detection under large facial shape variations due to significant facial expression and head pose changes. Wu and Ji [24] introduced a hierarchical BN shown in Fig. 5.29. The model consists of nodes from four layers. Nodes in the lowest layer represent the measurements of facial landmark locations for each facial component. Nodes in the second layer represent the true facial landmark locations for each facial component. In the third layer, latent nodes are introduced to represent the states of each facial component. The top layer contains two discrete nodes, representing the facial expression and the face pose.

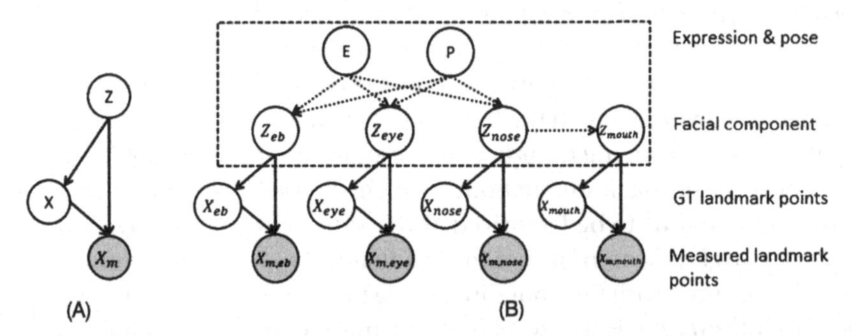

FIGURE 5.29 A hierarchical BN for facial landmark detection [24]. (A) Component model. (B) The proposed hierarchical model. Node connections marked as solid lines are fixed, whereas those with dotted lines are learned from the data.

This model captures two levels of shape information, including local shape information for each facial component and the global shape for the entire face, through the joint relationship among facial components, facial expression, and pose. The local shape information is captured by the nodes belonging to the same facial component. Based on mixture models, these nodes capture the variations of each facial component, including the eyebrow, eye, nose, and mouth, where Z is a discrete latent variable that captures the underlying state for each facial component. The global facial shape is captured by all the nodes inside the dotted rectangle, where the facial expression and face pose nodes are introduced to capture the spatial dependencies among different facial components as a function of facial expression and face pose variation. The model is parameterized by CPDs for each node. The E, P, and Z nodes are discrete, and their CPDs are specified by CPTs. $\mathbf{X}$ and $\mathbf{X}_m$ are continuous vectors. Whereas the CPD for $\mathbf{X}$ is specified by a multivariate Gaussian, the CPD for $\mathbf{X}_m$ is specified by a linear multivariate Gaussian. Given training data, the structure EM method (due to the presence of the latent variables Z) is employed to learn both the structure and parameters of the model. The learned model structure is shown in Fig. 5.29B. Given the learned model and the detected landmark position X_m, facial landmark detection can be formulated as a MAP inference problem through the junction tree method:

$$\mathbf{x}^* = \arg\max_{\mathbf{x}} p(\mathbf{x}|\mathbf{x}_m).$$

Compared to the three-way RBM discriminative model, the generative model shows better performance. However, the generative model requires additional information (the pose and expression labels) during training.

Rabinovich et al. [140] introduce a fully connected CRF model to perform multiobject detection and recognition (semantic segmentation) in a scene. The CRF model is used to capture the semantic relationships among object labels. Its unary function is derived from an independent object category recognition model using BoW features, and the pairwise potential captures the pairwise cooccurrences among the object labels. Given the labeled images, the pairwise potential function is learned for each pair of labels by maximizing their log-likelihood. To better generalize the label relationships to other datasets, Rabinovich et al. employ the Google Sets tool to generate a binary pairwise potential to capture the generic label relationships. Given the parameterized CRF model and a segmented image, MAP inference can be performed to jointly infer the labels of image segments and hence also the labels of the objects present in the scene. Their empirical evaluations on two benchmark datasets show that the captured label relationships can help improve object detection and object category recognition performance.

Galleguillos et al. [141] extend the semantic object relationships in [140] to also incorporate the spatial relationships of objects. A fully connected CRF model is constructed to capture both semantic and spatial relationships between objects. Likewise in [140], the unary potential function is provided by an independent recognition module. The pairwise potential captures both semantic and spatial relations among object labels. Specifically, the pairwise potential captures object cooccurrences with respect to each of the four spatial arrangements: above, below, inside, and around. The pairwise potential functions for the four spatial arrangements are learned jointly from the multiply labeled images by maximizing their joint log-likelihood. Experiments on the same two benchmark datasets show that their CRF model, which is based on combining both spatial and semantic object relationships, outperforms the CRF model in [140], which is based only on the semantic context. Strictly speaking, the CRF models in both [141] and [140] are not CRF models as their pairwise potential functions are independent of the image measurements. These models should thus be fully connected pairwise label-observation MRF models.

5.3.2 Scene recognition

Besides object recognition, scene recognition from images has been a major research topic in recent years. This task differs from object recognition in that scene classes are typically determined by the layout of the various scene objects instead of a single object of interest. Typical scene classes are related to environment, including outdoor views such as beach and street, and indoor views such as office and kitchen. In a more general setting, scene recognition also includes the categorization of other semantic environments such as sports venues and surveillance environments, and recognition of semantic events (e.g., sports, hiking, wedding), which can be characterized by various elements in an image. PGMs provide suitable tools for scene recognition due to their ability to characterize scene objects, their cooccurrences, and their spatial configurations. Two main groups of PGMs for scene modeling and recognition are latent topic models and CRF models.

5.3.2.1 Latent topic models

Two dominant latent topic models for scene recognition are the latent Dirichlet allocation (LDA) and the probabilistic latent semantic analysis (pLSA). Quelhas et al. [142] and Bosch et al. [143] used unsupervised pLSA or LDA models to compute the distribution $p(z|w)$ of the "hidden topics" (z in Fig. 3.47) given image features w extracted from the query image as the new features and then trained a discriminant model (e.g., SVM and KNN) based on the hidden topics for scene classification. Instead of training a separate classifier, Li and Perona [25] extended the traditional unsupervised LDA model to a supervised LDA model by considering the scene label of each image as shown in Fig. 5.30, where the shaded C node represents the object category label. The prior distribution parameters for the latent topic are now jointly determined by their hyperparameters θ and the scene label c, and hence θ is scene class dependent, that is, $\theta = \theta_c$. The prior distribution of the document class is specified by the parameter η. During training, given image feature $\mathbf{x}$ and document label c, the model parameters θ and β (η is fixed) are learned by maximizing the lower bound of the log-likelihood of the parameters through an EM procedure. Given the learned model and a query image represented by $\mathbf{x}$, its label is determined by identifying the document label c that maximizes $p(c|\mathbf{x}, \theta, \beta, \eta)$, which is approximately solved via variational inference.

Despite the success of pLSA and LDA models, they neglect the spatial correlations between hidden components and require the number of the components (topics) to be specified beforehand. To solve this problem, Sudderth et al. [144,145] employed a hierarchical Dirichlet process (HDP) and transformed the Dirichlet process[3] for scene recognition. Their models automatically learn the number of components and extend the traditional

[3] A Drichlet process is a random process whose realization is a random variable that follows a Dirichlet distribution.

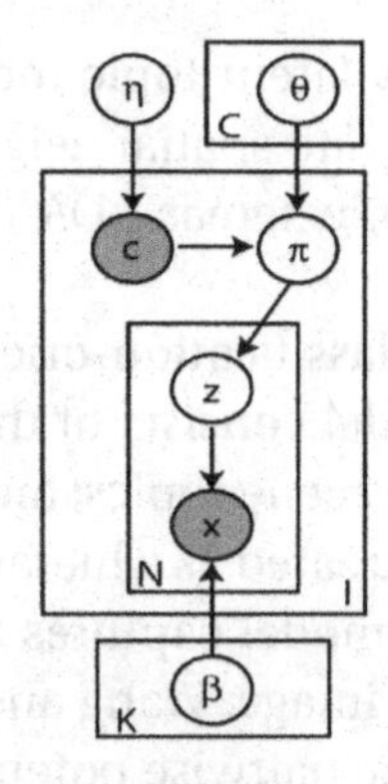

FIGURE 5.30 The supervised LDA model in [25], where the object label is represented by the shaded c node.

latent topic models by also modeling the spatial relations between object parts under multiple resolutions. As a cost, such models suffer from a computation burden much heavier than pLSA or LDA models.

Besides LDA models, Russell et al. [26] introduced a hierarchical directed model to capture the joint probability distribution of the scene objects, their appearance, and their spatial locations as shown in Fig. 5.31. The joint distribution can be factorized into products of object presence likelihood, object location likelihood, and object appearance likelihood. The parameters for object presence (θ_m) and object location likelihood ($\phi_{l,m}$) are learned online from training data. The parameters for the object appearance likelihood ($\eta_{m,l}$) are learned offline through an SVM classifier. Given the model and an input image, object detection and recognition are preformed by computing the presence probability for a particular object category in an image location.

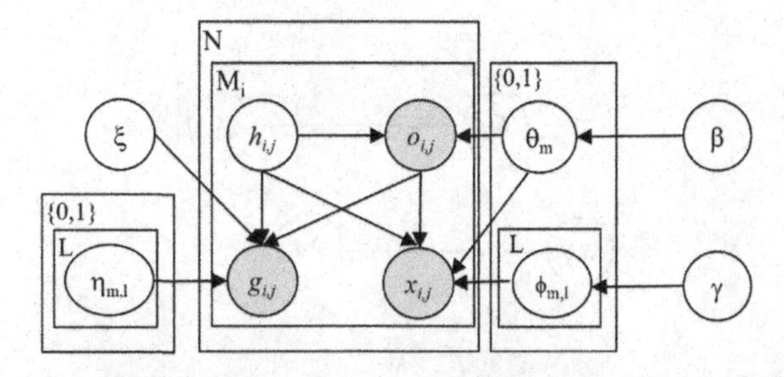

FIGURE 5.31 The hierarchical directed graphical model for modeling scene objects and their spatial appearance distribution [26], where o represents an image, g the object appearance, x the object location, and h indicates the presence or absence of an object in the image. The remaining symbols are the parameters of o, g, x, and h and their hyperparameters.

5.3.2.2 Conditional random field models

Another group of popular PGM models for scene understanding are the CRF models. CRF models differ significantly from LDA and pLSA models in several respects. First, CRF models are discriminative, whereas many latent topic models (e.g., pLSA and LDA) are generative. Second, CRF models incorporate spatial neighborhood interactions between the scene objects and the observed data, whereas LDA models neglect the spatial layout of scene objects.

Wang and Gong [27] proposed a classification-oriented CRF for natural scene recognition. As shown in Fig. 5.32, their model consists of three layers; the top layer is the class label; the middle layer represents the scene topics; and the bottom layer captures the image observations. This model can be treated as a hierarchical CRF model. The scene topics are produced by a pLSA model. The model captures the joint conditional distribution of the scene and latent topics given the image. Wang and Gong's model is parameterized by two unary potential functions and one pairwise potential function. The first unary potential function (appearance potential) captures the appearance compatibility of scene topics under scene label, whereas the second unary potential function (edge potential) captures the boundary (in terms of edges) interactions between neighboring scene patches under each scene label. The pairwise potential (spatial layout potential) captures the spatial compatibility between scene topics and scene label, that is, the spatial layout distribution of scene topic for each scene label. The pairwise potential is image independent. The quasi-Newton method was used during training to learn the model parameters by directly maximizing their log-likelihood. Given the learned CRF model, the scene label of a query image can be inferred by identifying the scene label that maximizes its posterior. However, during both training and testing, their method requires that a pLSA is first applied to identify the topics for each image. Jain et al. [146] proposed a selective hidden CRF to

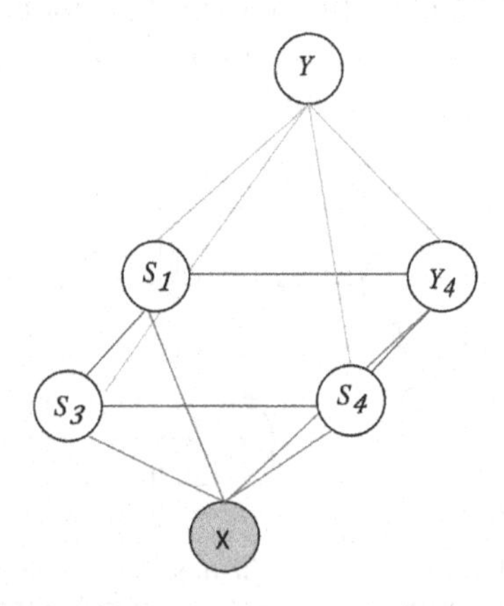

FIGURE 5.32 The classification orientation CRF model in [27], where y represents object class, s nodes represent the scene topics generated by a pLSA model, and x nodes represent image observations.

recognize sports scene categories. Both methods can be viewed as generalizations of the hidden CRF models [147,148], which introduced hidden variables into the original CRF model to capture the hidden topics/states of the image. For example, the CRF model in Fig. 5.32 becomes a hidden CRF model if the scene topics nodes S become latent, that is, their values are unknown during both training and testing. Many systems have subsequently been proposed to combine context with scene recognition. Some researchers employ global features to recognize scene category and local features for object classification, and then employ a graphical model to connect these two tasks. Such a graphical model can be generative [149,150] or discriminative [151,140,152]. In particular, Murphy et al. [16] presented a discriminative tree model for joint object detection and scene classification, as shown in Fig. 5.22. Their model captures the joint probability of scene label, object presence, and object locations.

5.3.3 Object tracking

Given a detected object, object tracking estimates the object position and its spatial scope over time. Object tracking requires detection of the object at each frame, and it includes two essential components: an object (appearance/shape) model and a dynamic model. The object model determines how to represent the target, whereas the dynamic model captures the dynamics of the target, and it decides where to search for the target in the forthcoming image frames. For the object (target) model, the same object models that we discussed in the previous section can be employed for object detection. The dynamic model characterizes the object state transition over time. Specifically, instead of performing object detection at each frame, object tracking exploits the dynamic transition of object states over time. Depending on the application, an object dynamic model can be manually specified or learned from training data. A manually specified dynamic model typically assumes local temporal smoothness without sudden changes. Tracking algorithms exploit either an explicit object dynamic transition model or the local smoothness assumption to restrict the search space in the upcoming frames.

Visual tracking of objects of interest, such as faces, has received significant attention in the CV community. It has been intensively studied for several decades, resulting in many algorithms. Despite these progresses, accurate, robust, and persistent target tracking remains challenging due to changes in target appearance as a result of variation in illumination, object pose, object shape, occlusion, camera parameters, and background clutters. Further details on them and methods to address them can be found in [153,154].

Like object detection and recognition, object tracking can be divided into a holistic approach and a part-based approach. The holistic approach assumes that an object as a whole is a rigid object and that every part of the object undergoes the same 3D motion. In contrast, part-based object tracking represents an object by its parts, and each part may undergo a different motion. Part-based tracking not only tracks all body parts but also maintains their spatio-temporal relationships. We further first discuss holistic object tracking, and then will discuss part-based object tracking in a separate section.

5.3.3.1 Holistic object tracking

Formally, the problem of object tracking can be stated as follows. Let $\mathbf{I}_t$ and $\mathbf{S}_t$ represent the image features and object state at time t, respectively. $\mathbf{S}_t$ can broadly capture the spatial, geometric, and appearance states of an object, which vary over time - including object position, size, orientation, speed, and so on. The initial state, $\mathbf{S}_0$, is either given manually or provided by an object detector. Probabilistically, tracking can be formulated as a temporal filtering problem that estimates $\mathbf{s}_t^*$ via $\mathbf{s}_t^* = \arg\max_{\mathbf{s}} P(\mathbf{s}_t|\mathbf{i}_{0:t})$ for $t = 1, 2, \ldots, T$.

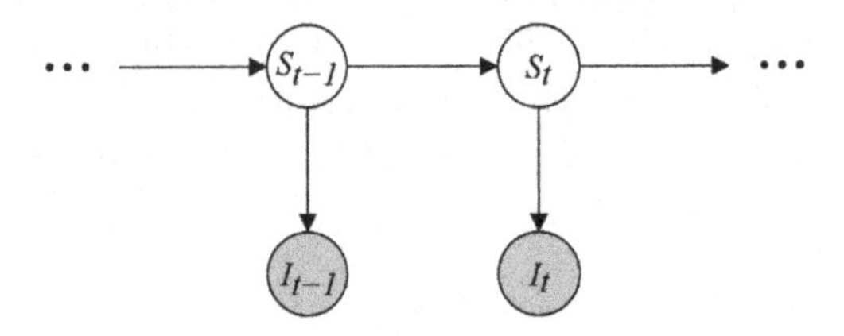

FIGURE 5.33 First-order linear dynamic system for tracking.

The PGM-based approaches often represent the joint distribution of $(\{\mathbf{I}_t\}, \{\mathbf{S}_t\})_{t=1}^T$ using dynamic models such as the linear dynamic system (LDS) model, as shown in Fig. 5.33, thereby reducing the tracking problem to the inference in the state-space model. The LDS model consists of two layers-a state layer at the top and the measurement layer at the bottom. It captures the joint probability distribution $(\{\mathbf{I}_t\}, \{\mathbf{S}_t\})_{t=1}^T$. Assuming the first-order Markov condition, that is, the object's current state S_t only depends on its state in the past frame $\mathbf{S}_{t-1}$, the joint probability of $(\{\mathbf{I}_t\}, \{\mathbf{S}_t\})_{t=1}^T$ for an LDS system can be written as

$$p(\mathbf{i}_0, \mathbf{i}_1, \ldots, \mathbf{i}_T, \mathbf{s}_0, \mathbf{s}_1, .., \mathbf{s}_T) = p(\mathbf{s}_0) \prod_{t=0}^{T} p(\mathbf{i}_t|\mathbf{s}_t) \prod_{t=1}^{T} p(\mathbf{s}_t|\mathbf{s}_{t-1}). \tag{5.19}$$

The LDS model consists of three components: the prior probability $p(\mathbf{S}_0)$, the measurement model $p(\mathbf{I}_t|\mathbf{S}_t)$, and the state transition model $p(\mathbf{S}_t|\mathbf{S}_{t-1})$. The prior model captures the initial (static) dependencies among the state variables. The measurement model captures the appearance/shape of an object under different states. The state transition model $p(\mathbf{S}_t|\mathbf{S}_{t-1})$ captures the object dynamics. Following the DBN definitions, the prior model can represent $p(\mathbf{S}_0)$, whereas the transition model can be used to represent $p(\mathbf{I}_t|\mathbf{S}_t)$ and $p(\mathbf{S}_t|\mathbf{S}_{t-1})$. In the absence of any knowledge about object dynamics, we can employ the local temporal smoothness model, which can be specified by a simple Gaussian smoothness model; that is,

$$p(\mathbf{s}_t|\mathbf{s}_{t-1}) \sim N(\mathbf{s}_{t-1}, \Sigma_t),$$

where the covariance matrix Σ_t can be learned from training data and is often assumed to be stationary over time. Following the DBN learning methods in Section 3.7.2.1, we can learn the parameters for each DBN component.

Given the DBN models, object tracking for a single object can be formulated as a filtering problem:

$$\mathbf{s}_t^* = \arg\max_{\mathbf{s}_t} p(\mathbf{s}_t|\mathbf{i}_{0:t}). \tag{5.20}$$

Various DBN inference methods introduced in Section 3.7.2.2 can be used to solve for the filtering problem recursively. In particular, for a simple chain structure like the LDS model, the efficient forward–backward inference method in Algorithm 3.12 can be used.

In case of linear and Gaussian dynamic and measurement models, the LDS system becomes a Kalman filter [155]. For complex state transition and measurement models, analytic inference methods may not be possible. Approximate inference methods, such as sampling-based methods, are often used for object tracking. Furthermore, sampling-based methods also allow one to relax the unimodality assumption with state distribution for simultaneous tracking of multiple objects. In particular, the sequential importance sampling is often used to approximately solve this problem. It represents the conditional density $P(\mathbf{S}_{t-1}|\mathbf{I}_{0\dots t-1})$ at time $t-1$ by a set of n weighted particles $\{(w_{t-1}^{(i)}, \mathbf{S}_{t-1}^{(i)})\}_{i=1}^{n}$. Given an image $\mathbf{I}_t$ at time t, to estimate the target state at t, we first sample $P(\mathbf{S}_t|\mathbf{I}_{0\dots t-1})$ to generate new samples $\mathbf{S}_t^{(i)}$, $i = 1, 2, \dots, n$, and then we compute their weights w_t^i, which are the normalized likelihood of $\mathbf{S}_t^{(i)}$, that is,

$$w_t^{(i)} = \frac{P(\mathbf{i}_t|\mathbf{s}_t^{(i)})}{\sum_{i=1}^{n} P(\mathbf{i}_t|\mathbf{s}_t^{(i)})}. \tag{5.21}$$

The target state distribution $P(\mathbf{S}_t|\mathbf{I}_{0\dots t})$ at time t can be represented by a set of n weighted particles $\{(w_t^{(i)}, \mathbf{S}_t^{(i)})\}_{i=1}^{n}$. As tracking continues, this sample and reweight process repeats. At each time t, the target state can be computed either as the mode of the particles or as their weighted average

$$\mathbf{s}_t = \sum_{i=1}^{n} w_t^{(i)} \mathbf{s}_t^{(i)}.$$

This is the so-called particle filter (PF) method or, in particular, the well-known condensation algorithm [156]. Despite its merits, the PF method is known to suffer from performance degradation when the dimensionality of the state increases, as sampling and search in high-dimensional space can be inefficient. Methods addressing this challenge include the variational and Blackwellization methods. The variational approximation method was introduced in [157] to approximate the posterior distribution of the state variables, often with factorizable distribution, yielding an efficient sampling method. The unscented PF [158] and the Rao–Blackwellization [28,159,160] methods aim at finding a proposal distribution that reduces the particle search space. In particular, for efficient inference with DBN, Doucet et al. [28] proposed the Rao–Blackwellized algorithm, which reduces the particle sample space by analytically marginalizing out a subset of variables. Specifically, their method assumes that the state variables can be divided into two sets and that the posterior probability of the first set of state variables given the observations and the second set of variables can be computed analytically. As a result, following the chain rule, the posterior probability of state variables can be written as the product of two terms. The first term contains the posterior probability of the first set of state variables, which can be computed analytically, and the second term is the posterior of the second set of state variables. A PF can be first applied to compute the posterior probability of the second set of state variables via sampling. Then, given the samples of the second set of state variables, one can analytically compute the posterior probability of the first set of state variables. To evaluate the performance of their method, Doucet et al. applied it to robot tracking and localization with a factorial HMM model as shown in Fig. 5.34, where the grid color state variables $M_t(i)$ can be analytically computed, whereas the location state variables L_t can be computed by particle sampling. Experiments show that their method can achieve inference accuracy close to the exact method.

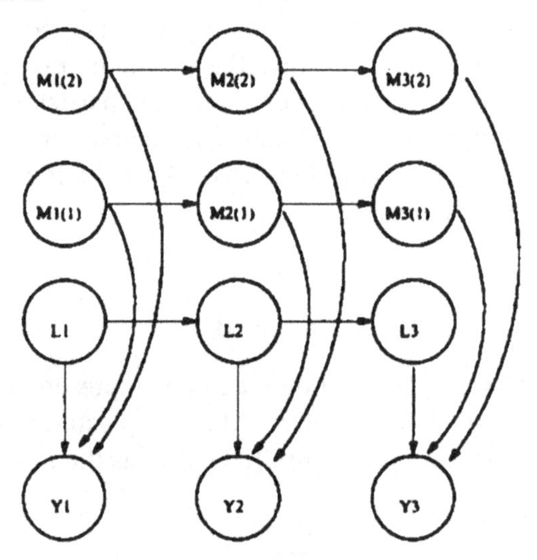

FIGURE 5.34 A factorial HMM with three hidden chains for robot localization [28], where the state variables consist of color variables $M_t(i)$ and location variables L_t Given the topology, $p(M_t, L_t|\mathbf{Y})$ can be decomposed into a product of $p(L_t|Y)$ and $p(M_t|L_t, \mathbf{Y})$, where $p(M_t|L_t, \mathbf{Y})$ can be computed analytically, whereas $p(L_t|\mathbf{Y})$ can be computed using the PF method.

In addition to generative PGM models such as LDS, Kalman filtering, and DBNs, discriminative models such as CRFs have also been widely applied to dynamic modeling in general and object tracking in particular. Unlike generative tracking, discriminative tracking directly optimizes the prediction accuracy of a tracker - namely $P(\mathbf{S}|\mathbf{I})$; this would be a preferred approach. Fig. 5.35A shows a CRF model for object tracking, where the first layer represents the object state S, and the second layer represents their image observations I.

While similar in structure to the corresponding HMM in Fig. 5.35C, the fully connected CRF model in (A) differs from HMMs in two respects. First, the CRF model is constructed by an undirected graph, whereas the corresponding HMM is formed by a directed graph. Moreover, the observation layer is fully connected to the state layer, whereas the observations for HMM are only connected to the corresponding states. Second, the state transition for the CRF model is parameterized by the pairwise conditional potential function $\Psi(S_t, S_{t-1}|\mathbf{I}_{1:t})$, and the emission probability is quantified by the conditional unary potential function $\Phi(S_t|\mathbf{I}_{1:t})$. More importantly, instead of modeling the joint probability of

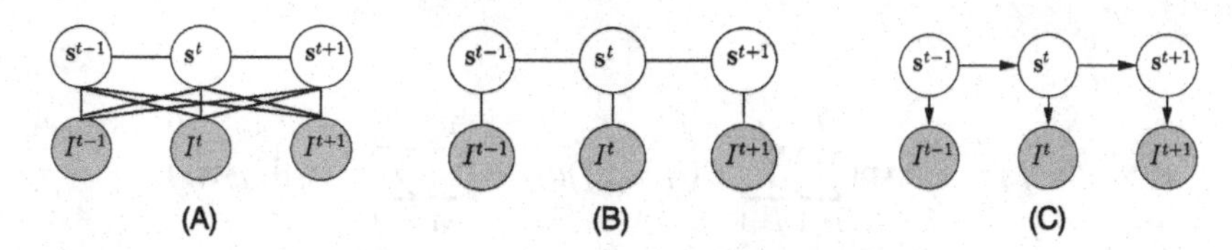

FIGURE 5.35 CRF models for object tracking. (A) Fully connected CRF model, (B) partially connected CRF model, and (C) corresponding HMM.

states $\mathbf{S}_{1:T}$ and their image observations $\mathbf{I}_{1:T}$, the CRF models the conditional distribution of states given the image observations:

$$p(\mathbf{s}_{1:T}|\mathbf{i}_{1:T}) = \frac{1}{Z}\prod_{t=1}^{T}\Psi(s_t, s_{t-1}|\mathbf{i}_{1:t})\Phi(s_t|\mathbf{i}_{1:t}), \tag{5.22}$$

The fully connected CRF model is complex in learning and inference. In practice, it is often simplified to the CRF model in Fig. 5.35B, where the observation nodes are not fully connected to the state layer. Instead, each observation node is only connected to the corresponding state node, effectively resulting in a pairwise label observation CRF. As a result, the pairwise and unary potential functions are changed to $\Psi(S_t, S_{t-1}|\mathbf{I}_t, \mathbf{I}_{t-1})$ and $\Phi(S_t|\mathbf{I}_t)$, producing the simplified state conditional distribution

$$p(\mathbf{s}_{1:T}|\mathbf{i}_{1:T}) = \frac{1}{Z}\prod_{t=1}^{T}\Psi(s_t, s_{t-1}|\mathbf{i}_t, \mathbf{i}_{t-1})\Phi(s_t|\mathbf{i}_t), \tag{5.23}$$

where the unary potential measures the compatibility between object state and its observations, and the pairwise potential function captures the object transition or dynamics conditioned on their observations. Given the definition of the state distribution in Eq. (5.23), the object tracking $p(S_t|\mathbf{I}_{1:t})$ can be implemented recursively:

$$p(s_t|\mathbf{i}_{1:t}) \propto \Phi(s_t|\mathbf{i}_t)\int \Psi(s_t, s_{t-1}|\mathbf{i}_t, \mathbf{i}_{t-1})p(s_{t-1}|\mathbf{i}_{1:t-1})ds_{t-1}.$$

For object tracking with CRF, we can use Eq. (5.24) to perform filtering:

$$\mathbf{s}_t^* = \arg\max_{\mathbf{s}^t} p(\mathbf{s}_t|\mathbf{i}_{1:t}). \tag{5.24}$$

Taycher et al. [161] proposed using the simplified CRF model for human tracking. They use similarity-preserving binary embedding for modeling the unary potential function, and the grid-based discrete PF was used during tracking. Evaluation of both synthetic and real data demonstrates that their CRF model outperform other tracking algorithms, including the condensation algorithm. Ross et al. [162] presented the CRF model with latent variables

($\mathbf{u}$ and $\mathbf{v}$)

$$p(\mathbf{s}_{1:T}|\mathbf{i}_{1:T}) = \frac{1}{Z}\exp(\sum_{t=1}^{T}\sum_{j=1}^{J} E(s_t, s_{t-1})u_{tj} + \sum_{t=1}^{T}\sum_{k=1}^{K} E(s_t|\mathbf{i}_{1:t})v_{tk}). \tag{5.25}$$

Instead of using all features to compute the unary and pairwise potential functions at each time, they introduce latent binary switching variables (u_{tj} and v_{tk}) to select a subset of the most relevant features at each time to compute the unary and pairwise functions during tracking as shown in Eq. (5.25), where $E(S_t, S_{t-1})$ and $E(S_t|\mathbf{I}_{1:t})$ are the pairwise and unary energy functions. Consequently, their model can flexibly switch features on and off at different times so that only the most relevant features at different times are selected for tracking. The contrastive divergence method was used to learn the model parameters by marginalizing out the switching variables. During inference, the state variables and switching variables are inferred alternately through either MCMC sampling methods or message passing. One of the key advantages of Ross et al.'s approach is that they have complete flexibility to choose the observation and dynamics features to maximize the tracking performance. Their model was applied to tracking the position of a basketball that follows a complicated trajectory in a video. Experiments show that their approach can handle missing and erroneous data and performed significantly better than the traditional Kalman filter.

5.3.3.2 Part-based object tracking

In addition to holistic object tracking, part-based object tracking is also widely employed in CV. Part-based object tracking represents the object by parts and tracks the movements of all parts. Part-based object tracking requires tracking not only each object part but also capturing the dynamic dependencies among the object parts. Compared to holistic object tracking, part-based object tracking can be more robust to object occlusion. In addition, as each object part may undergo different 3D motions, part-based object tracking can be applied to articulated object tracking.

Dynamic graphical models such as the dynamic BNs (DBNs) are often used to capture the spatial-temporal relationships among the object parts for part-based object tracking. Let $\mathbf{X} = \{X_i\}_{i=1}^{N}$ represent the N object parts, and let $\mathbf{I} = \{\mathbf{I}_i\}$ represent their image measurements. According to the DBN formulation shown in Fig. 5.36, a DBN model may consist of the prior model G_0 and the transition model $G_{\rightarrow}$. Each node X_i in the prior and transition model represents the state of an object part (i.e., its position and/or orientation), and $\mathbf{I}_i$ represents its image measurements for object part X_i. The links in the prior model capture the spatial dependencies among object parts, whereas the links in the transition network capture the dynamic transition for each object part and the dynamic dependencies among object parts.

Following the DBN parameter and structure learning methods discussed in Section 3.7.2.1, we can learn the parameters and the structures for both the prior and transition models. Given the learned DBN model, part-based object tracking can be performed

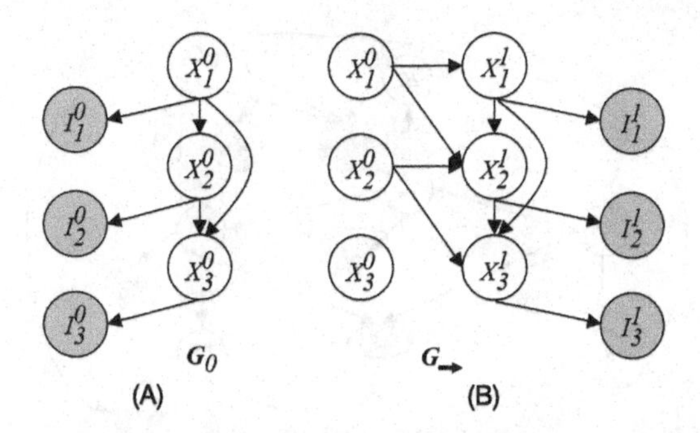

FIGURE 5.36 A DBN model for part-based object tracking, where (A) is the prior model, and (B) is the transition node. Node X_i represents the ith body part, and I_i represents its image measurements.

using the DBN filtering method to recursively estimate the part variables $\mathbf{X}^t$ at time t given their image measurements in the past:

$$\mathbf{x}^{*t} = \arg\max_{\mathbf{x}^t} p(\mathbf{x}^t | \mathbf{i}^{1:t}).$$

PGM-based tracking has received significant attention in the articulated object tracking literature, such as for people tracking and, in particular, body pose tracking. The goal of body pose tracking is to track each body part (joint) position (or joint angles) over time in a video. Thus, it is also referred to as *human motion estimation.* Various PGMs have been proposed for human body motion and pose tracking. These models vary in the architecture of their PGM models, in the types of image measurements, and in performing either 2D or 3D body pose tracking. Different PGM architectures include the kinematic tree representations (in 2D [163], 2.5D [30], and 3D [164,165]), the 3D loose-limbed model [31], and the 3D motion capture joint angles. The types of image features used may include various shape and texture features [166–168], outputs from individual part detectors [169,32, 170], and, more recently, deep model learned features.

Zhang and Ji [29] proposed a DBN model for upper body tracking. Based on physical (topological) arrangement among upper-body parts, a tree BN (kinematic chain) is first manually constructed as shown in Fig. 5.37, where the white circle nodes represent the position or joint angles for each body part, and the shaded circle nodes represent the image measurements for corresponding body parts.

To capture all physically feasible body motions instead of only the body motions present in the training data, Zhang and Ji incorporate various physical, anatomical, and biomechanical constraints on the relationships among body parts into the parameters of their model. As a result, unlike the existing body model, their model can track any kind of body motion without being restricted to typical movements in the training data. Based on the anatomical knowledge and physical constraints, both the BN structure and parameters are manually specified. The BN model is then extended to a DBN model by assuming smooth transition between body parts over time. The standard DBN inference method is

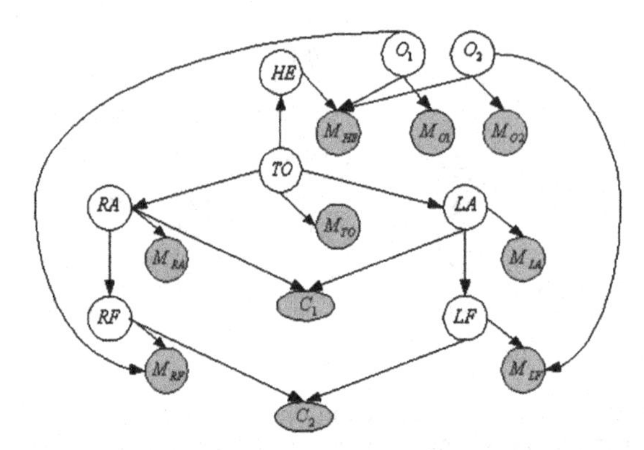

FIGURE 5.37 Upper body BN [29], where the white nodes represent body parts, the shaded M nodes represent their image measurements, and the shaded C nodes represent upper-body constraints.

then used to perform body pose tracking via Bayesian filtering. Their model can be applied to both 2D and 3D body tracking.

In [30] Palovic et al. introduced a DBN-based switching linear dynamic system (SLDS) for modeling complex and rich dynamic behavior such as the human figure motion. As shown in Fig. 5.38, the SLDS model consists of three layers. The top layer is the switching variable layer $\mathbf{H} = \{h_1, \ldots, h_T\}$, which represents the underlying dynamic types (e.g., jogging or walking). The middle layer is the state layer, and it captures the body position transitions over time. The bottom layer represents the image measurements of the body joints over time. The bottom two layers form an HMM, whereas the top layer allows state switching between different dynamic processes. The state transition at the second layer is now determined by both the previous state and the value of the current switching variable, that is, the transition probability changes from $p(\mathbf{S}_t|\mathbf{S}_{t-1})$ to $p(\mathbf{S}_t|\mathbf{S}_{t-1}, h_t)$. The state switching variables allow for automatic switching between different types of body motions, which, in turn, provides more accurate dynamic modeling during tracking. Palovic et al.'s model thus overcomes the unary dynamics assumption with the conventional HMM and allows the model to model a complex dynamic process that consists of different types of simple dynamic processes. They use a conditional linear multivariate Gaussian to parameterize the state transition and a multivariate Gaussian to parameterize the image observations. Due to the presence of the latent variables, the SLDS model is learned via an EM algorithm from image data. For inference, they need perform the joint posterior probability $P(\mathbf{S}, \mathbf{H}|\mathbf{X})$, for which exact inference becomes intractable. They introduced three different approximation schemes. The Viterbi approximation decouples $P(\mathbf{S}, \mathbf{H}|\mathbf{X})$ into the product of $p(\mathbf{H}|\mathbf{S}, \mathbf{X})$ and $p(\mathbf{S}|\mathbf{X})$. Viterbi decoding is first applied to solve $p(\mathbf{S}|\mathbf{X})$, yielding $\mathbf{S}^*$. Viterbi decoding is then applied to solve $p(\mathbf{H}|\mathbf{S}^*, \mathbf{X})$. The variational inference decouples the SLDS model into a HMM model (for the upper layer) and a linear dynamical system (for the lower layer). Finally, the generalized pseudo-Bayesian scheme attempts to collapse the model into a mixture with a reduced (and fixed) number of components. The SLDS was shown to provide more robust tracking performance than simple dynamic models. In addi-

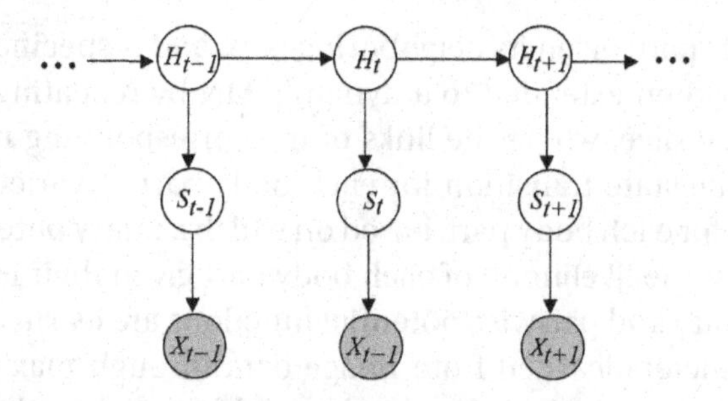

FIGURE 5.38 Switching linear dynamical system (SLDS) [30], where the nodes in the top layer represent the state switching variables, the nodes in the middle layer represent the hidden state variables, and the nodes in the bottom layer the image observations.

tion, experiments also show that the SLDS model outperforms the HMM in simultaneous segmentation and action classification of videos consisting of different types of primitive actions (e.g., jogging and running).

Aside from DBNs, undirected dynamic models have also been used for 3D human body pose estimation and tracking. Following the body kinematic chain, Sigal et al. [31] proposed using an MN to represent the body as a collection of 10 loosely connected body parts (including head, torso, upper/lower-left/right-arm/leg) as shown in Fig. 5.39, where each node represents a body part, and the state of each body part is measured by six spatial parameters, capturing its geometric dimension and its spatial relationships with the neighboring body parts. The short solid links capture the spatial dependencies among the body

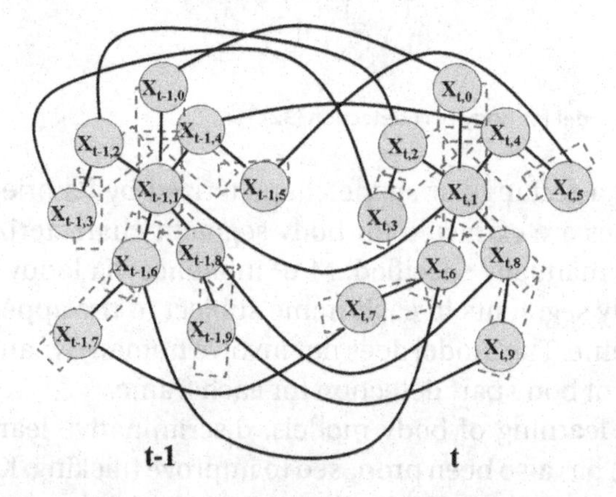

FIGURE 5.39 The graphical model for the loose-limbed human model [31], where the human body is composed of 10 parts (head, torso, and upper/lower-left/right-arm/leg) depicted by red dashed rectangles. Each body part (a node in a cyan circle) is represented by a six-dimensional vector of 3D position and orientation. The short solid black edges capture the spatial dependency between adjacent parts (spatial association), and the long solid edges from time $t-1$ to t capture temporal dependencies among part parts.

parts. Each link is quantified by a pairwise potential function that measures the compatibility of two body parts. The pairwise potential function is quantified by the conditional probability of the body part, given its neighboring part, and is specified as a Gaussian mixture. The static MN is then extended to a dynamic MN by repeating the structure of the static MN at each time slice, where the links of the corresponding nodes at two consecutive times capture the state transition for each body part. A variety of low-level image features are extracted for each body part, based on which a unary potential function is constructed that measures the likelihood of each body part given their image measurements. Parameters for the unary and pairwise potential functions are learned separately, with the unary potential parameters learned from image data through maximum entropy learning, while the pairwise potential functions are learned from the motion capture data. Body part tracking is performed via a MAP inference of each body part, given its measurements. For tractable inference, Sigal et al. employed the nonparametric belief propagation (i.e., approximating messages by sampled particles) [171] via the PF method to combine the unary and pairwise potential functions to estimate the 3D body part pose in each frame. The main advantage of this approach is that it reduces the complexity of search space to be linear in the number of body parts by using the bottom-up information from the part detectors.

Ramanan and Forsyth [32] used a nine-segment tree Markov model to capture nine body parts (consisting of the torso plus the left/right upper/lower arms/legs) and their kinematic dependencies for 2D body pose estimation as shown in Fig. 5.40. Each body

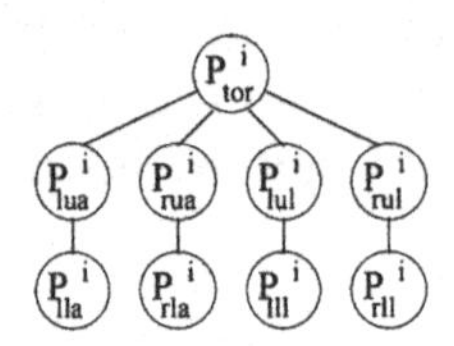

FIGURE 5.40 A tree Markov model for body part detection [32].

part is represented by a rectangular shape characterized by its orientation and centroid position. Image features are extracted for body segment characterization and detection. Model parameters are manually specified. MAP inference via loopy belief propagation is used to locate the body segments in each frame subject to the appearance constraint for each body part over frame. The model does not involve human dynamics modeling; hence, it performs independent body part detection for each frame.

Besides generative learning of body models, discriminative learning (i.e., maximum conditional likelihood) has also been proposed to improve tracking. Kim and Pavlovic [172] showed that even with a simple model like linear LDS, sophisticated discriminative learning objectives can significantly improve the tracking accuracy. Two learning algorithms are introduced, the conditional maximum likelihood and the slicewise conditional maximum likelihood. They evaluate the generalization performance of their methods on the 3D human pose tracking problem from monocular videos.

To reduce the complexity associated with using the generative 3D model for inference, Sminchisescu et al. [33] proposed extending the directed conditional model of [173] to the continuous state space for 3D pose estimation. By reversing the link directions between state and observation to point from observation to state as shown in Fig. 5.41, the directed conditional model is parameterized by $p(\mathbf{x}_t|\mathbf{r}_t)$ for image observation and by $p(\mathbf{x}_t|\mathbf{x}_{t-1}, \mathbf{r}_t)$ for state transition. Both observation and transition models are conditioned on the image observations $\mathbf{r}_t$.

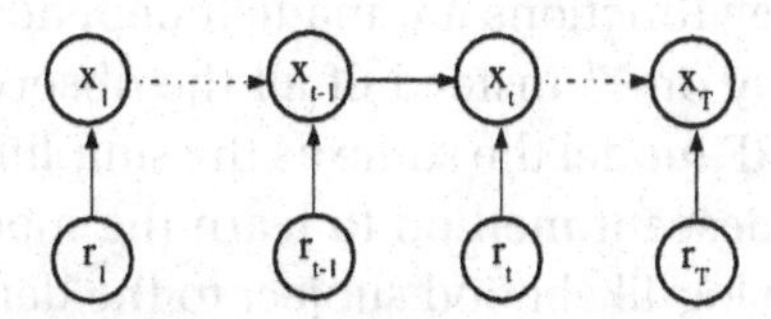

FIGURE 5.41 The directed conditional model [33].

With the directed conditional model, it is easy to show that the filtering density function can be written recursively as

$$p(\mathbf{x}_t|\mathbf{r}_{1:t}) = \int_{\mathbf{x}_{t-1}} p(\mathbf{x}_t|\mathbf{x}_{t-1}, \mathbf{r}_t)p(\mathbf{x}_{t-1}|\mathbf{r}_{1:t-1})d\mathbf{x}_{t-1}.$$

Instead of parameterizing $p(\mathbf{x}_t|\mathbf{x}_{t-1}, \mathbf{r}_t)$ and $p(\mathbf{x}_{t-1}|\mathbf{r}_{t-1})$ using a linear Gaussians (as discussed in Section 3.2.3.2), they proposed to parameterizing them using a Bayesian mixture of experts, that is, mixtures of Gaussians, with mixture weights specified as a function of $\mathbf{r}$. They further introduced a Bayesian EM method to approximately learn Bayesian mixture of experts. Then given the learned model, they use the filtering density function for 3D body pose tracking.

Fully discriminative models have also been employed for multitarget tracking. Taycher et al. [161] applied a CRF-like discriminative model to human pose tracking, where they discretize the continuous state space into grids. The proposed grid filter algorithm results in accurate tracking with almost real time. However, this grid-based approach generally requires a huge number of poses to be known a priori to achieve good approximation. In [34], Kim and Pavlovic presented a discriminative undirected graphical model, called the conditional state-space model (CSSM), for body pose tracking. As shown in Fig. 5.42, the CSSM model is similar to the fully connected CRF model in Fig. 5.35A in both structure and modeling. It conditions the states **X** on the entire measurement sequence **Y** while exploiting the sequential structure of the problem. By directly modeling the conditional distribution $p(\mathbf{X}|\mathbf{Y})$ its modeling matches well with the prediction (tracking) task. Kim and Pavlovic proposed representing $p(\mathbf{X}|\mathbf{Y})$ with a Gibbs distribution as

$$p(\mathbf{x}^{1:T}|\mathbf{y}^{1:T}) \propto \exp(-[\sum_{t=1}^{T} \mathbf{x}^t S\mathbf{x}^t + \mathbf{x}^t Q\mathbf{x}^{t-1} + \mathbf{x}^t E\phi(\mathbf{y}_t)),$$

where the first and second terms in the exponent, together, form the pairwise energy function, $\mathbf{X}^t S\mathbf{X}^t$ is the spatial energy function, which captures the spatial dependencies among elements of $\mathbf{X}^t$, and $\mathbf{X}^t Q \mathbf{X}^{t-1}$ is the temporal energy function , and it assumes linear Gaussian dynamics. It captures the temporal dependencies (transition) between consecutive state vectors. $\mathbf{X}^t E\phi(\mathbf{Y}_t)$ represents the unary energy function, and $\phi(\mathbf{Y}_t)$ is a feature vector derived from observations at time t only. The unary energy function measures the correlation between the state vector $\mathbf{X}^t$ and the observations $\mathbf{Y}^t$. Note that to simplify the computations, the pairwise energy functions are made independent of $\mathbf{Y}$, whereas the unary energy function depends only on $\mathbf{Y}^t$ instead of all the observations $\mathbf{Y}$. These simplifications effectively make this CRF model the same as the simplified CRF model in Fig. 5.35B. They introduced a gradient descent method to learn the model parameters S, Q, and E by minimizing their negative log-likelihood subject to the density integrability. Given the learned model parameters, they introduced a recursive inference method based on message passing to perform the decoding inference:

$$\mathbf{x}^{*1:t} = \arg\max_{\mathbf{x}^{1:t}} p(\mathbf{x}^{1:t} | \mathbf{y}^{1:t}).$$

Due to the recursion, their inference algorithm is much faster than Kalman filtering since it allows the model to have a large number of measurement features. They applied their model to 3D human pose tracking, where $\mathbf{X}$ represents the 3D joint angles. Experiments show that their discriminative model produced significantly lower estimation errors than the corresponding generative models.

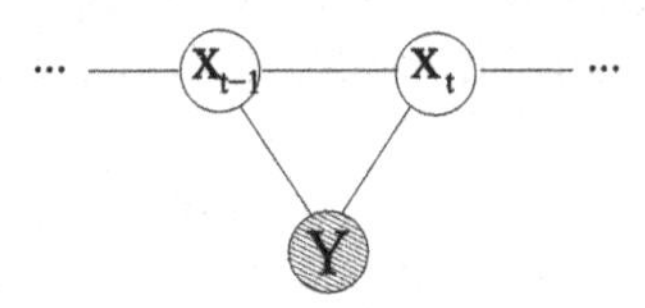

FIGURE 5.42 Conditional SSM (CSSM) [34].

Part-based object tracking can be further extended to multiobject tracking, where each object may be treated as a rigid object part. Compared to part-based single object tracking, multiobject tracking is generally more challenging since we do not have the corresponding measurements for each object. Hence, we need to simultaneously estimate targets' states and the target-measurement association. In addition, the spatial-temporal relationships among object parts can be arbitrary, and they are much harder to capture. In PGM-based approaches, besides modeling the tracking state and observation variables, the model should also represent the association variables for each tracker. Assume that at time t, there are M trackers $\mathbf{x}_t = (x_{1,t}, \ldots, x_{M,t})$, where the ith tracker has a tracking state $x_{i,t}$ at time t, m_t measurements $\mathbf{y}_t = (y_{1,t}, \ldots, y_{m_t,t})$, where $y_{k,t}$ is the kth measurement at time t, and a latent association vector $\mathbf{a}_t = (a_{1,t}, \ldots, a_{M,t})$, where variable $a_{i,t} \in \{0, \ldots, m_t\}$ associates the tracker i with a particular measurement (0 indicates object disappearance). Given the observations $\mathbf{y}_t = (\mathbf{y}_1, \mathbf{y}_2, \ldots, \mathbf{y}_t)$, the multitarget tracking can be formulated as a maximum a posteriori (MAP) inference problem;

$$\begin{aligned} \mathbf{x}_t^* &= \arg\max_{\mathbf{x}_t} \log p(\mathbf{x}_t|\mathbf{Y}_t) \\ &= \arg\max_{\mathbf{x}_t} \log \sum_{\mathbf{a}_t} p(\mathbf{x}_t, \mathbf{a}_t|\mathbf{y}_t) \\ &= \arg\max_{\mathbf{x}_t} \log \sum_{\mathbf{a}_t} p(\mathbf{x}_t, |\mathbf{a}_t, \mathbf{y}_t) p(\mathbf{a}_t|\mathbf{y}_t), \end{aligned} \tag{5.26}$$

where the first term $p(\mathbf{x}_t, |\mathbf{a}_t, \mathbf{Y}_t)$ can be formulated as a traditional multitarget tracking with known measurement-tracker association. The second term captures the association vector distribution given observations so far. Solving Eq. (5.26) requires enumerating all possible configurations of $\mathbf{a}_t$, which may become intractable for a large number of trackers and measurements. It can be solved approximately by replacing $\sum_{\mathbf{a}_t} p(\mathbf{x}_t, |\mathbf{a}_t, \mathbf{y}_t)p(\mathbf{a}_t|\mathbf{y}_t)$ with either $\frac{1}{S}\sum_{s=1}^{S} p(\mathbf{x}_t, |\mathbf{a}_{s,t}, \mathbf{y}_t)$ or $p(\mathbf{x}_t, |\mathbf{a}_t^*, \mathbf{y}_t)$, where $\mathbf{a}_{s,t}$ is a sample from $p(\mathbf{a}_t|\mathbf{y}_t)$, and $\mathbf{a}_t^*$ is the mode of $p(\mathbf{a}_t|\mathbf{y}_t)$.

Following the same problem definitions, Yu et al. [174] formulated the multitarget tracking as the following a MAP estimation problem of the data association variables:

$$\mathbf{a}_t^* = \arg\max_{\mathbf{a}_t} \log p(\mathbf{a}_t|\mathbf{y}_t).$$

This model can be decomposed into the association prior model $P(\mathbf{a}_t|\mathbf{x}_t)$ and the likelihood model $P(\mathbf{y}_t|\mathbf{x}_t, \mathbf{a}_t)$. A fully connected pairwise MRF was used to represent $P(\mathbf{a}_t|\mathbf{x}_t) \propto \prod_{i,j} \psi_{i,j}(a_{i,t}, a_{j,t}|\mathbf{x}_t)$. They further assume that the pairwise potential for the association prior is independent of tracker state $\mathbf{x}_t$, leading to the exclusive-OR indicator potential, that is, $\psi_{i,j}(a_{i,t}, a_{j,t}|\mathbf{x}_t) = \delta(a_{i,t} \neq a_{j,t})$. Given the data association, the likelihood model is further factorized into pertracker emission models, that is, $P(y_t|x_t, a_t) = \prod_i P(y_{a_{i,t},t}|x_{i,t})$. During the optimization, they regard $\mathbf{x}_t$ as hidden variables and maximize $P(\mathbf{a}_t|\mathbf{Y}_t)$ with respect to $\mathbf{a}_t$ by a (variational) EM-like iteration. Interestingly, the state posterior in the E-step; namely, $P(\mathbf{x}_t|\mathbf{Y}_t)$ becomes the desired quantity in tracking. This framework is computationally attractive as it allows the optimization to be done in a distributed manner.

Khan et al. [175] use the joint PFs of [176], where they construct MRF models on the fly to model the spatial interaction among the objects nearby. Specifically, instead of assuming that trackers are spatially independent, they assume that they are related at each time, and they use $\psi_{i,j}(x_{i,t}, x_{j,t}) \propto \exp(-g(x_{i,t}, x_{j,t}))$ to capture the pairwise interaction between two trackers i and j at time t. Here, $g(x_{i,t}, x_{j,t})$ is the number of pixels overlapping between two trackers i and j at time t, which penalizes (or avoids) collapsing objects, a property observed in social agents (e.g., insects). The pairwise interaction potential is incorporated into the tracker transition model and is used jointly with weight measurements of each sample during tracking. For efficient sampling over all trackers, Khan et al. introduced an MCMC sampling method that samples the state for one tracker at a time.

5.3.4 Three-dimensional reconstruction and stereo vision

Three-dimensional reconstruction involves estimating the 3D geometric properties of an object from its 2D image. The 3D geometric properties include 3D coordinates, the 3D orientation, and the depth for each 3D point. Three-dimensional coordinate (3 degrees of freedom) reconstruction is also referred to as full reconstruction as it completely captures the 3D geometry of the object. Three-dimensional orientation estimation is also referred to as 3D shape reconstruction, as it only recovers the three-dimensional orientation (2 degrees of freedom) for each 3D point on the object. The size of the 3D shape remains unknown. Finally, depth reconstruction is the simplest 3D reconstructions as it only recovers the z coordinate (one degree of freedom) of each 3D point. Three-dimensional reconstruction techniques in CV can be classified into 3D reconstruction from monocular single images (shape from X), 3D reconstruction from two images (passive or active stereo), and 3D reconstruction from a sequence of images (structure from motion). PGMs have been applied to each type of 3D reconstructions, in particular, to passive stereo. Although we will focus on the application of PGM to passive stereo, we will also discuss PGM applications to 3D reconstruction from monocular images.

5.3.4.1 PGM for passive stereo

Three-dimensional reconstruction with passive stereo typically involves two-steps: point matching and 3D reconstruction of the matched points via triangulation. The challenge lies in the point matching step, which identifies the corresponding points in two images. Two points from two images match and become corresponding points if they are generated by the same 3D point. Point matching is time-consuming and ambiguous as the search space is often large and there is often not enough information to uniquely match two points despite the use of various geometric constraints such as the epipolar constraint, spatial order constraint, and so on. One solution to this challenge is performing point matching jointly for all image points instead of matching one at time, and imposing the local or global structural constraints on the geometry recovered from the matching points. PGMs like MNs are often employed to perform joint point matching and impose structural constraints.

In CV, point matching is often reformulated as a disparity estimation. For a rectified stereo images (i.e., corresponding points are located on the same image rows), disparity is defined as the absolute horizontal (column) difference between two matched points. Given the estimated disparity, the depth for each pixel can be recovered since disparity is inversely proportional to the depth. Hence disparity is often used as a subrogate measure of the pixel depth. Disparity estimation methods can be divided into local and global approaches. The local approach estimates the disparity for each pixel individually, whereas the global approach simultaneously estimates the disparities for all pixels. PGM models such as MRFs are widely employed by the global approach.

Specifically, given two images of the same scene (the reference image on the left and the matching image on the right), disparity estimation can be stated as follows. Let $\mathbf{X} = \{X_1, X_2, \ldots, X_N\}$ be the N pixels on the reference image, and let $\mathbf{D} = \{D_1, D_2, \ldots, D_N\}$ be the corresponding disparity map for each pixel with $D_n \in [1, K]$, where K is the number of pixels in a row. We can further define $\mathbf{Y} = \{Y_1, Y_2, \ldots, Y_N\}$ as being the pixels on the matching image. Mathematically, the disparity estimation can be stated as follows: given **X** and **Y**, find **D** that best match pixels in **X** with those in **Y**. We can formulate this as the MAP inference problem

$$\mathbf{d}^* = \arg\max_{\mathbf{d}} p(\mathbf{d}|\mathbf{x}, \mathbf{y}). \tag{5.27}$$

The problem can be formulated as an MAP-MRF inference problem. We can construct a pairwise grid-like two-layer label-observation MRF model to capture the relationships between **X**, **Y**, and **D**, as shown in Fig. 5.43, where each node d_n in the label layer represents the disparity measurement for the nth pixel. The corresponding node in the bottom observation layer is the image measurements $I_n = f(X_n, Y_{n+D_n})$ derived from X_n in the reference image and from Y_{n+D_n} in the matching image. The links among the label nodes capture the local disparity smoothness.

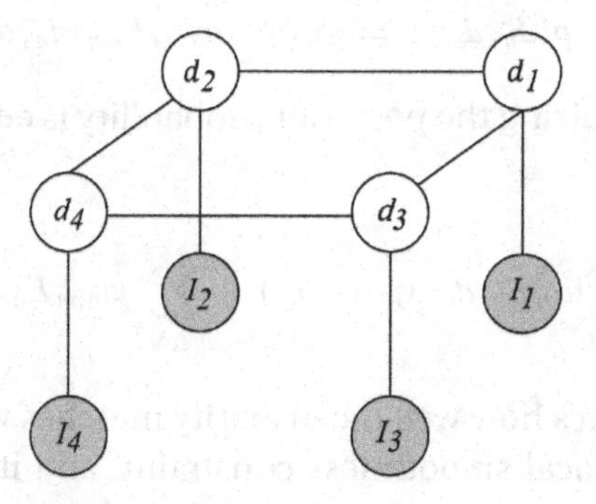

FIGURE 5.43 A label-observation model for image disparity estimation.

Given the MRF model, $p(\mathbf{D}|\mathbf{X}, \mathbf{Y})$ can be rewritten as follows:

$$\begin{aligned} p(\mathbf{d}|\mathbf{x}, \mathbf{y}) &= p(d_1, d_2, \ldots, d_N | x_1, x_2, \ldots, x_N, y_1, y_2, \ldots, y_N) \\ &\propto p(d_1, d_2, \ldots, d_N) p(x_1, x_2, \ldots, x_N, y_1, y_2, \ldots, y_N | d_1, d_2, \ldots, d_N) \\ &= \prod_{n=1}^{N} \prod_{m \in \mathcal{N}_n} p(d_n, d_m) p(x_n, y_{n+d_n} | d_n), \end{aligned} \tag{5.28}$$

where $\mathcal{N}_n$ is the neighbor for pixel n. The global disparity prior $p(d_1, d_2, \ldots, d_N)$ is approximated by products of local pairwise disparity prior $p(d_n, d_m)$. The local prior can be used to implement a local smoothness constraint. $p(x_n, y_{n+d_n} | d_n)$ measures the likelihood of d_n for pixel X_n and the corresponding pixel Y_{n+d_n} in the matching image. Within the MRF framework, $p(d_n, d_m)$ can be specified by a pairwise energy function $E_{nm}(d_n, d_m)$, and $p(x_n, y_{n+d_n} | d_n)$ can be specified by a unary energy function $E_n(x_n, y_{n+d_n}, d_n)$. Specifically, the pairwise energy function $E_{nm}(d_n, d_m)$ encodes the local disparity smoothness. We may simply use the Potts model to specify the pairwise energy function. In practice, the pairwise energy function is typically a function of the disparity differences among neighboring pixels in the matching image, that is, $E_{nm}(d_n, d_m) = \rho(d_n - d_m)$, where ρ is an increasing function, for example, the quadratic function.

The unary energy function $E_n(x_n, d_n, Y_{d_n+n})$ is quantified by the negative cross-correlation between X_n and Y_m with the corresponding pixel at $n + d_n$ on the matching image:

$$E_n(x_n, y_n + d_n, d_n) = -CC(x_n, y_{n+d_n})$$

Aside from negative cross-correlation, other matching costs- including a sum of squared differences and absolute intensity difference - can also be used to quantify the unary energy function. Given the energy functions, we can then construct $p(d_n, d_m)$ and $p(X_n, Y_{n+d_n}|d_n)$ as Gibbs distributions:

$$\begin{aligned} p(x_n, y_{n+d_n}|d_n) &= \exp(-\alpha_n E_n(x_n, y_{n+d} \\ p(d_n, d_m) &= \exp(-w_{n,m} E_{nm}(d_n, d_m)). \end{aligned}$$

We can show that maximizing the posterior probability is equivalent to minimizing the total energy function

$$E(\mathbf{d}) = \sum_{n=1}^{N} \{\alpha_n E_n(d_n, x_n, y_{n+D_n}) + \sum_{m \in \mathcal{N}_n} w_{n,m} E_{nm}(d_n, d_m)\},$$

where the first term measures how well the disparity matches with the pair of images. The second term encodes the local smoothness constraint, and it measures disparity differences among local pixels. To avoid imposing the smoothness constraint along the object boundary, where the depth could change significantly, a discontinuity-preserving smoothness constraint based on robust statistics has been proposed. One such solution is to make the local smoothness constraint, that is, the pairwise energy term, depend on the intensity differences. For example, a local penalty function in terms of image gradient is often used to penalize the local smoothness assumption for boundary pixels with larger gradients. The advantages in using an MRF for passive stereo are twofold. First, it allows one to systematically encode the local structure smoothness constraint through pairwise energy function of MRF. Second, through MAP inference, it allows one to jointly infer the disparities for all pixels simultaneously subject to the local smoothness constraint. Finally, it allows one to leverage the latest MRF learning and inference methods.

MRF learning can then can be used to learn the MRF parameters $(\alpha_n, w_{n,m})$. We can thus employ various MRF MAP inference methods to solve for Eq. (5.27) to obtain the disparity map $\mathbf{d}^*$. A variety of algorithms discussed in Section 4.6 can be used to find a (local) minimum, including belief propagation, graph-cuts, sampling methods, variational methods, and simulated annealing. In addition, discrete optimization methods, including linear programming and dynamic programming, are also applied to stereo matching [177,178].

Given two corresponding rows of two rectified images, dynamic programming constructs a matrix of all pairwise matching costs between pixels on the two rows. It then recursively finds a minimal cost path through the matrix of all pairwise matching costs. Although dynamic programming can find the global minimum for independent rows in polynomial time, it has difficulty with enforcing interrow consistency.

Tappen and Freeman [35] applied the above formulation for disparity estimation, where they use the Birchfield–Tomasi matching cost to specify the unary energy function and a variant of the Potts model to specify the pairwise energy function. They applied both the belief propagation and graph cuts methods to perform the MAP inference. For belief propagation, they used both synchronous BP and accelerated BP methods. Their study shows that the graph-cuts method can produce the best result. Fig. 5.44 shows the estimated disparity maps obtained by different inference methods.

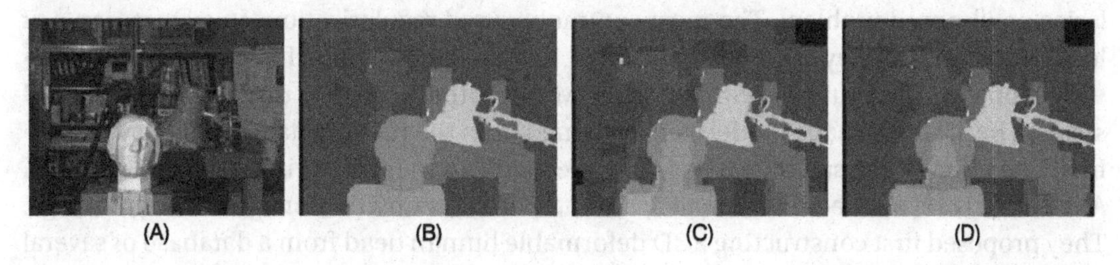

FIGURE 5.44 Disparity estimation on the Tsukuba image obtained by different methods [35]. (A) Tsukuba Image. (B) Graph-cuts. (C) Synchronous BP. (D) Accelerated BP.

The full grid-based MRF models are computationally expensive. Different simplified MRF models have been introduced, including the tree model in [178], where the authors employ the minimum spanning tree method to prune the full grid-based model into a tree and employed dynamic programming to perform MAP inference on the tree model. Their model was evaluated on the Middlebury database. Although their model did not achieve state-of-the-art in accuracy, their method was very fast (a fraction of a second for one image) and achieved a good trade off in terms of accuracy and computational efficiency.

To handle image occlusion and discontinuity, Sun et al. [179] proposed using three coupled MRFs with belief propagation to perform stereo matching. The three coupled MRFs model a smooth field for depth/disparity, a line process for depth discontinuity, and a binary process for occlusion. During inference, instead of doing joint optimization, they breakdown the optimization into three sub-optimizations. They first estimate depth discontinuity and then estimated binary occlusion. Given the estimated depth discontinuity and occlusion, they finally apply belief propagation to obtain the MAP estimation of the smooth disparity map. They also incorporate additional visual cues derived from segmentation, edges, and corners into the unary energy function to improve stereo matching. Finally, they extended their two-frame stereo matching to multiview stereo. Scharstein and Szeliski [180] provided a summary and evaluation of different MRF inference techniques for establishing two-frame stereo correspondences. They provided a detailed summary of the taxonomy of several dense two-frame stereo correspondence methods according to the matching costs, aggregation methods, and optimization techniques. They also quantitatively evaluated the performance of these methods on the Middlebury dataset, using different performance metrics. Their evaluation results show that the global methods based on MRF formulation using either graph-cuts or belief propagation generally outperform other methods.

5.3.4.2 Three-dimensional reconstruction from monocular images

In CV, 3D reconstruction from monocular images is collectively referred to as shape from X. This type of 3D reconstruction includes shape from shading, shape from texture, shape from focus, photometric stereo, and shape from geometry. Three-dimensional reconstruction reconstruction from a single image is challenging and is an ill-posed problem as multiple 3D geometries can produce the same image. Additional cues are needed to regularize an ill-posed problem. The most commonly used regularizations involve variants of local smoothness. They are local and cannot capture global shape information. One possible global regularization is employing PGM to capture the prior distribution of the 3D shape. The 3D prior shape model can be learned from the training data or manually specified. The prior model can then be used as a regularization term during 3D reconstruction. Atick et al. [181] applied this idea to 3D human head reconstruction from a single image. They proposed first constructing a 3D deformable human head from a database of several hundred laser-scanned heads. The deformable model is then incorporated into the shape from shading equation. By solving the constrained shape from shading equation they can derive the coefficients for the deformable model and, hence, the 3D face from a single face image. Although interesting and novel, their method used only a geometric shape model and did not involve a graphical model. In contrast, Saxena et al. [36] proposed using an MRF model to capture the object 3D shape for 3D depth estimation from a single still image. Their model employs a hierarchical MRF prior model at different scales as shown in Fig. 5.45 to capture the global shape prior of the scene. They use a 3D scanner to collect training data, consisting of scene images and their corresponding depth map. Using the collected training data, they learn a multiscale conditional MRF to capture the conditional distribution of the depth given corresponding image features. Specifically, for depth estimation, they first divide an image into rectangular patches and estimate the single depth measurement for each patch. Image features that capture textural variations, texture gradients, and color are extracted at three spatial scales. To capture contextual and global information, image features at nearby patches are also extracted. These features are combined to form a large feature vector of 646 dimensions for each patch. Both relative depth and absolute depth are used to characterize the depth for each patch. They then construct a multiscale Gaussian MRF model as shown in Fig. 5.45 to capture $p(\mathbf{d}|\mathbf{x})$, where $\mathbf{d}$ is the depth map for an image, and $\mathbf{x}$ is the image features. As shown in Eq. (5.29), $p(\mathbf{d}|\mathbf{x})$ consists of two terms. The first term is the unary energy function that is learned from the training data that captures the relationship between depth and image features through a linear regression. The second term captures the local smoothness of depth at different scales.

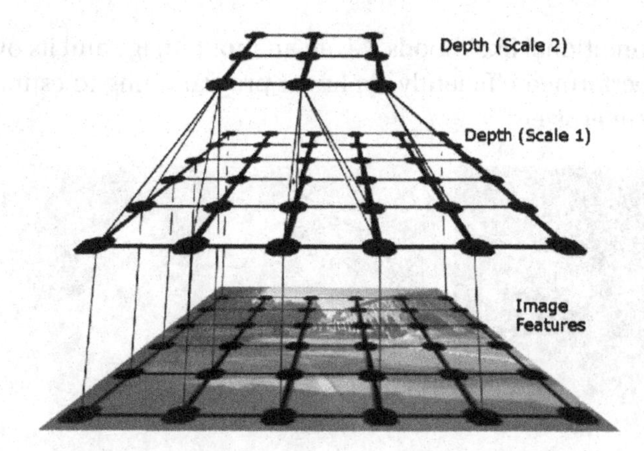

FIGURE 5.45 The multiscale conditional MRF for depth estimation from a single image [36]. The top layers capture the 3D depths in multiple scales, and the bottom layer captures the relationships between 3D depths and their image observations.

$$p_G(\mathbf{d}|\mathbf{x},\theta,\sigma)=\frac{1}{Z_G}\exp\left(-\sum_{i=1}^{M}\frac{(d_i-\mathbf{x}_i^\top\theta_r)^2}{2\sigma_{1r}^2}-\sum_{s=1}^{2}\sum_{i=1}^{M}\sum_{j\in N_s(i)}\frac{(d_i(s)-d_j(s))^2}{2\sigma_{rs}^2}\right). \tag{5.29}$$

The maximum likelihood method is used to independently learn the parameters for the unary and pairwise energy functions. Given the MRF model and an image, depth estimation can be formulated as a MAP inference by maximizing Eq. (5.29). As $p(\mathbf{d}|\mathbf{x})$ is quadratic in $\mathbf{d}$, its maximum is easily found in closed form. Besides modeling $p(\mathbf{d}|\mathbf{x})$ as Gaussian, they also propose modeling $p(\mathbf{d}|\mathbf{x})$ as a Laplacian distribution for relative depth estimation, as relative depth distribution more closely follows Laplacian than Gaussian distributions. They applied their models to reconstructing different outdoor scenes in a variety of environments and compared their methods with a stereo method. Their results show the improved reconstruction accuracy over the stereo method for scenes similar to the training images. For environments different from the training images, the stereo method still outperforms their methods.

Saxena et al. [182,37] proposed a similar learning approach for 3D reconstruction from a single image, where they assume that a scene image is made up of superpixels, and superpixels are generated by the projection of 3D planar surfaces in the scene. As shown in Fig. 5.46, their method started with the oversegmentation of an input image into superpixels. They then use an MRF to encode the parameters (3D location and 3D orientation) of the corresponding 3D planes, their spatial relationships, and the relationships between the 3D planes and superpixels. Specifically, the unary potential of MRF model captures the relationships between the image features extracted from within the superpixels and the corresponding 3D plane parameters, whereas the pairwise potential encodes information about the local connectedness, coplanarity, and colinearity between local 3D planes. To simplify the training, they employ multiconditional learning to train their MRF model. Multi-conditional learning [183] approximates the joint likelihood through a product of several marginal conditional likelihoods. Given an input image and its oversegmentation, MAP inference is performed efficiently via linear programming to estimate the 3D plane parameters of the superpixels.

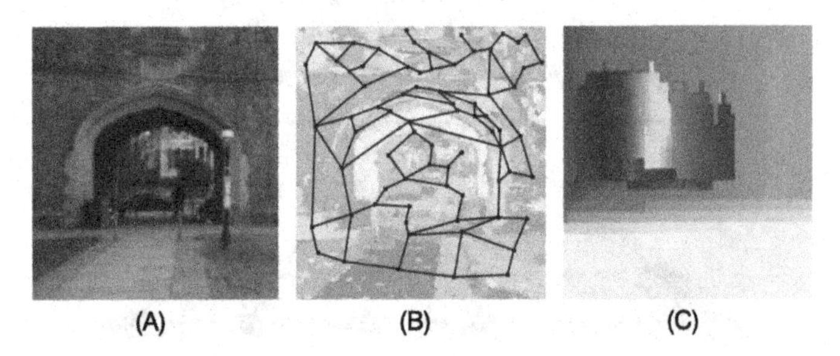

FIGURE 5.46 The superpixel MRF model from [37]. (A) An original image, (B) the over-segmented image in the background, and the corresponding MRF model shown in black edges overlaid, and (C) the estimated depth map for the original image.

Liu et al. [184] proposed a method for 3D depth estimation from a single image using semantic labels. Unlike from other methods that perform depth estimation directly from the image, their method first performs a semantic image segmentation to obtain the labels for objects in the scene. Depth estimation is then performed on the semantically segmented image for each pixel, given their semantic labels. The incorporation of semantic class label information into depth estimation allows one to leverage geometric constraints/priors for each object type to improve depth estimation. For semantic image segmentation, they employed a standard MRF model to perform multiclass (seven classes) labeling of each image pixel. The unary functions are learned using boosted decision tree classifiers over a standard set of 17 filter response features computed in a small neighborhood around each pixel. The pairwise potential captures the mean square-difference between color vectors for neighboring pixels. For the depth estimation, they construct pixel-based and a superpixel-based conditional MRFs, where both the unary and pairwise energy functions are conditioned on the semantic labels. Eq. (5.30) shows the total energy function for the pixel-based MRF, where p refers to a superpixel. A separate unary function is learned for each class. The unary term (first term) measures the compatibility between depth for each pixel and its pointwise depth estimated by a loglinear function of the image features for each semantic class. The standard smoothness prior (the second term) is used to impose the colinearity constraints on the neighboring superpixels (p, q, and r). Liu et al. also include additional energy terms (the third, fourth, and fifth terms) to capture certain class-specific geometric constraints on the depth estimates. These geometric constraints, derived separately for each class, can limit the range of their depth estimates and their orientations. All constraints are encoded into the energy terms as a Huber loss function[4], and standard learning, and inference methods are applied to both MRF models to learn model parameters and to perform MAP inference of the depth estimates. Experiments for depth

[4] A robust regression loss function that reduces the outlier effects.

estimation for outdoor scenes show the improved performance of their method compared to related methods as a result of their use of semantic information, which allows one to produce additional geometric constraints not possible with other methods.

$$
\begin{aligned}
\mathbf{E}(\mathbf{D}, \mathcal{I}, \mathbf{L}) &= \underbrace{\sum_{p} \phi_d(D_p)}_{\text{data term}} + \underbrace{\sum_{pqr} \psi_{pqr}(D_p, D_q, D_r)}_{\text{smoothness}} \\
&+ \underbrace{\sum_{p} \psi_{pg}(D_p, D_g) + \sum_{p} \psi_{pb}(D_p, D_b) + \sum_{p} \psi_{pt}(D_p, D_t)}_{\text{geometry}}. \qquad (5.30)
\end{aligned}
$$

Delage et al. [38] introduced a DBN model for autonomous 3D reconstruction from a single indoor image. The DBN is used to capture some prior knowledge about the scene and resolves some of the ambiguities that are inherent to monocular 3D reconstruction. Their model assumes a "floorwall" geometry consisting of a flat floor and straight vertical walls on the scene and is used for 3D reconstruction from a single image. They further assume that the Y axis of the camera is orthogonal to the floor plane and the camera is located at known height above the floor. Specifically, as shown in Fig. 5.47, a conditional DBN (more precisely, an extended BN) was constructed specifically for more accurate detection of the floor boundary. The DBN captures the conditional joint probability of the floor boundary pixel locations, their image orientations, and their image features given the color of the floor. Each slice of the DBN captures the image data in a column of an image. Given the manually constructed DBN model, the conditional probability for each node is specified as either Gaussian or mixture of Gaussians. The DBN model parameters are learned from training data through a maximum likelihood estimation. Given the model, MAP inference via the junction tree algorithm is employed to find the most likely locations of the floor boundary pixel locations and their orientations. Then given the detected floor boundaries, they apply a projection model plus geometric constraints (planar plane, vertical and parallel lines) to analytically recover the 3D coordinates for both the floor pixels and vertical wall pixels. Experiments with real indoor scenes show robustness and accuracy of their method in reconstructing the 3D indoor scene from a single image. However, their method has some strong assumptions, including the scene consisting of a flat floor and vertical walls, a calibrated camera, known camera position, and so forth.

5.4 PGM for high-level computer vision tasks

In this section, we focus on the application of PGMs to high-level CV tasks. High-level CV tasks focus on interpretation and understanding of events or activities in the images/videos. Typical high-level CV tasks include facial expression recognition, human action/gesture recognition, and complex human activity recognition.

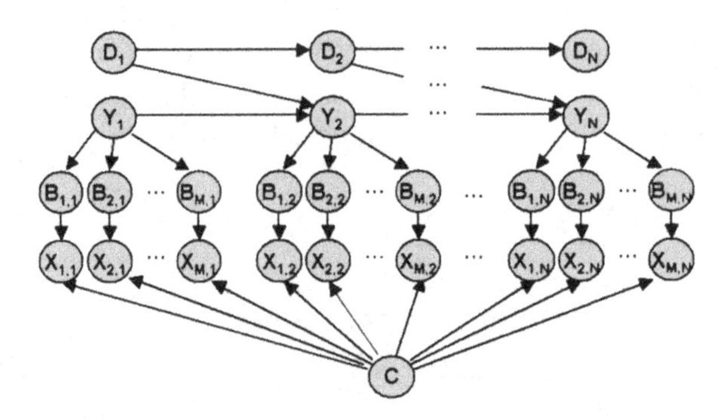

FIGURE 5.47 The DBN model for floor boundary detection [38], where C represents the floor intensity, Y_i denotes the floor boundary position in the ith column of the image, D_i represents the image orientation of the ith boundary, B_{ij} is a binary variable indicating the presence or absence of a boundary pixel at image position (i, j), and $X_{i,j}$ are image features.

5.4.1 Facial expression recognition

Facial expressions are formed by facial muscle contractions. According to the facial action coding system (FACS) [185], there are a total of 44 facial muscles, and their contractions produce different facial expressions. Facial expression recognition can be classified into local facial action recognition and global facial expression recognition. Whereas the local facial action unit (AU) recognition involves detecting the presence (contraction) of a facial muscle near a facial component (e.g., an eye brow) and its degree of contraction (intensity), the global facial expression recognition typically involves recognizing the overall facial expression as a result of simultaneous contraction of multiple AUs. Facial expressions typically include six prototype facial expressions-namely, happy, surprise, fear, sad, disgust, and angry. PGMs may be used to capture dependencies among different facial muscles for both facial action and facial expression recognition.

Recognizing the fact that AUs depend on each other due to both the underlying facial anatomy and the need to form a meaningful facial expression, AUs should be recognized jointly instead of individually to exploit their dependencies. Tong et al. [39] introduced a DBN to capture and encode such AU dependencies and to leverage them for improved AU recognition. As shown in Fig. 5.48, the static part (right side of the figure) of their DBN model captures the spatial dependencies among AUs, where the nodes represent AUs, and the directed links capture the AU dependencies. The dynamic part (left side of the figure) of the model is represented by the self-point arrows and the transition links between nodes in t and $t-1$. The dynamic part captures dynamic dependencies among AUs. Both the DBN structure and parameters are learned using the score-based hill-climbing method introduced in Algorithm 3.7. To incorporate human knowledge about the relationships among AUs, they add a prior structure term to the Bayesian information criterion (BiC) score, as shown in Eq. (5.31), where the first term is the prior structure term, the second term is the likelihood term, and the third term is the penalty term:

$$score(\mathcal{G}) = \log p(\mathcal{G}^0) + \log p(\mathbf{D}|\mathcal{G}, \Theta^*) - \frac{d}{2}\log(N), \tag{5.31}$$

where d is the degrees of freedom of the structure $\mathcal{G}$, and N is the number of training samples. The prior structure term measures the similarity between the learned structure $\mathcal{G}$ and a manually specified prior structure G^0. An appearance-based method is then used to perform individual AU detection to obtain initial AU measurements. Then they attach the image measurements to the corresponding AU nodes in the DBN, as shown by the shaded nodes in Fig. 5.48. Given the initial AU measurements, a MAP inference is then performed through the DBN model to determine the optimal state for each AU. The DBN model allows systematic integration of the AU relationships and AU measurements, leading to more robust and accurate AU recognition.

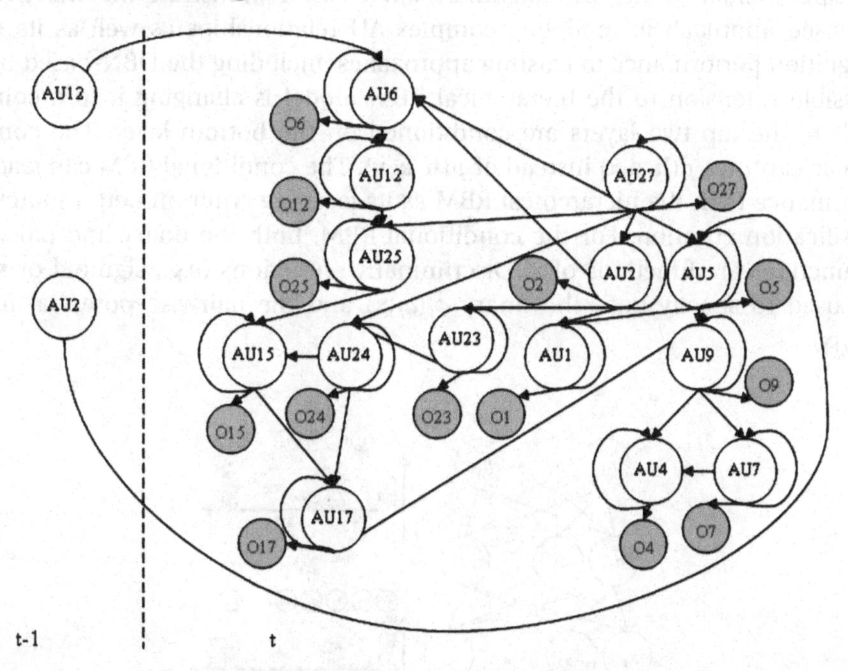

FIGURE 5.48 The DBN for AU modeling and recognition [39]. The nodes represent AUs, and the directed links between nodes capture AU dependencies. The self-arrow for each AU node and the transition links between nodes at time $t-1$ and time t capture the AUs self-temporal evolution and temporal dependencies between different AUs. The shaded nodes represent the image measurements for AUs produced by an independent AU detector.

The DBN models in [39] capture the AU dependencies; typically, they capture the local pairwise dependencies. They fail to capture long-range and global AU dependencies between AUs. To overcome this limitation, Wang et al. [40] proposed a hierarchical RBM model that adds another visible layer at the bottom as shown in Fig. 5.49, where the bottom layer represents the AU image measurements, the middle level represents the groundtruth AUs, and the top level consists of binary hidden nodes. The top two layers form an RBM model. They capture both local and global relationships among the AUs. The bottom two layers form a pairwise label-observation MN that captures relationships between groundtruth AUs and their measurements. The model thus combines the bottom-level image features and the top-level AU relationships to jointly recognize AUs in a principled manner. The total energy function for the hierarchical model, besides the unary and pairwise energy terms for the RBM model, contains an extra pairwise term to capture the interactions between the nodes in the AU measurement layer and AU groundtruth layer. Due to the presence of latent nodes, the parameter learning for the hierarchical RBM model is performed by maximizing the marginal loglikelihood via the contrastive divergence method. During AU recognition, an image-based method was first used to perform AU measurements. An MAP inference is then performed through the model to infer the optimal AU values, given the AU measurements. Experimental results on benchmark databases demonstrate the effectiveness of the proposed approach in modeling complex AU relationships as well as its superior AU recognition performance to existing approaches, including the DBN-based methods. One possible extension to the hierarchical RBM model is changing it to a conditional RBM, where the top two layers are conditioned on the bottom layer. The conditional RBM hence captures $p(\mathbf{h}, \mathbf{a}|\mathbf{x})$ instead of $p(\mathbf{h}, \mathbf{a}, \mathbf{x})$. The conditional RBM can lead to better performance than the hierarchical RBM as its learning criterion better matches with the classification criterion. For the conditional RBM, both the unary and pairwise potential functions are functions of $\mathbf{x}$. Discriminative functions (e.g., sigmoid or softmax) may be used to specify both the unary $\phi_i(a_i|\mathbf{x})$ and the pairwise potential functions $\psi_{ij}(a_i, a_j|\mathbf{x})$.

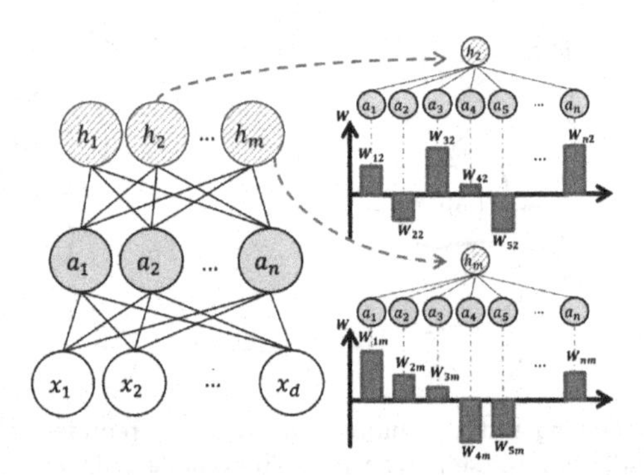

FIGURE 5.49 The hierarchical RMB model for joint facial action unit modeling and recognition [40]. Left: graphical depiction of the model, where the first layer represents the AU measurements, the second layer represents the groundtruth AUs, and the top layer includes hidden nodes that capture global AU dependencies. Right: the captured AU combination patterns of two latent units implied by their parameters.

Besides capturing dependencies among AUs for AU recognition, PGMs can also be used to capture relationships between AUs and facial expressions for global facial expression recognition. Zhang and Ji [41] introduced a hierarchical DBN for facial expression recognition. As shown in Fig. 5.50, the model consists of multilevels, with the top level representing the global facial expression, middle level representing the local facial actions, and the bottom level representing the image measurements of the local facial motions. The structure of the DBN model is manually specified based on Ekman's facial action coding system (FACS) [185]. Specifically, the AUs in the middle level are further divided into primary AUs and axillary AUs, and they are connected to the corresponding facial expressions in the top layer. The image measurements for different facial regions in the bottom layer are connected to the corresponding AUs. The model thus encodes the spatio-temporal relationships between the global facial expressions, local AUs, and their image measurements. Given the structure of the model, maximum likelihood estimation was used to learn the model parameters. Given the DBN model, MAP inference is performed to infer the most likely facial expression given the image measurements of the local facial actions.

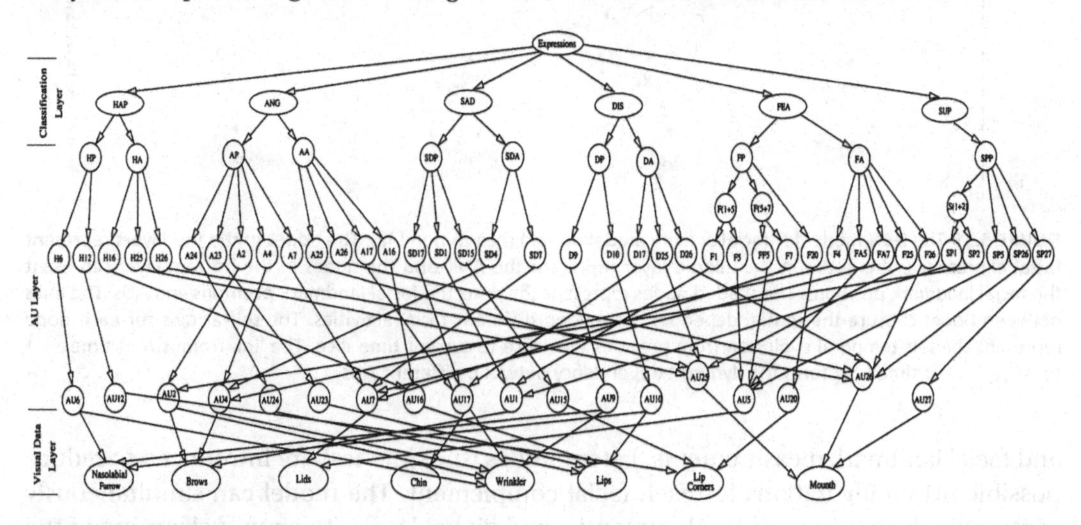

FIGURE 5.50 The DBN model of six basic expressions recognition [41]. The top layer represents the six prototype facial expressions, the middle layer represents the local facial actions units associated with each facial expression, and the bottom layer represents image measurements for different facial regions.

A similar hierarchical DBN model for joint facial activity modeling, tracking, and recognition was proposed in [42]. As shown in Fig. 5.51, their DBN model is organized into three levels. The top layer consists of six nodes that represent the six prototype facial expressions (happy, sad, disgust, angry, surprise, and fear). Each node is binary and represents the presence or absence of the facial expression. With six binary nodes instead of one multivalue node, as in [41], their model does not assume that the six expressions are mutually exclusive but rather that they can coexist. The top layer thus captures the overall facial shape. The nodes in the middle layer represent the AUs as defined in the FACS. Each node is binary, representing whether the AU is present or not. The presence of an AU represents the contraction of the corresponding facial muscle such as lid tightener, eyebrow raiser, and so on. The nodes in the middle layer jointly encode facial shapes for each facial component, that is, eye, eyebrow, mouth, and so on. The nodes in the bottom layer represent the positions of facial landmark points. The shaded nodes represent the AU

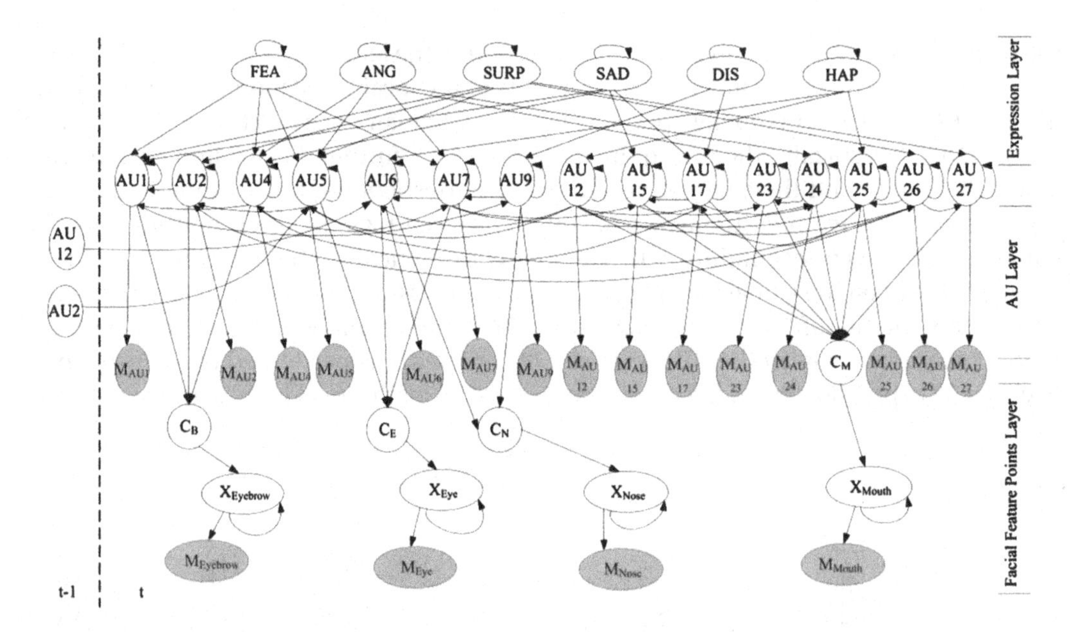

FIGURE 5.51 The DBN model for facial activity modeling and recognition [42]. The nodes in the top layers represent facial expressions, the nodes in the middle layer represent the AUs, and the nodes in the bottom layer represent the facial landmark positions. The shaded nodes represent the detected facial landmark positions and AUs. The links between nodes capture the spatial dependencies among different facial activities. The self-arrows for each node represent the self-temporal evolution from previous time slice to current time slice. The link from AU_i at time $t-1$ to $AU_j (j \neq i)$ at time t captures the dynamic dependency between different AUs.

and facial landmark measurements. Latent nodes (the C nodes) are introduced to reduce possible AU configurations for each facial complement. The model can simultaneously capture the facial shape at local, semantic, and global levels. To simplify learning of the model, part of the model structure is manually specified, whereas other parts are learned from data. Specifically, the structure that relates an expression to AUs, their groundtruth landmarks, and their landmark measurements is manually specified, whereas the semantic and dynamic relationships among AUs are learned automatically from data using the hill-climbing method. Given the structure, parameters are learned by maximizing the likelihood separately for each part. For the part with latent nodes, the EM method was used. During recognition, independent facial landmark point detection and AU detection are first applied to obtain the initial positions of the facial landmark points and initial AU measurements. The detected facial landmark positions and AUs are then used as evidence for the DBN model to perform a joint MAP inference of all three levels of facial activities, recognizing facial expressions, AUs, and tracking facial landmark points.

5.4.2 Human activity recognition

As CV is being increasingly applied to humans, human activity recognition has become an active area of research. Like object recognition, human activity recognition can also be classified into two major approaches, feature-based and model-based. Feature-based human action recognition involves extracting appearance and geometric features, and then performing activity recognition using the extracted features. Appearance features include 2D image features, such as SIFT, HoG, and GIST features extracted from each frame, and 3D space-time image features, such as 3D interest points (e.g., SIFT features) extracted from an image sequence. Geometric features are typically extracted from human silhouettes or skeletons [186,69,187]. They include centroid position, area, filling ratio, and first-order moments. In addition, various dynamic features can also be extracted from the human motion trajectory - including velocity, moving direction, and optical flow - to capture the motion feature at pixel level. Instead of extracting handcrafted features, recent developments involve directly learning spatio-temporal features through deep learning models. Given the features, standard classifiers such as SVM or logistic regressions are often used for activity classification. The feature-based approach to human activity recognition remains a dominant approach in CV. A detailed review of this topic may be found in [188].

The feature-based methods are purely bottom-up and data-driven. They ignore the underlying temporal structures of human activity. In contrast, the model-based approach constructs a model for each activity and then performs activity recognition using the models. Like the object model for object recognition, an activity model captures the underlying spatio-temporal structures of the activities. It represents an activity as a sequence of spatial/appearance transitions over time. PGMs are the dominant approach to model-based activity recognition. A review of model-based human activity recognition can be found in [189,190]. In addition, Herath et al. [191] provide the latest comprehensive review of current research in human activity recognition, including research on the use of both handcrafted features and deep model learned features.

Human activity can be defined at different levels. The lowest level is human gesture, which involves upper-body (hand-arm) motion patterns. The middle level involves full-body motion patterns, and it is referred to as human action/event recognition. Each action pattern is typically composed of a simple cyclic motion such as walking, running, jumping, and so on. Human events also involve interaction between a human and an object. The highest level of human activity is complex human activity, which typically involves executing certain simple human actions in a certain order or in parallel, over a period of time, such as picking up a person or dropping a package. Human activity at each level is composed of essential elements (a.k.a. atomic action or primitives), and the spatio-temporal dependencies among them over time that compose the human activity. For example, for a body gesture, the essential elements are the joint positions of the hand and arm, and their interactions over time to form a specific gesture. For human action, the essential elements consist of some key body poses and their interactions over time. For complex human activity, the essential elements are primitive human actions whose interactions over a period of time form a complex human action. As PGM models can effectively capture the spatio-temporal dependencies among the essential elements, PGMs (varying from static probabilistic models to different types of dynamic probabilistic models) have been widely applied to human activity modeling and recognition at each level.

5.4.2.1 Body gesture recognition

Body gesture can be represented by the interaction among upper-body parts over time. A PGM gesture model should capture such body parts and their interactions. A gesture can be modeled holistically or locally. For holistic modeling, the model consists of hidden state variables that characterize the underlying upper-body shapes and their image measurements. Hence, the model captures the dynamics of the human gesture in terms of their hidden states transition over time and their image measurements. HMMs are ideally suited to holistic gesture modeling. For local gesture modeling, the model may explicitly represent body parts (e.g., upper-body joint positions or their angles), the dependencies of the body parts over time, and their corresponding image measurements. As all variables are observed during training, the DBNs may be used for local gesture modeling. The dynamic models for part-based object tracking discussed in Section 5.3.3.2 can be employed for body gesture recognition. In this section, we will focus on holistic gesture recognition using HMM and its variants.

Before we discuss application of HMMs to gesture recognition, we briefly summarize HMMs. Details may be found in Section 3.7.3.1. An HMM is a special kind of DBN, and it consists of two layers: the top hidden layer $\mathbf{s}$ and the bottom observation layer $\mathbf{o}$ as shown in Fig. 5.52. The hidden layer consists of a sequence of t hidden nodes $\mathbf{s}_t$ representing the

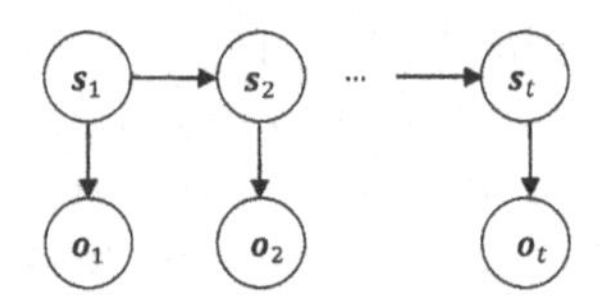

FIGURE 5.52 Hidden Markov model for human action modeling and recognition.

hidden state and the dynamic system, and it starts from $t = 0$. Each hidden node $\mathbf{s}_t$ can take one of K possible values. To each hidden state node, the corresponding image features $\mathbf{o}_t$ are connected, which are extracted from the tth image frame and may be continuous or discrete. An HMM $\mathcal{G}$ can be parameterized by $\mathbf{\Theta} = \{\boldsymbol{\pi}, \mathbf{A}, \mathbf{B}\}$, where $\boldsymbol{\pi} = \{\pi_k = p(s_0 = k)\}_{k=1}^{K}$ represents the prior probability of node s_0. $\mathbf{A} = \{a_{kj} = p(\mathbf{s}_t = k|\mathbf{s}_{t-1} = j)\}$ represents the state transition probability, and $\mathbf{B} = \{b_k = p(\mathbf{o}|s = k)\}$ represents the emission probability.

Given M training image sequences $\mathbf{I}_1^k, \mathbf{I}_2^k, \ldots, \mathbf{I}_M^k$ for the kth human action, where $k \in \mathbf{C}$ with $\mathbf{C}$ representing the classes of human action, we can learn the kth HMM model parameters $\mathbf{\Theta}_k$ using the Baum–Welch algorithm introduced in Section 3.7.3.1.3. During learning, due to the EM nature of the learning, $\mathbf{\Theta}_c$ need be initialized. A clustering method (such as K-means) is often used to determine the number of hidden states and to initialize emission probability. The prior probability and transition probabilities can be initialized to the uniform distribution. This process repeats for each human action class. Given K learned HMMs $\mathbf{\Theta}_k$ and a query video $\mathbf{I}_q$, feature extraction is first performed to generate observation sequence $\mathbf{o}_q$ from the query video. Then the action for a query video can be recognized by evaluating the model likelihood against the query video and identifying the model that produces the highest likelihood:

$$k^* = \arg\max_{k} p(\mathbf{o}_q|\Theta_k). \tag{5.32}$$

The forward and backward method introduced in Section 3.7.3.1.2 can be used to efficiently compute the likelihood.

HMM-based human gesture recognition typically includes three steps: image feature extraction, vector quantization, and model likelihood estimation. Image feature extraction performs feature extraction to produce image features for the body. Vector quantization groups the image features into codes (clusters), which are then served as observations for the HMM. Likelihood estimation identifies the HMM model that produces the highest likelihood. In one of the early studies using HMMs, Yamato et al. [43] introduced an HHM for recognizing six tennis stroke actions (backhand volley, backhand stroke, forehand volley, forehand stroke, smash, and serve). As shown in Fig. 5.53, their method starts with a preprocessing to binarize the image sequences. Then body shape (mesh) features are extracted from the binary images to characterize the body shape. Finally, vector quantization is performed on the mesh features to generate a codeword $\mathbf{o}_t$ for each image at time t.

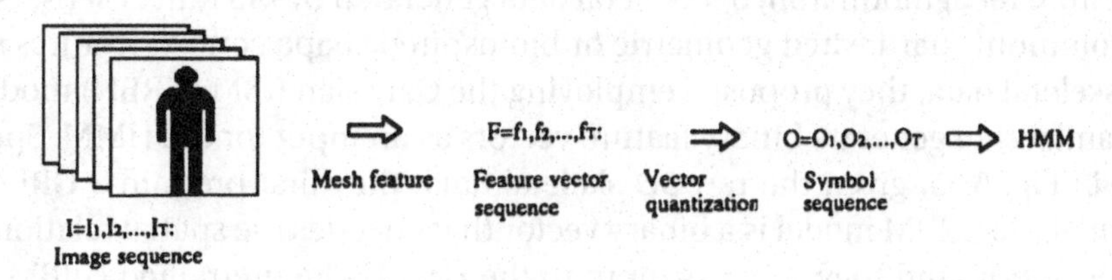

FIGURE 5.53 The processing flow for the tennis action recognition in [43].

The number of hidden states is set to 36. The Baum–Welch algorithm was used to learn six classes of tennis actions, one for each class. The forward and backward procedure was used during recognition to compute the likelihood for each model given a query video. Experiments on a few human subjects show that their method achieves high performance for within-subject action recognition, but its performance drops significantly for across subject action recognition. This may be due, in part, to the limited power of HMMs in capturing the intraclass variation between subjects or to model overfitting as a result of using a large number of hidden states (36) with their model.

In [44] the authors first proposed using a standard HMM to model three Tai Chi gestures: single whip, cobra, and brush knee. As shown in Fig. 5.54A, two three-state HMMs were used to model single whip and cobra gestures, whereas a four-state HMM was used to model brush knee gestures. Geometric shape features are extracted as state measurements. The standard HMMs achieved an average recognition accuracy of 69%. Then they introduced a nonsymmetric coupled HMM (CHMM;Fig. 5.54B) to separately model the left and right arms and to capture their interactions. The best number of hidden states for each HMM in the CHMM was empirically determined for each gesture. The experiments show that, because of explicit modeling of the left and right arm interactions, the nonsymmetric CHMM achieved an average recognition accuracy of 94% - a significant improvement over the standard HMM.

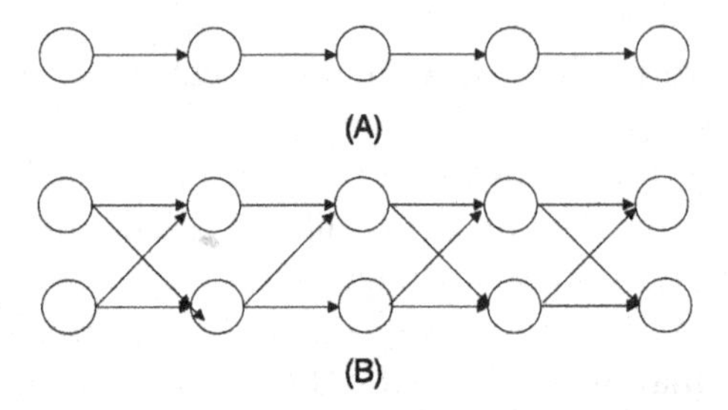

FIGURE 5.54 The standard HMM (A) and the coupled HMMs (B) for Tai Chi gesture recognition [44], where the top layer models the left arm temporal process, and the bottom layer models the right arm temporal process. Note that, for clarity, the figures only show the hidden state nodes and the corresponding observation nodes are omitted.

Besides extracting gesture features from RGB data, features are also extracted from 3D skeletal data for gesture recognition. Wu et al. [45] propose an HMM-based approach to human gesture recognition from 3D skeletal data generated by MS Kinect sensors. Instead of using commonly hard wired geometric or bioinspired shape context features extracted from the skeletal data, they proposed employing the Gaussian RBM (GRBN) model (as the vector quantizer) to generate binary feature vectors as an input for the HMM. Specifically, as shown in Fig. 5.55, given the raw 3D skeletal data, they first pretrain a GRBN model. The output of the GRBM model is a binary vector that encodes the spatial relations among the joint positions, and they serve as input to the HMM. The pretrained GRBM model is then fine tuned by jointly training with the HMM for each action class. Experiments on benchmark datasets show improved gesture and action recognition performance.

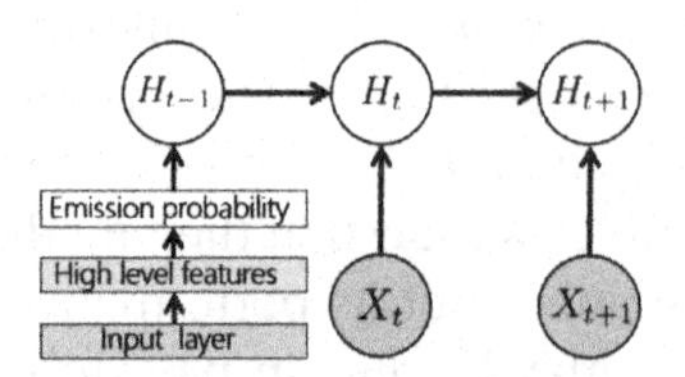

FIGURE 5.55 Combining an HMM with a GRBN for gesture recognition from skeleton data [45].

HHMs have been applied to hand gesture recognition, in particular, for sign language recognition. Sign language recognition focuses on finger gesture recognition. Finger gestures are used to spell words letter by letter. In [46] the authors introduce an HMM to model hand gestures for recognizing sentence-level American Sign Language. Their HMM consists of four states, as shown in Fig. 5.56. Preprocessing is first performed to extract hand blobs, from which 16 geometric features (consisting of 2D coordinates of the hands, axis angle of least inertia, and the eccentricity of the bounding ellipse) are extracted to characterize the shape of each hand. To model long-range temporal dependencies, their model also includes second-order temporal links. The Viterbi algorithm, combined with statistical grammar, is used during recognition.

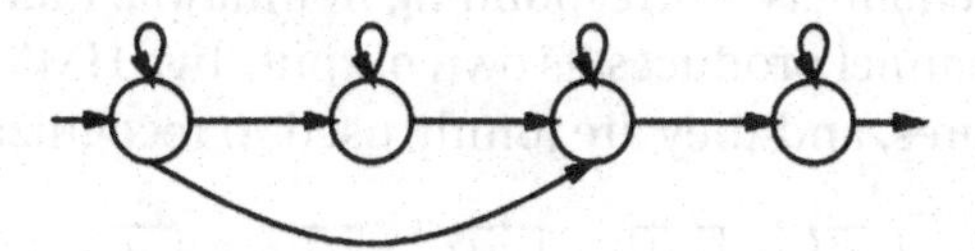

FIGURE 5.56 The four-state second-order HMM used for American sign language recognition [46]. For clarity, only the state nodes are shown; their corresponding observations are omitted.

Yoon et al. [192] introduced an HMM-based method for recognizing planar hand gestures generated in front of the camera. Hand tracking is first performed to generate hand movement trajectories, from which image features (in terms of combined weighted location, angle, and velocity) are extracted. This is followed by C-means clustering to generate the codebook (vector quantization) as input to the 10-state left-to-right HMMs for recognizing hand gestures, consisting of 12 graphic elements (circle, triangle, rectangle, arc, horizontal line, and vertical line) and 36 alphanumeric characters (10 Arabic numerals and 26 alphabets). A similar HMM was proposed in [47] for single-handed dynamic hand gesture modeling and recognition. Kalman filtering was used to track the hand shape over time. Hand shape features extracted from the hand tracker are coded via vector quantization and fed into HMM models with four states for hand gesture recognition. Their model was demonstrated to recognize five hand gestures for robot control in real time. Fig. 5.57 shows their method for hand gesture recognition.

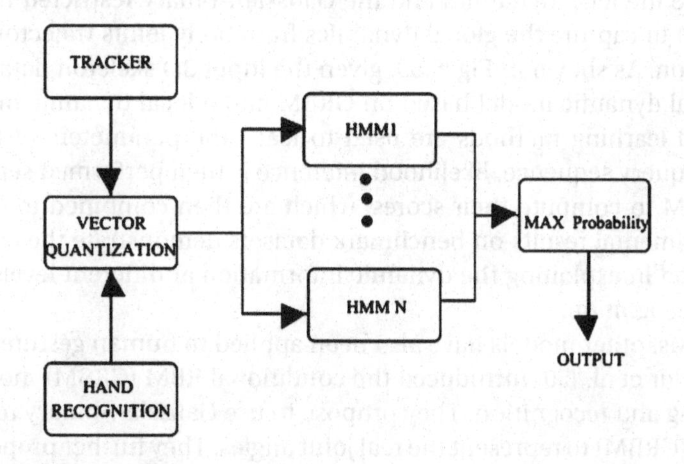

FIGURE 5.57 Hand gesture recognition method in [47].

Conventional HMMs are local and can only model one dynamic process. Sign language typically consists of parallel and interacting processes. To model interacting dynamic processes and to reduce the number of combination states, Vogler and Metaxas [48] introduced the parallel HMMs (PaHMMs), as shown in Fig. 5.58, for American sign language recognition. As an extension to the multiple state multiple observation HMMs (MSMOHMMs) and to the factorial HMMs, PaHMMs factorize the hidden state space into two independent "state channels" corresponding to multiple independent temporal processes, and each state channel produces its own output. Two HMMs are used to model the left and right hand gestures, and they are jointly used to recognize sign language. During

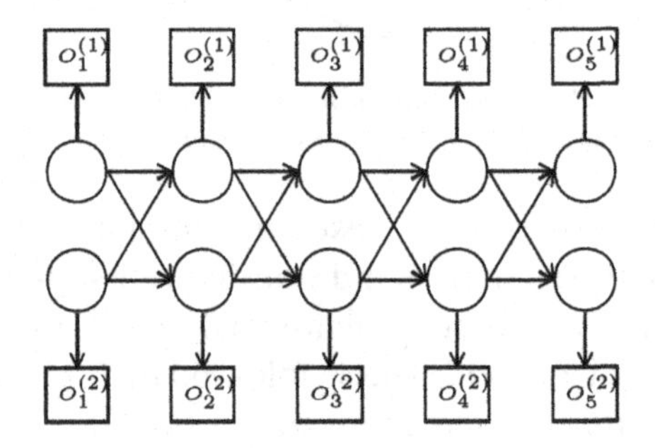

FIGURE 5.58 An Parallel HMM [48], where separate HMMs are used in parallel to model the left and right hand gestures.

recognition, the probabilities from two HMMs are combined by their product as the two HMMs are assumed to be independent of each other.

The conventional HMM only captures the local dynamics for human action modeling and recognition. To overcome this limitation, Nie et al. [49] presented a hybrid dynamic model for human gesture and action recognition. Specifically, they proposed using an HMM to capture the local dynamics and the Gaussian-binary restricted Boltzmann machine (GB-RBM) to capture the global dynamics from body joints trajectories for human action recognition. As shown in Fig. 5.59, given the input 3D skeleton data, their method includes a global dynamic model based on GRBM and a local dynamic model based on HMM. Standard learning methods are used to learn the parameters of the HMM and GRBM. Given a query sequence, likelihood inference is then performed separately for the HMM and GRBM to compute their scores, which are then combined to produce the final score. Experimental results on benchmark datasets demonstrate the capability of the proposed method in exploiting the dynamic information at different levels for improved human action recognition.

Besides HMMs, other models have also been applied to human gesture modeling and recognition. Taylor et al. [50] introduced the conditional RBM (CRBM) model for human motion modeling and recognition. They propose to use Gaussian-binary restricted Boltzmann machine (GRBM) to represent the real joint angles. They further proposed modeling the dynamics by allowing the latent and visible variables at each time step to receive directed connections from the visible variables at the last few time-steps as shown in Fig. 5.60. Specifically, they add two types of connections: autoregressive connections between the visible layers in last few steps and the current visible layer, and the connections from the past visible layers to the current hidden layers. These new connections capture the dynamically changing bias that can influence current dynamics. They call such a high-

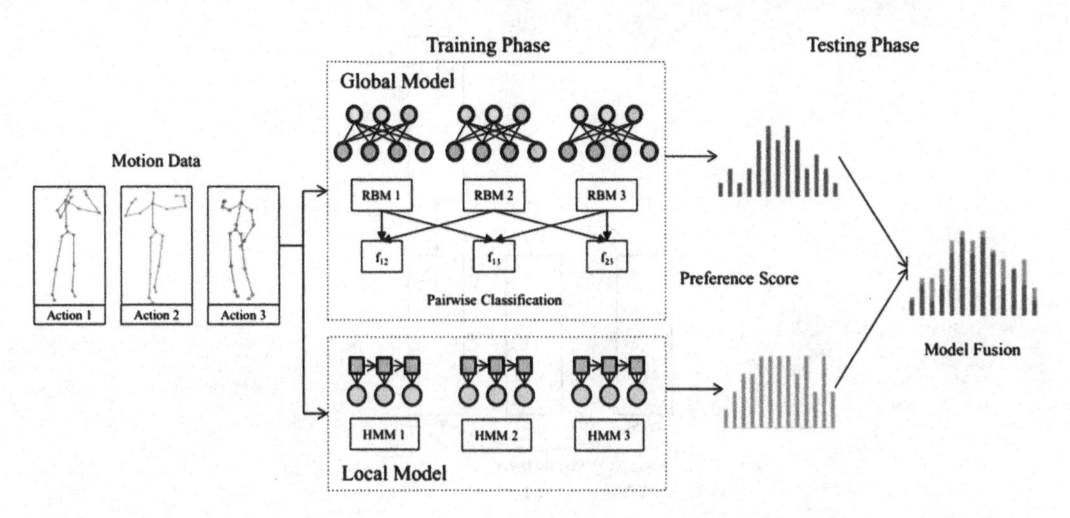

FIGURE 5.59 The hybrid dynamics models [49]. Framework of the proposed method. For each class of action, one GRBM and one HMM are trained to represent the global and local dynamics. The preference scores of the GRBM and HMM are combined to produce the final score.

order autoregressive RBM model a conditional RBM.[5] As a result of these new connections, the energy functions for the CRBM will also be augmented with additional pairwise terms to account for interactions between the current observation layer and past observation layers and between past observation layers and the current hidden layer. Taylor et al. use the contrastive divergence method by maximizing the marginal likelihood to train the CRBM model. Their model was evaluated in terms of human motion synthesis and filling of missing data.

To model the spatial interactions in high-dimensional input data, Nie et al. [51] propose theoretically extending conventional RBMs to explicitly capture the local spatial interactions in the input data. As shown in Fig. 5.61, the visible nodes in the bottom layer of the RBM model at each time slice are represented by an undirected subgraph that captures the spatial dependencies among elements of the input vector. As a result, a new pairwise energy term is introduced to quantify the interactions among the elements of the input vector. The new energy function captures two kinds of data interactions, direct interactions among elements of the input vectors at the same time and indirect interactions through the latent node for input vectors at two consecutive times. A contrastive divergence based learning method was then employed to learn one model for each class. Experimental results based on benchmark databases demonstrate the improved performance of Nie et al.'s algorithm for human action recognition.

[5] Strictly speaking, it should not be called a conditional RBM, as it still captures the joint distribution of latent and observed variables instead of the conditional distribution of latent variables, given the observed variables.

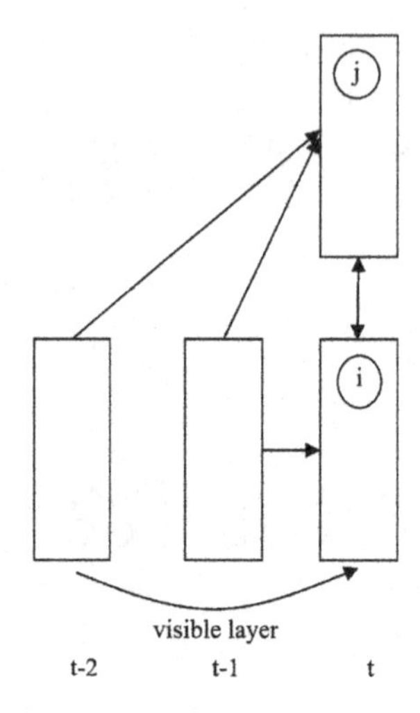

FIGURE 5.60 The proposed CRBM model [50], where the visible nodes in the past time slices are directly connected to both the visible nodes and latent nodes at the current time.

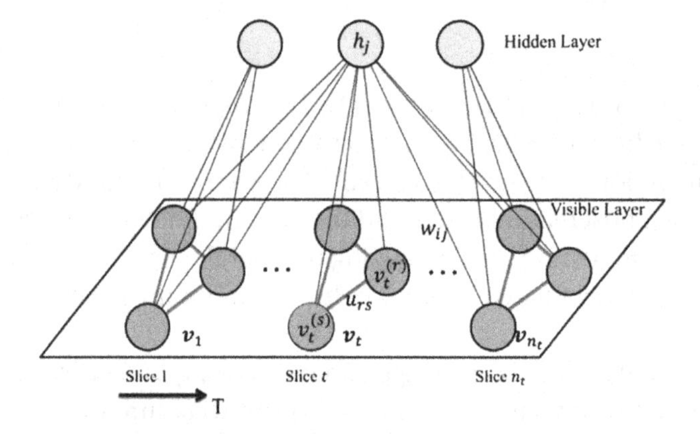

FIGURE 5.61 The extended RBM model [51]. Each hidden unit connects to each visible unit to capture the global temporal patterns (some links are omitted for brevity). The nodes and links in each visible unit capture the elements of the visible vector and their spatial interactions.

5.4.2.2 Human action/event recognition

Whereas gesture recognition focuses on upper-body actions, human action/event recognition focuses on whole-body actions. Human actions also include interactions between humans and between a human and an object. For human action recognition, graphical models may be used to capture the underlying motion structure in terms of spatiotemporal dependencies among action primitives (key body poses). For example, the key body poses for a walking action include leg crossing: right leg forward, and left leg forward. It is their repeated execution in order that generates the walking action. Among different graphical models, HHMs remain the most widely used PGMs models for human action modeling and recognition. The states of the hidden nodes in an HMM represent action primitives that compose the action, whereas the transition probability among hidden states incorporates the action dynamics. The observation node provides the measurement for the action primitives with image features. The power of an HMM lies in its hidden states and its explicit modeling of the state transition and state observations. Methods for human action recognition using HMM mainly vary in the feature representation they use to represent image sequence. The learning and inference methods are more or less the same. As HMMs for whole-body action recognition are similar to those of human gesture recognition, our further discussion will focus on recognizing human actions that involve interactions between humans or between humans and objects.

To represent interactions among multiple entities, Oliver et al. [52] used HMMs to model and recognize the interactions among people. Similar to their work in [44], they proposed using both a standard HMM and a CHMM to model human interactions. Fig. 5.62 shows their standard HMM and the CHMM. Human detection and tracking was used to detect humans as blobs and then extract geometric and dynamic features from the detected human blobs. These image features serve as input to the HMMs. For the CHMM, each chain represents one person. Three or five hidden states are selected for both the HMM and CHMM for different types of interactions, depending on their complexity. During HMM inference, the standard forward–backward method is used to compute the likelihood for a given query video. For CHMM inference, they proposed a revised forward–backward method for likelihood computation. The performance of their models was eval-

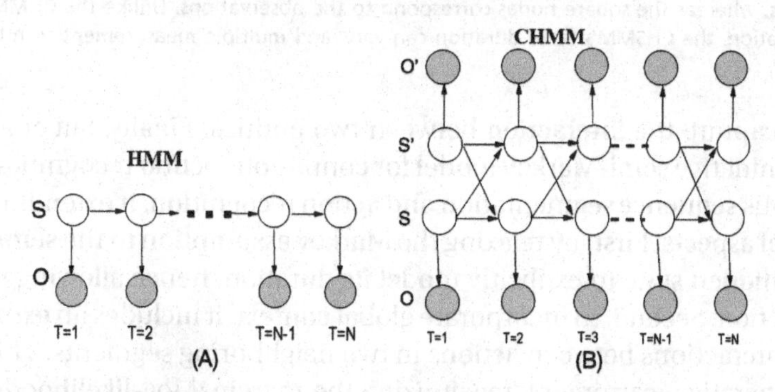

FIGURE 5.62 The standard HMM (A) and the CHMM (B) for modeling and recognizing human interactions [52], where **S** and **S′** represent the hidden states for two interacting dynamic entities, and **O** and **O′** are their image observations.

uated on both synthetic and real pedestrian data. The synthetic data involves five types of interactions between two agents, including: 1) follow; 2) approach, meet, and go separately; 3) approach, meet, and go together; 4) change direction to meet and go together; and 5) change direction to meet separately. The real data consist of three types of pedestrian interactions, including 1) follow, 2) meet and continue together, and 3) meet and split. Their evaluations on synthetic data show that the CHMM achieves perfect recognition of the five interactions, whereas the HMM achieves about 87% recognition accuracy. Similar performance was achieved when evaluating their model on the real data.

To overcome the constant state duration assumption with HMMs, semi-HMMs have been used for action modeling and recognition. Tang et al. [193] adopted a variable-duration HMM, which includes a latent duration variable that models durations of states in addition to the transitions between states. They further employed the max-margin framework during training, which allows for simultaneous discovery of discriminative and interesting segments of video and performing event recognition. In [53], the authors combined the CHMM with semi-Markov models to produce the coupled hidden semi-Markov models (CHSMM) for recognizing human activities involving interactions among different agents. As shown in Fig. 5.63, CHSMM can simultaneously allow variable duration for each

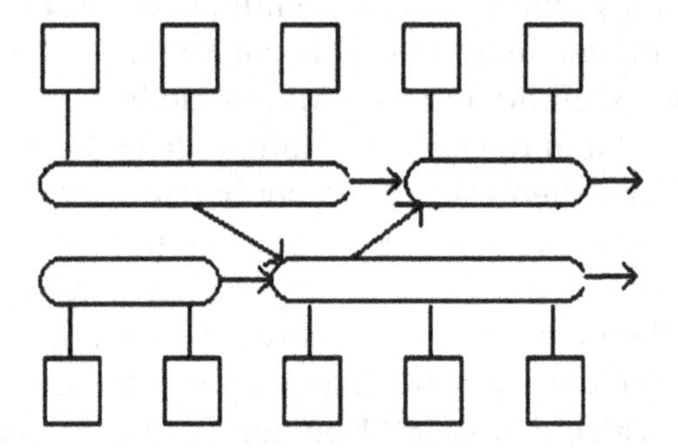

FIGURE 5.63 The CHSMM [53], where the long elliptical nodes of different lengths (different time durations) represent state variables, whereas the square nodes correspond to the observations. Unlike the CHMM, which involves constant state duration, the CHSMM's state duration can vary, and multiple measurements can be produced at a state.

state and can capture the interaction between two entities. Finally, Shi et al. [194] introduced a discriminative semi-Markov model for continuous action recognition by performing simultaneous sequence segmentation and action recognition. It extends the traditional HMM in several aspects. First, by relaxing the Markov assumption to the semi-Markov one, it allows each hidden state to explicitly model its duration, hence allowing variable duration for each action. Second, to incorporate global context, it includes an extra energy term to model the interactions between actions in two neighboring segments. Third, instead of performing generative learning by maximizing the marginal log-likelihood, Shi et al. introduce a discriminative learning method that maximizes the conditional log-likelihood, which is implemented as maximizing the margin between the groundtruth label and incorrect labels. Finally, for inference, they introduce a Viterbi-like dynamic programming algorithm that simultaneously finds the labels for all segments by maximizing the conditional loglikelihood. They evaluated the performance of their methods on different human action datasets, including segment-based datasets such as KTH and MoMo and their own continuous dataset WBD.

Besides HMMs and their variants, other generative models such as DBNs [54,195,55, 196] have also been applied to human action modeling and recognition. Compared to HMMs, DBNs can model more general temporal relationships among different entities of the activity. In addition, with their flexible structure, DBNs are more expressive in representation and afford greater model design flexibility. Luo et al. [54] employed a discrete-state DBN model unrolled in two time slices, as shown in Fig. 5.64 for sports event modeling and recognition. At each time slice, there are five state nodes and four observation

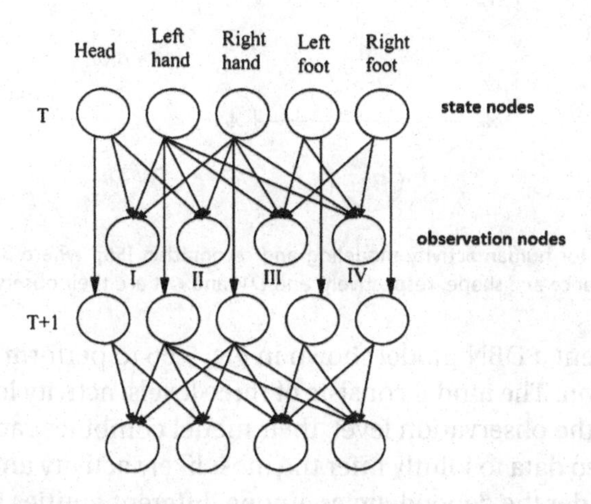

FIGURE 5.64 A DBN model (unrolled in two time slices) for human body motion modeling and recognition [54]. Each time slice has five state nodes to represent the states of different body parts (head, hands, and feet) and four observation nodes to represent the image measurements of the state variables.

nodes. The state nodes respectively represent states of head, hands, and feet. Each state node has four values representing their position relative to the center of gravity: upper-right, upper-left, lower-left, and lower-right. The four observation nodes are shared by the state nodes and represent mean x and y coordinates for each of the four states. Different from a CHMM, which models interactions among body parts directly through state variables, Luo et al.'s model captures the body part interactions indirectly through shared observations. Their model can thus be treated as a variant of a CHMM. A Gaussian distribution is used to model the CPD for each observation node. The maximum likelihood method is used to learn the DBN model parameters. For efficient inference, they converted the DBN model to the corresponding HMM and use the HMM forward–backward procedure to perform both likelihood and decoding inference. Their method was evaluated on sports actions, including bowling, downhill skiing, golf swing, pitching, and ski jumping.

Wang and Ji [55], proposed using a DBN model, as shown in Fig. 5.65, for human action recognition. The model consists of two levels, the feature level and the state level. Whereas the feature level encodes the observations from the images, the state level abstracts the basic states of the activity. It includes two hidden states V and A to model the underlying shape and appearance dynamics, respectively. The two hidden states have their own observations OV and OA, which are causally coupled, effectively forming a CHMM. Instead of learning the DBN model generatively by maximizing the likelihood for each model independently, they proposed learning the DBN models jointly and discriminatively by maximizing their joint conditional likelihood. Their experiments on the KTH dataset show the superior performance of the discriminatively learned DBN models to the generatively learned DBN models.

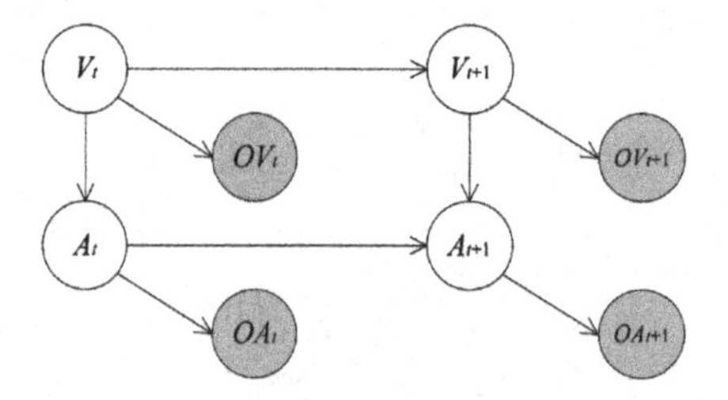

FIGURE 5.65 A DBN model for human activity modeling and recognition [55], where A and V are state variables representing object appearance and shape, respectively, and OV and OA are their observations.

Wu et al. [56] present a DBN model shown in Fig. 5.66 to perform simultaneous object and activity recognition. The model consists of three levels: activity level, object level, and observation level. At the observation level, their model combines radio-frequency identification data and video data to jointly infer the most likely activity and object labels. Their model does not consider the dependencies among different entities at the same level and assumes that the object level nodes are latent. Their model can be viewed as a layered HMM with multiple observations, where the bottom two layers are treated as an HMM with multiple observations, and the top layer provides the activity label. The EM algorithm is used to learn the parameters for the bottom two layers, and the parameters for the top layer are manually specified. With the top label layer, their model can recognize multiple activities with one model.

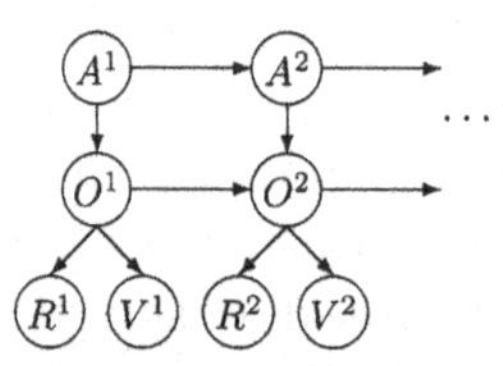

FIGURE 5.66 A DBN model for simultaneous object and activity recognition [56]. The model consists of three levels: activity level (A), object level (O), and observation level.

Zhang and Ji [57] introduced a dynamic chain graph for modeling and recognizing complex human actions, involving interactions between two different subjects. As an extension to DBNs, which include only directed links, the dynamic chain graph consists of both directed and undirected links. Specifically, the dynamic chain graph consists of a static chain graph as shown in Fig. 5.67A, and a transition chain graph as shown in Fig. 5.67B. The static chain graph contains both directed and undirected links, which are used to capture both causal and noncausal spatial/semantic relationships among elements of a human action at a time. Specifically, the static chain graph is composed of two mixed graphs (i.e., a graph with both directed and undirected links), one for each subject. Within each mixed graph, the subject action (a_1 or a_2) is represented by its states, including shape s, appearance p, and motion m. The interactions among state variables are captured by undirected links, whereas the interactions between state variables and their measurements M are captured by directed links. Furthermore, the interactions between the two mixed graphs are captured by undirected links to represent their mutual dependencies. The transition chain graph, consisting of a two-slice chain graph at two consecutive times, models the dynamic evolution of an activity. The evolution is captured by the directed temporal links between corresponding nodes at time t and $t-1$ to capture the temporal causality. Learning and inference methods were introduced to learn the dynamic chain graph. For human action recognition, they learn one dynamic chain graph for each class of activity, and the activity for a query video is determined by finding the chain model with the highest likelihood.

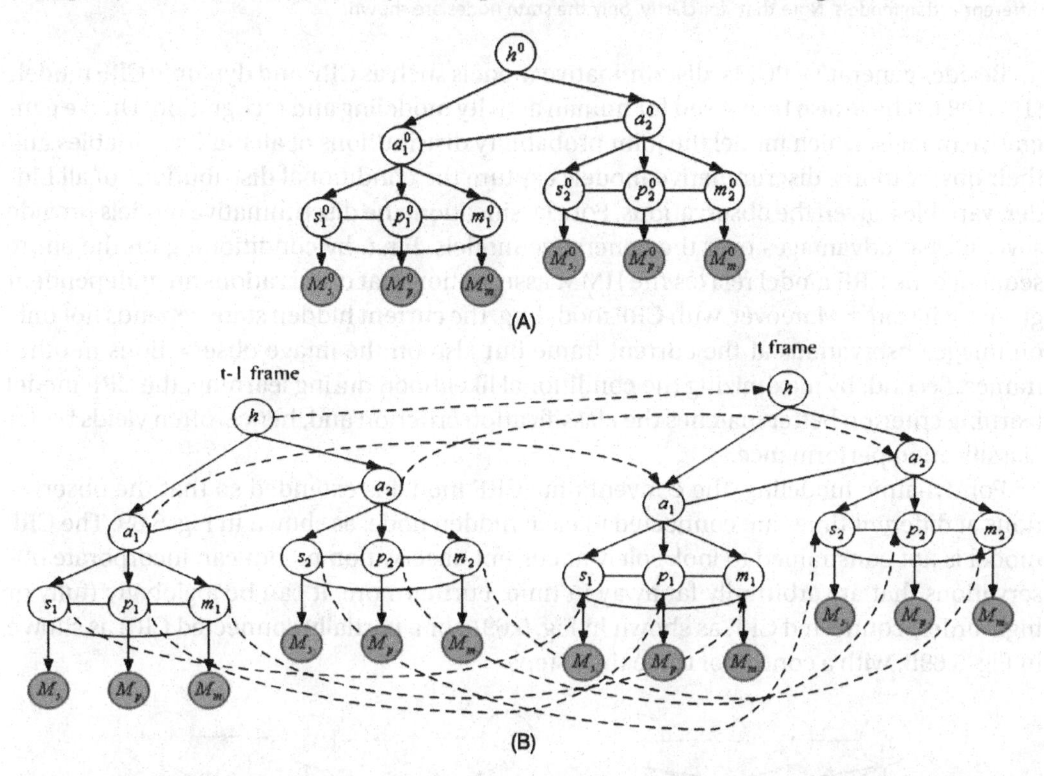

FIGURE 5.67 A dynamic chain graph for human activity modeling and recognition [57]. (A) Structure of the static (prior) model and (B) structure of the transition model.

Lv et at. [58] introduced a two-layer DBN action model to model a single human action. Like an HMM, an action model has two layers: the top layer consists of state nodes that represent key body poses that make up an action, as shown in Fig. 5.68A, and the bottom layer's nodes represent their corresponding image measurements (note: the bottom layer is omitted from the figure). The temporal links between the key body pose nodes represent the temporal transition from one key body pose to another over time. Unlike an HMM, where the hidden nodes have no semantic meaning, the state nodes in Lv et at.'s action model represent key body poses. They further propose combining action models together to form an action net (Fig. 5.68B) for recognizing continuous human action or complex human action that is composed of several primitive human actions.

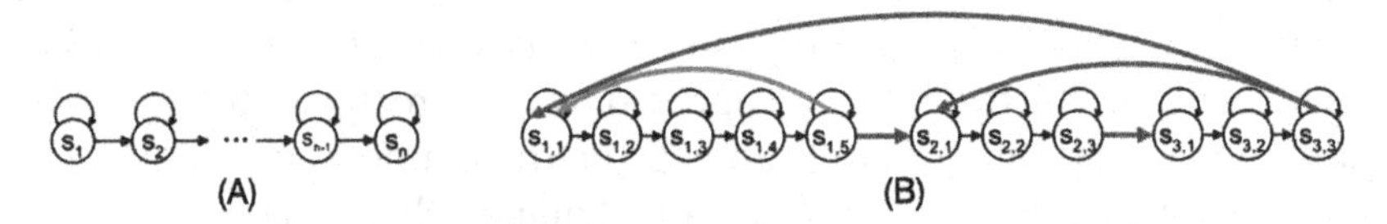

FIGURE 5.68 Action graph models [58]. (A) The action model and (B) the action net model consisting of three action models connected in series [58], where the blue (dark gray in print version) links represent the transition between different action models. Note that, for clarity, only the state nodes are shown.

Besides generative PGMs, discriminative models such as CRF and dynamic CRF models [197,198,61] have also been used for human activity modeling and recognition. Unlike generative models, which model the joint probability distributions of all hidden variables and their observations, discriminative models capture the conditional distributions of all hidden variables, given the observations. For classification, the discriminative models provide several clear advantages over the generative models. First, by conditioning on the entire sequence the CRF model relaxes the HMM assumption that observations are independent given their states. Moreover, with CRF modeling, the current hidden state depends not only on image observations at the current frame but also on the image observations in other frames. Second, by maximizing the conditional likelihood during learning, the CRF model learning criterion better matches the classification criterion and, hence, often yields better classification performance.

For dynamic modeling, the conventional CRF model is extended so that the observations at different times are connected to each hidden node, as shown in Fig. 5.69. The CRF model is not constrained to look solely at current observation $\mathbf{o}_t$, but can incorporate observations that are arbitrarily far away in time. Furthermore, it can be a globally (fully or high-order) connected CRF, as shown in Fig. 5.69A, or a partially connected CRF, as shown in Fig. 5.69B, with a context of three time steps.

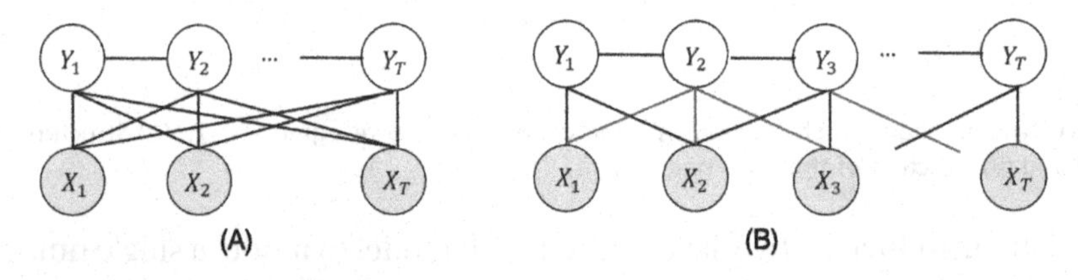

FIGURE 5.69 Dynamic CRF models. (A) A fully connected CRF, and (B) a locally connected dynamic CRF with a context of three time steps, as shown by the red (dark gray in print version) links.

Through either full connections or partial connections, CRFs overcome the HMM assumption that observations at different times are independent, given their state by allowing observations at different times to influence the state at each time. Specifically, the global CRF models allow one to use the observations for the entire sequence, whereas the local CRF models exploit local neighboring observations. For human activity understanding, the additional contextual information provided by observations from other times is critical for correctly resolving locally ambiguous classes.

Vail et al. [59] used a linear chain CRF (Fig. 5.70A) for action recognition. Their CRF model captures the conditional distribution of labels **y** given their observations **x**, which can be parameterized as

$$p(\mathbf{y}|\mathbf{x}) = \frac{1}{Z}\sum_{t=1}^{T}\exp(\mathbf{w}^\top E(y_t, y_{t-1}, \mathbf{x})), \tag{5.33}$$

where **w** is the weight vector, and $E(y_t, y_{t-1}, \mathbf{x}))$ is the energy function, which can be further divided into pairwise and unary potential functions. Given the parameterizations, model parameters **w** are learned by maximizing Eq. (5.33) through gradient ascent. As a discriminative counterpart of a standard HMM, a linear chain CRF is advantageous when the features violate the independence assumption. Inference is performed by either filtering, (i.e., estimating the action label y_t at time t given all previous observations $x_{1:t}$) or decoding (i.e., obtaining the labels **y** for a query sequence **x**).

To model the interactions among multiple objects, Wang et al. [60] introduced a factorial CRF (FCRF) model. As shown in Fig. 5.70B, like a factorial HMM, an FCRF has multiple linear chains of hidden nodes with connections between temporal nodes to represent distributed hidden states and their interactions. However, unlike a factorial HMM, an FCRF, by connecting hidden states to not only current observations but also to past observations, does not assume that observations are independent, given the hidden states. Following the temporal CRF model parameterizations in Eq. (5.33), the conditional distribution of the label for the FCRF model can be parameterized as

$$\begin{aligned} p(\mathbf{y}|\mathbf{x}) &= \frac{1}{Z}(\sum_{t=1}^{T}\sum_{l=1}^{L}\exp(\mathbf{w}_l^\top E(y_{t,l}, y_{t-1,l}, \mathbf{x})) \\ &\quad + \sum_{t=1}^{T}\sum_{l=1}^{L-1}\exp(\mathbf{w}_l'^\top E(y_{t,l}, y_{t,l+1}, \mathbf{x})), \end{aligned} \tag{5.34}$$

where the first term captures the potential within the same state chain l, and the second term captures the potential between two state chains. The parameters **w** and **w**′ can be learned by maximizing the conditional likelihood. Action recognition can then be performed by labeling an input sequence through the MAP inference, that is, $\mathbf{y}^* = \arg\max_{\mathbf{y}} p(\mathbf{y}|\mathbf{x}, \mathbf{w}, \mathbf{w}')$.

To further improve the CRF model representation power for human actions, the authors in [62,199] embed latent variables layer **h** into the temporal CRF model in Fig. 5.70A, yielding the so-called hidden CRF (HCRF) for human action recognition, as shown in Fig. 5.71. With the introduction of the latent variables **h**, the label conditional distribution can be

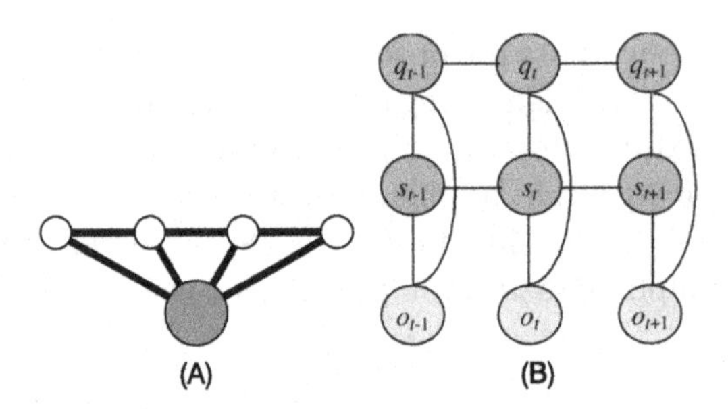

FIGURE 5.70 (A) linear chain CRF [59], where the hidden states (white nodes) in the top layer are connected to all observations (the shaded node); (B) Factorial conditional random field [60,61], which consists of two interacting hidden state layers (blue (mid gray in print version) nodes). Note that, for clarity, only state links to their current observations (green (light gray in print version) nodes) are shown.

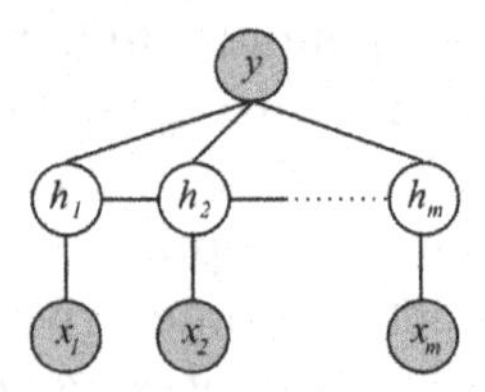

FIGURE 5.71 A hidden (hierarchical) CRF model [62], where the white nodes represent the hidden states, and shaded nodes represent the observations and output labels.

expressed as

$$p(\mathbf{y}|\mathbf{x},\boldsymbol{\theta}) = \sum_{\mathbf{h}} p(\mathbf{y}|\mathbf{x},\mathbf{h},\boldsymbol{\theta})p(\mathbf{h}|\mathbf{x},\boldsymbol{\theta}). \tag{5.35}$$

By assuming disjoint sets of hidden states associated with each class label, $p(y = j|\mathbf{x}, \mathbf{h}) = 0$ for $\mathbf{h} \notin \mathbf{H}_j$, and Eq. (5.35) is simplified to

$$p(\mathbf{y}|\mathbf{x},\boldsymbol{\theta}) = \sum_{\mathbf{h}\in\mathbf{H}} p(\mathbf{h}|\mathbf{x},\boldsymbol{\theta}), \tag{5.36}$$

where $\mathbf{h}_j$ are the hidden states belonging to action label j, and $\mathbf{H}$ represents the union of hidden states for all action classes. Then they follow Eq. (5.33) to specify $p(\mathbf{h}|\mathbf{x})$. The parameters $\boldsymbol{\theta}$ are learned by maximizing the posterior distribution of the parameters, whereby a Gaussian prior is used to specify the parameter prior. For action recognition, given a query sequence $\mathbf{x}$, MAP inference is performed to estimate the most probable label sequence $\mathbf{y}^*$ that maximizes $p(\mathbf{y}|\mathbf{x},\boldsymbol{\theta})$. Wang and Mori [200] further improved the HCRF model's performance by introducing a max-margin learning method, leading to the so-called max-margin hidden conditional random field (MMHCRF).

Aside from dynamic models, static models have also been used for human action recognition. Filipovych et al. [63] use a BN to capture the relationship between actor and object, as shown in Fig. 5.72. The model consists of four types of nodes: temporal state node M, static state nodes S, and their observations O_M and O_S. In this model, different static states of the actor and objects interacted with each other through the temporal state node. One BN model is learned for each dynamic scenario. Given an image sequence, scenario recognition is formulated as a detection problem by identifying the spatio-temporal location in the video sequence that maximizes the posterior probability of an action model.

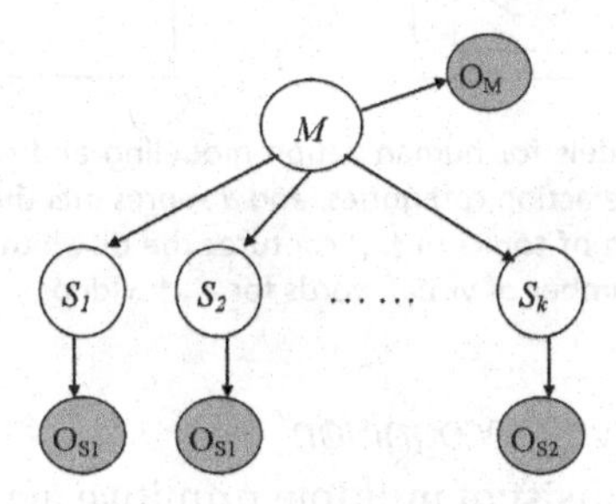

FIGURE 5.72 BN structure of the actor-object model, where the S and M nodes represent the static states and dynamic states of the actor [63].

Hierarchical Bayesian models, including the latent Dirichlet allocation (LDA) and the probabilistic latent semantic analysis (pLSA), have also been employed for unsupervised human action learning and recognition. Niebles et al. [64] proposed using both the pLSA and LDA for human action modeling and recognition. Their goal is to automatically learn different classes of actions from a set of unlabeled videos and employ the learned model to perform action categorization and localization in the query video sequences. They assume that the human actions are composed of a set of spatio-temporal visual words, which can be learned using bag-of-word method. The distribution of these visual words form different human actions (topics). Specifically, as shown in Fig. 5.73, the pLSA model takes bag of words w as inputs and assumes that the word distribution in a video sequence d can be represented by a weighted combination of word distributions for each topic $p(w|z)$ with the weight computed by $p(z|d)$. Given the unlabeled video sequences, EM is used to learn the number of topics (the number of states for z), the word distribution for each topic $p(w|z)$, and the mixture weight (the topic distribution for each document) $p(z|d)$. Each of the latent topics corresponds to one human action. During recognition, the action for a query sequence can be recognized as the action topics in z (a state of z) that produces the highest probability $p(z|d)$. The LDA extends the pLSA by treating the parameter θ of topic distribution in documents ($p(z|\theta)$) as a random variable with distribution $p(\theta|\alpha)$ parameterized by hyperparameters α. As a result, unlike the pLSA, which captures the joint distribution $p(d, z, w)$, the LDA captures the joint distribution of $p(\theta, z, w)$. Given a set of unlabeled training sequences, learning is performed to obtain the hyperparameters α and the word-topic distribution parameter β. LDA learning is often computationally expensive, and approximate methods, such as the variational method are often used. During recognition, given a testing image sequence, the LDA first estimates the parameter θ, based on which it can identify the topic (action) with the highest probability $p(z|\alpha, w)$.

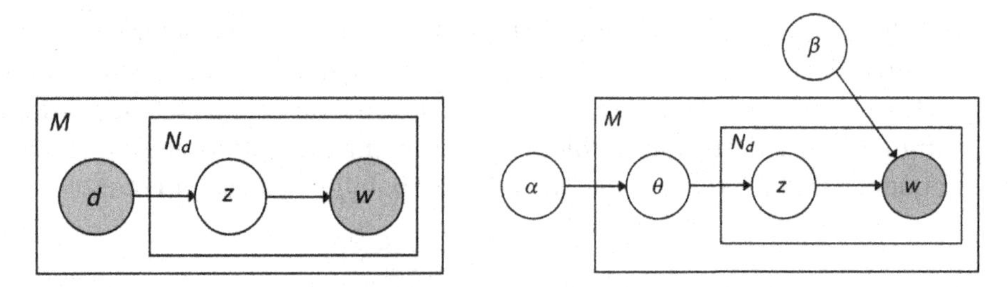

FIGURE 5.73 Hierarchical Bayesian models for human action modeling and recognition [64]. (A) The pLSA model, where w is the input visual words, z are action categories, and d represents the video sequence. (B) The LDA model, where θ parameterizes the distribution of topics in z, β captures the distribution of words for each topic in z, M is the number of videos, and N_d is the number of visual words for each video.

5.4.2.3 Complex human activity recognition

Complex activities typically consist of multiple primitive actions happening in parallel or sequentially over a period of time. Complex activities are hence structurally defined by the temporal relationships among activity primitives. Understanding complex activities requires recognizing not only each individual action primitive but, more importantly, capturing their spatio-temporal dependencies over different time intervals. Hence the key in complex activity recognition is modeling the structural relationships among primitive actions over time.

Hongeng and Nevatia [65] employ a hierarchical representation to model and recognize composite actions (e.g., theft in a phone booth, and attack and chase). As shown in Fig. 5.74, they assume that a complex action consists of interacting primitive actions over a period of time, whose temporal relationships can be modeled by a semi-HMM. By em-

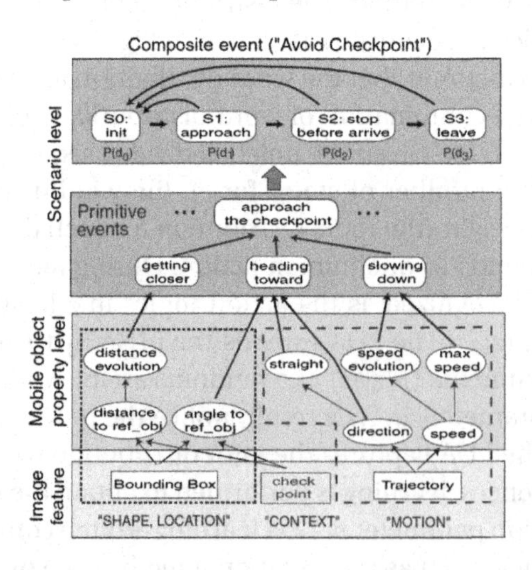

FIGURE 5.74 Hierarchical representation of a composite action "avoid checkpoint" [65], where Bayesian Networks are used to recognize primitive human actions and SHMMs are used to capture the interactions between primitive actions for modeling and recognizing complex activities.

ploying the semi-HMM, they can explicitly model the state duration variations for different primitive actions. Their method starts with image feature extraction in the bottom layer, based on which motion properties of objects in the video are estimated. The properties of the detected moving objects are then fed into a BN that performs primitive action recognition. The output of the BN serves as the inputs for the semi-HMM at the top to model complex human actions. One such hierarchical model is constructed for each complex action. Complex actions are recognized by identifying the model that produces the highest likelihood. Sun and Nevatia [201] introduce a similar structure for complex activity modeling and recognition. They assume that a complex activity is composed of primitive activity concepts and their interactions over time. The hidden node is used to represent the primitive activity. Their model was evaluated on the TRECVID MED 11 Event Kit dataset [202] for recognizing 15 complex human actions/events, and it outperformed the state-of-the-art methods.

Zhang et al. [66] proposed a two-layer HMM for unusual event detection. As shown in Fig. 5.75, the model consists of two levels. The basic events (primitive actions) are modeled by sub-HMMs. Upper level complex activity is modeled as an ergodic K-class HMM, where the hidden states are the basic events (primitive actions).

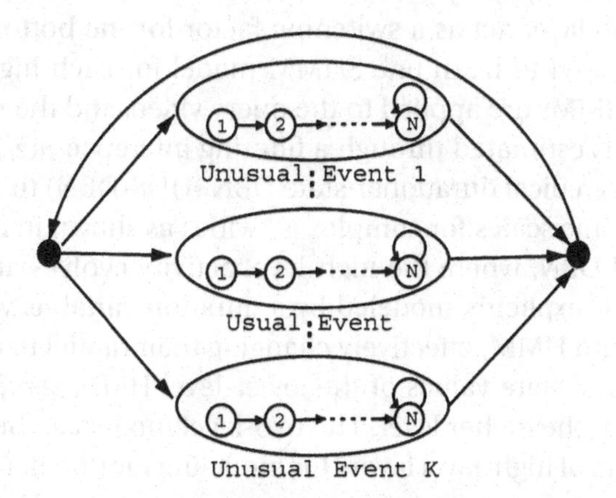

FIGURE 5.75 A two-layer HMM for unusual event detection [66], where the lower-level HMM is used to recognize primitive human actions. The higher-level HMM, taking outputs from the lower-level HMM, models the complex activities.

Duong et al. [67] introduced the switching hidden semi-Markov model (S-HSMM) for modeling and recognizing complex office activities. As shown in Fig. 5.76, the model consists of two layers. The top layer nodes Z_t represent six high-level activities, including 1) entering-the-room and making breakfast, 2) eating-breakfast, 3) washing-dishes, 4) making-coffee, 5) reading-morning-newspaper and having coffee, and 6) leaving-the-room. The complex activities are composed of atomic activities. For example, activity "at-stove" is an atomic activity for complex activity making-coffee. The x_t nodes in the bottom layer represent the atomic activities and their duration ϵ_t using hidden semi-Markov

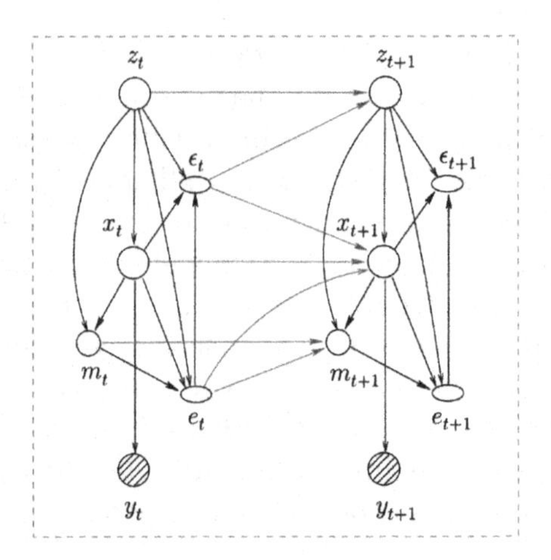

FIGURE 5.76 A DBN representation of the switching hidden semi-Markov model [67], where $\mathbf{z}_t$, $\mathbf{x}_t$, and y_t represents complex activities, the current atomic activities, and the image measurements of x_t, respectively. ϵ_t represents the state duration.

models with the duration explicitly modeled by Coxian distribution.[6] Hence the complex activity node in the top layer act as a switching factor for the bottom-layer model. DBN learning methods are used to learn one SHMM model for each high-level activity. During recognition, the SHMMs are applied to the query video, and the most likely high-level activity at each time t is estimated through a filtering inference $p(z_t|y_{1:t})$. Similarly, Du et al. [68] proposed a hierarchical durational-state DBN (HDS-DBN) to model two stochastic processes at different time scales for complex activities as shown in Fig. 5.77A. The model consists of a two-level DBN, where the high-level activity evolves at a slower time scale, and its state duration is explicitly modeled by a duration variable, whereas the low-level activity uses the standard HMM, effectively changing their model to a hierarchical hidden semi-Markov model. The state values of the lower-level HMM serve as the observations for the state variables at the higher level. The two-level model can be further extended by adding additional layers of high-level layers for modeling motion details at different scales as shown in Fig. 5.77B.

Xiang et al. [69] propose a dynamic multilinked hidden Markov model (DML-HMM) for modeling complex activities that involve interactions among multiple primitive dynamic events. As shown in Fig. 5.78, DML-HMM consists of a set of interconnected HMMs, with one HMM for each dynamic event. Unlike the CHMM, which manually connects the hidden states among different HMMs, the DML-HMM employs DBN structure learning to automatically discover the necessary temporal links among the constituent HMM processes, resulting in sparse and more optimized connections among a subset of relevant hidden state variables across multiple temporal processes (HMMs). This property provides DML-HMMs with more efficient computations during both learning and inference.

[6] A probability distribution consists of a mixture of exponential distributions of different rates.

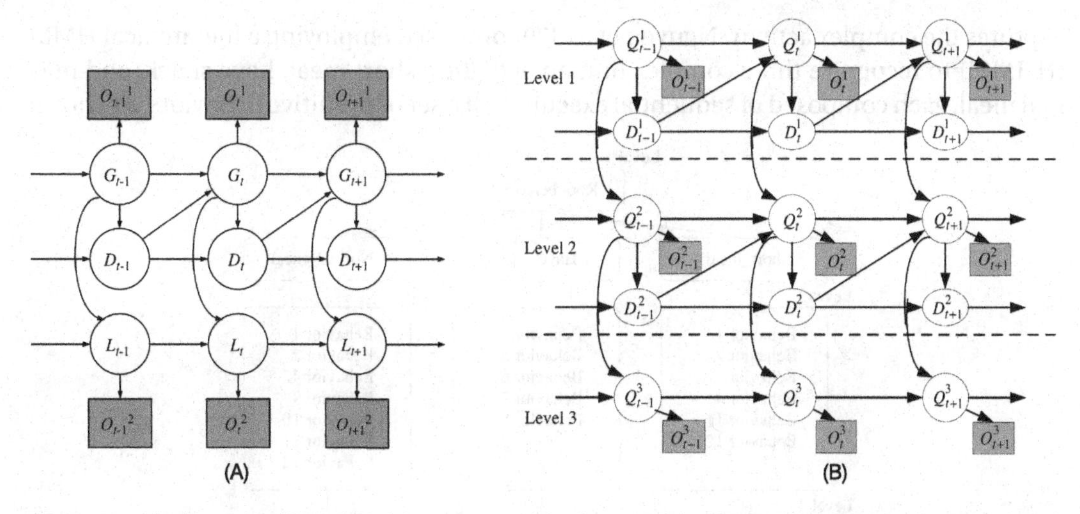

FIGURE 5.77 The hierarchical DBN [68]. (A) An HDS model, where G and L are the global and local state variables, respectively. D is the duration variable for the global state, and Os are the observations. (B) The extension of HDS to three levels of motion details.

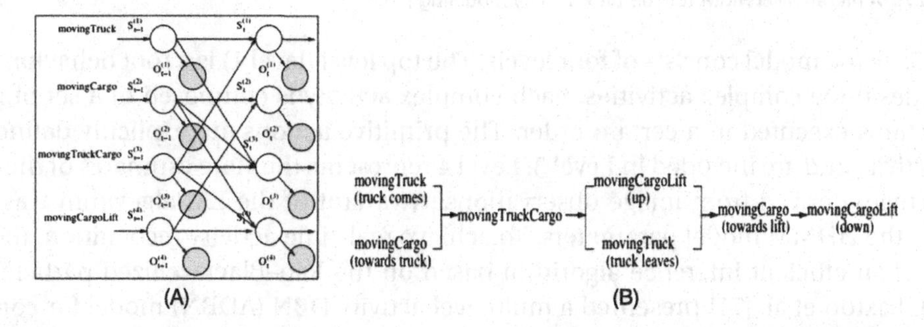

FIGURE 5.78 (A) A DML-HMM for modeling airport cargo loading and unloading activity and (B) the temporal relationships among the primitive events [69].

Xiang et al. demonstrated the performance of DML-HMM for airport cargo loading and unloading activity recognition. Besides using only HMMs, hybrid two-level models have been proposed for complex activity modeling and recognition. Ivanov and Bobick [203] introduced a two-level model for complex activity recognition. The lower level consists of HMMs to represent activity primitives. The higher level employs stochastic context-free grammars (SCFG) to capture the temporal configurations among primitives. Truyen et al. [204] introduced the boosted Markov random forests (AdaBoost+MRF) and applied them to multilevel activity recognition. The bottom layer of their model captures the basic actions, and the higher-level activities are captured by the upper layer. A boosting algorithm is employed to learn the model parameters.

Instead of using a two-level model, some complex activity models employ a multilevel structure, where the lower levels capture the activity primitives, whereas the higher level captures the complex activity. Nguyen et al. [70] proposed employing a hierarchical HMM (HHMM) to recognize three complex human activities: short meal, have snack, and normal meal, each composed of sequential execution of a set of primitive behaviors. As shown

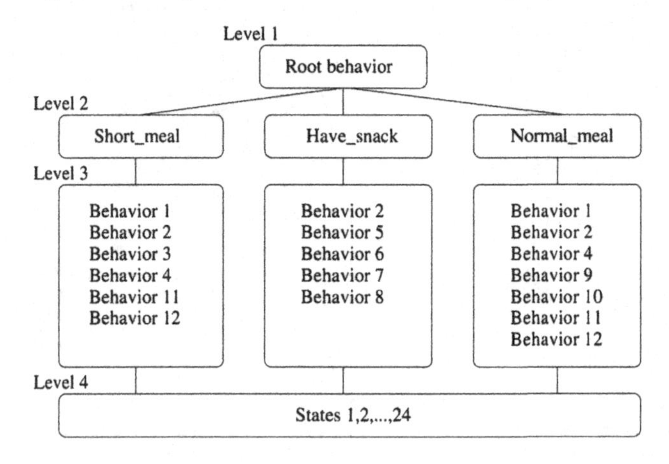

FIGURE 5.79 A hierarchical HMM for complex activity modeling [70].

in Fig. 5.79, the model consists of four levels. The top level (level 1) is a root behavior. Level 2 includes three complex activities. Each complex activity is composed of a set of primitive actions executed in a certain order. The primitive actions are explicitly defined for each activity and are included in Level 3. Level 4 represents the state estimates of the complex activity derived from image observations. A variant of the EM algorithm was used to learn the HHMM model parameters. To achieve real-time activity recognition, they introduced an efficient inference algorithm based on the Rao–Blackwellized particle filter (RBPF). Laxton et al. [71] presented a multilevel activity DBN (ADBN) model for complex activity modeling and recognition. As shown in Fig. 5.80, a complex activity (such as brewing coffee) is made up of a sequence of subactivities (e.g., grind coffee), which, in turn, can be further decomposed into a set of atomic activities (open grinder, pour beans, etc.). Each level can be modeled by a BN with its nodes corresponding to activities in the lower level. Each node has three possible states: waiting, active, and finished. In the lowest level, the atomic activity nodes are connected to the corresponding image measurements of the atomic activity. During recognition, given a query video, the most likely states for the activities at each level are estimated over time through probabilistic filtering using a Viterbi-like algorithm.

Lillo et al. [72] introduced a hierarchical MRF model for simultaneously recognizing complex human activities and primitive human actions from human skeleton video data. As shown in Fig. 5.81, their model consists of four levels: the complex activity level at the top, the primitive actions at the second level, the body poses at the third level, and the video features at the bottom level. The model is quantitatively captured by a total energy function consisting of energy terms for each level and an energy term for pose transition over time. They propose using a max-margin method to simultaneously learn all energy

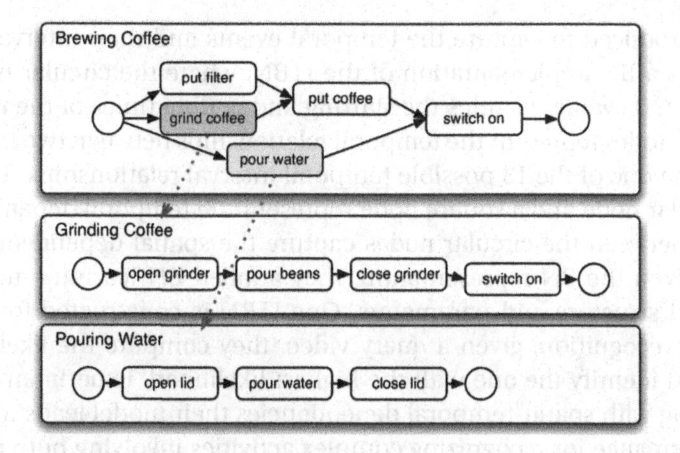

FIGURE 5.80 One slice of the multilevel ADBN model that captures a complex activity at three levels [71]. Each level can be modeled by a BN with its nodes corresponding to activities in the lower level.

parameters at different levels. The max-margin optimization is solved using the concave-convex procedure [205]. During recognition, given the query video, they perform inference to simultaneously recognize the complex activity and its composing actions.

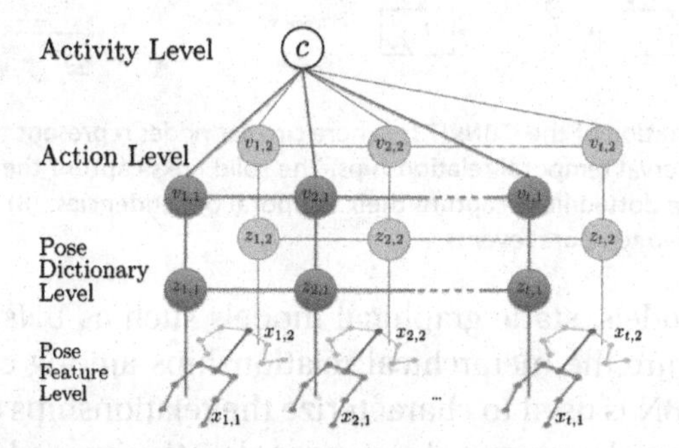

FIGURE 5.81 The hierarchical model for recognition of composable human activities. At the top level, activities are compositions of actions that are inferred at the intermediate level. These actions are in turn compositions of poses at the lower level, where pose dictionaries are learned from data [72].

The graphical model-based approaches discussed previously have two major limitations. First, time-sliced graphical models such as HMMs and DBNs are typically based on points of time; hence, they can only capture three temporal relations: precedes, follows, and equals. Second, HMMs are probabilistic finite-state machines that grow exponentially as the number of parallel events increases. To address these issues, Zhang et al. [73] introduced the interval temporal BN (ITBN), a novel graphical model that combines a BN with the interval algebra to explicitly model the temporal dependencies among primitive temporal events over time intervals. According to interval algebra, a complex activity consists of temporal events and their temporal relationships over time intervals. There are 13 such temporal interval relationships between any two temporal events as shown in Fig. 5.82B.

The ITBN is introduced to capture the temporal events and their interval relationships. Fig. 5.82A shows a BN implementation of the ITBN, where the circular node represents a temporal event. Its value includes the starting and ending times of the temporal event. The rectangular nodes represent the temporal relationships between two temporal events. They can assume one of the 13 possible temporal interval relationships. The dotted links between a circular node and a square node represent the temporal dependence, whereas the solid links between the circular nodes capture the spatial dependencies among the event nodes. Given the BN representation, they applied BN learning methods to learn the ITBN model structure and parameters. One ITBN is constructed for each complex activity. During recognition, given a query video, they compute the likelihood for each ITBN model and identify the one with the highest likelihood. Experimental results show that by reasoning with spatio-temporal dependencies their model leads to a significantly improved performance for recognizing complex activities involving both parallel and sequential events.

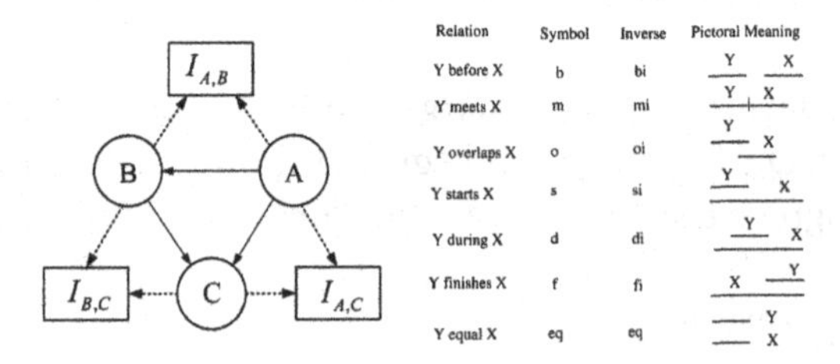

Relation	Symbol	Inverse	Pictoral Meaning
Y before X	b	bi	Y X
Y meets X	m	mi	Y X
Y overlaps X	o	oi	Y X
Y starts X	s	si	Y X
Y during X	d	di	Y X
Y finishes X	f	fi	X Y
Y equal X	eq	eq	Y X

FIGURE 5.82 (A) BN implementation of the ITBNs [73], where circular nodes represent the temporal events, and rectangular nodes capture the interval temporal relationships. The solid links capture the spatial dependencies among temporal entities, whereas the dotted links capture their temporal dependencies. (B) The 13 possible interval temporal relationships between two temporal events.

Besides dynamic models, static graphical models such as BNs or their variants have also been used to capture the hierarchical relationships among elements of an activity. Typically, a three-level BN is used to characterize the relationships among image features, atomic activities, and complex events. Hongeng et al. [74] proposed a hierarchical naive BN to model both single- and multistate scenarios (a scenario is equivalent to a complex activity). As shown in Fig. 5.83, a naive BN is used to model the relationship between dynamic image features, properties of moving objects, and subscenarios. Another BN is used to capture the relationships between subscenarios and scenarios. Combining the two naive BNs yields the hierarchical naive BN. Given an image sequence, dynamic features are extracted to obtain the dynamic properties of moving objects, which are then used to infer the probabilities for the subscenarios. Finally, the measurements of the subscenarios are used as evidence to infer the most likely scenario.

Aside from the BNs, hierarchical Dirichlet process (HDP) models, which are widely used for natural language processing (NLP), have also been used in complex activity modeling and recognition. Since the structures of complex activities are easily represented by a language scheme, with the analogy of action primitives as document topics and an activity as a document, these models can be naturally extended to activity modeling and

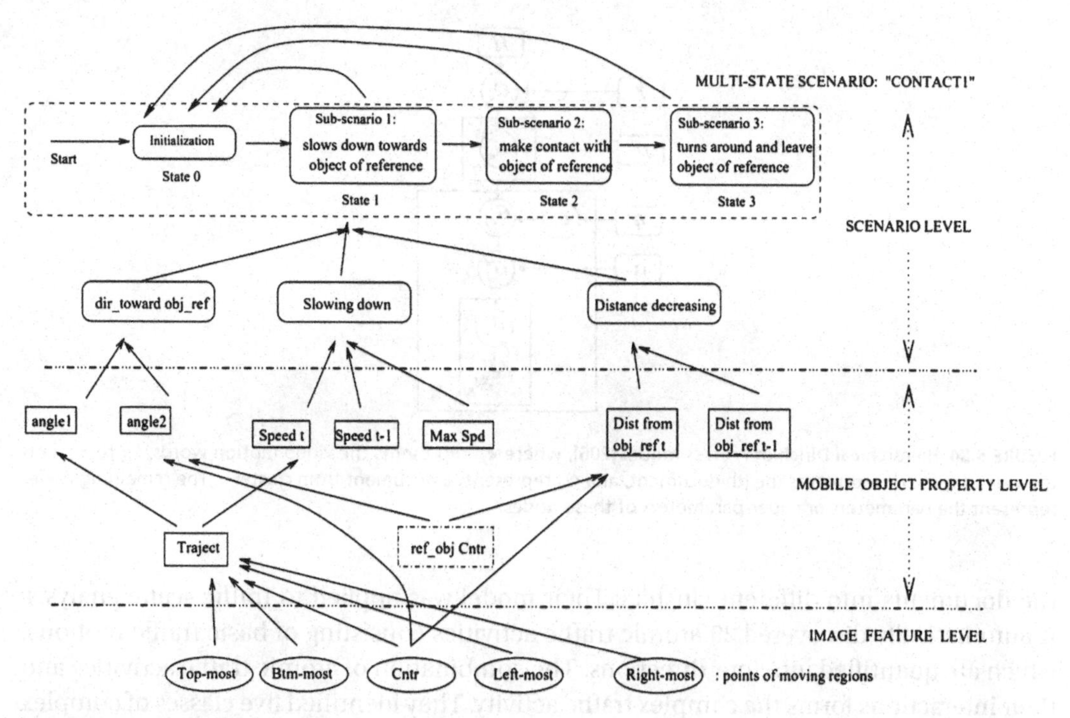

FIGURE 5.83 Representation of activity "contact" defined as: a person walks toward another person, makes contact with them, turns around, and walks away. Image features are shown in ovals, mobile object properties in rectangles, scenarios in rectangles with round corners, and context in dotted rectangles [74].

recognition. Wang et al. [206] introduced a hierarchical Dirichlet process (HDP) model to capture the relationships between different activity levels: visual feature, atomic activities, and complex activities, as shown in Fig. 5.84, where $\mathbf{x}_{ij}$ represent the ith video word in the jth document, θ_{ji} represents a document topic in the jth document with word i, G_j^D represent the jth document, and G_c represents a document from cluster c of L clusters (documents). The remaining nodes represent the parameters (base probability) or hyper-parameters (concentration parameters) for each of these variables. For example, the jth document G_j^D follows a Dirichlet process with base probability G_{c_j} and concentration parameter α, that is, $G_j^D \sim DP(\alpha, G_{c_j})$. Following the plate representation, the innermost plate represents N words, the big plate captures M complex activities, and the plate over G_c captures L clusters of complex activities. Each video sequence represents a complex activity from one of the L clusters, which is, in turn, is composed of a distribution of atomic activities θ_{ji}, which is represented by the distribution of video words extracted from the video frames. The topics of words are actually a summary of typical atomic activities in the scene. Each topic follows a multinomial distribution over words.

Compared to the traditional LDA model, the HDP model allows one to automatically decide the number of topics, instead of manually specifying them. Furthermore, to model different types of complex activities, Wang et al. introduced a new cluster variable to group

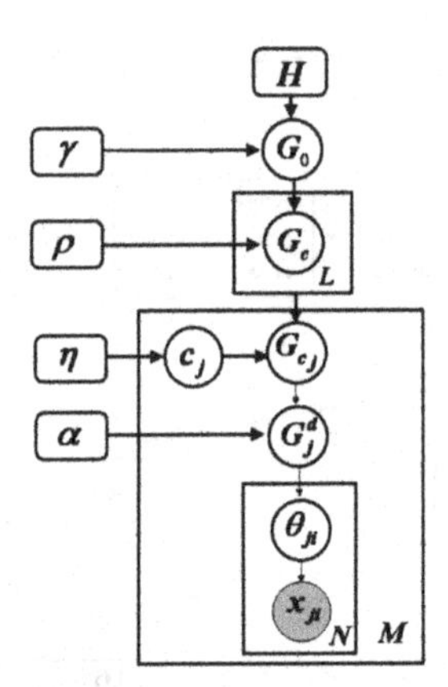

FIGURE 5.84 Hierarchical Dirichlet Process model [206], where $\mathbf{x}_{ji}$ represents the video motion words, θ_{ji} represents the video topic, G_j^D represents the jth document, and G_c represents a document from cluster c. The remaining nodes represent the parameters or hyper-parameters of these nodes.

the documents into different clusters. Their model was applied to traffic scene analysis. It automatically discovered 29 atomic traffic activities consisting of basic traffic motions, which are quantified into four directions. The combination of atomic traffic activities and their interactions forms the complex traffic activity. They identified five classes of complex traffic activities and applied their model to complex activity classification, achieving an average accuracy of 85.74% for 540 video clips. They can also compute the likelihood of a query video for abnormality detection. Finally, they show that their model can be used to perform a semantic query to detect a certain activity of interest based on a particular combination of atomic activities and their interactions.

5.4.3 Capturing contexts for human activity recognition

Besides target information, contextual information can often provide additional information on the type of human action or activity. Gupta et al. [75] proposed combining human motion understanding and object recognition with a BN to model human–object interactions. As shown in Fig. 5.85, the model includes four types of nodes: the object class node denoted by O, the reach motion node denoted by M_r, the manipulation motion node denoted by M_m, and the object reaction node denoted by O_r. The shaded nodes are the corresponding measurements. The object node is the root node of the model. The M_r, M_m, and O_r nodes are its child nodes. Also, the manipulation motion node M_m is the child node of the reach motion node M_r, and the object reaction node O_r is the child node of node M_m. During recognition, Gupta et al. employed HOG-based detector for object and human body part detection and HMMs for motion (reach, reaction, manipulation motions) recognition, yielding their image measurements e_{O_r}, e_{M_m}, e_{M_r}, and e_O, respectively. Given the object and motion measurements, they then employed the belief propagation algorithm to perform simultaneous recognition of an object and six human actions. By joint modeling of human motion and object and exploiting their interactions both the object recognition and human motion understanding tasks can be improved.

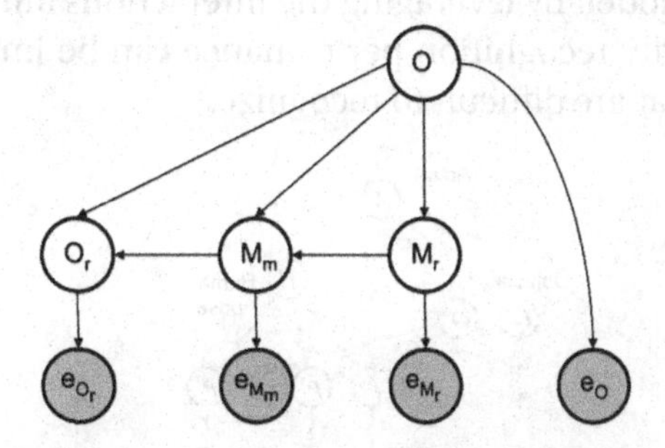

FIGURE 5.85 The BN in [75] for modeling human–object interaction contexts in action videos.

Gupta et al. [76] further proposed a model to divide the objects into two types, scene and manipulable objects. The scene objects are mostly located in the scene, and the manipulable objects are those manipulated by humans. In the example of a tennis game the net would be a scene object, and the ball and racket would be manipulable objects. Fig. 5.86 shows the model that captures the interactions between the scene objects (SO), the manipulable objects (MO), the scene (S), and the human (H). The shaded nodes are their corresponding measurements.

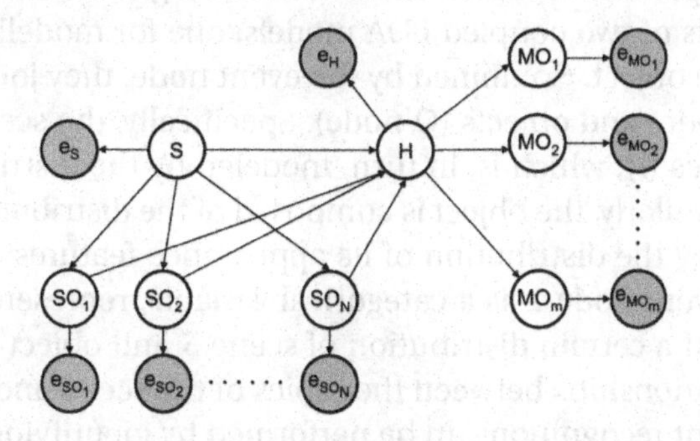

FIGURE 5.86 The BN in [76] for modeling human–object interaction contexts in static images.

Besides the semantic context of human and object, Yao et al. [77] went one step further by studying the semantic relationships between human pose and object. As shown in Fig. 5.87, they proposed an MRF model to encode the cooccurrence/mutually exclusive relationships of objects, human poses, and activities in static images, where the A node represents the activity, H represents the type of human pose, P represents the body part, and O represents different objects. With the mutual context model, the activity, human pose, and objects can benefit the recognition of one another. The weighting coefficients in the pairwise and unary potential terms are the only parameters to be learned. These weighting coefficients are estimated using the maximum likelihood estimation. During testing, the individual measurements of activity, objects, and human poses are all taken as inputs to the context model to simultaneously infer the states of activities, objects, and human poses through the model. By leveraging the interactions among object, human pose, and activity, human activity recognition performance can be improved, in particular, for those human activities that are difficult to recognize.

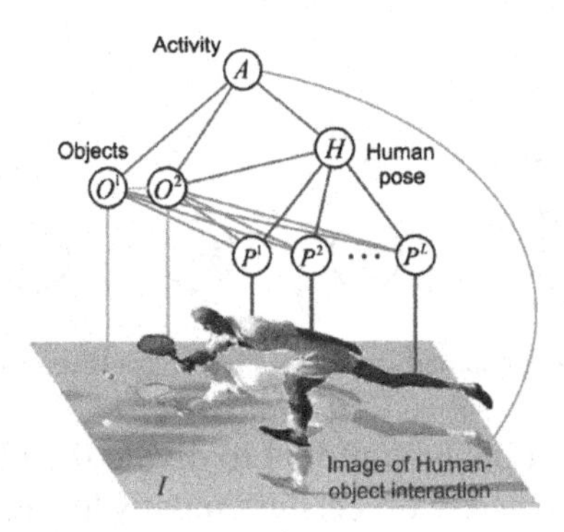

FIGURE 5.87 The MRF model proposed by Yao et al. [77] for encoding the mutual context of objects, poses, and activities.

To capture the contextual information at the event, scene, and object levels, Li et al. [78] proposed a hierarchical probabilistic model for event classification by integrating the classification of scene patches and objects. As shown in Fig. 5.88, this model is also a latent topic model. It consists of two coupled LDA models, one for modeling the scene and the other for modeling the object. Combined by the event node, they jointly model the distributions of scene (S node) and objects (O node). Specifically, the scene is modeled by the distribution of M topics (t), which is, in turn, modeled by the distributions of image appearance feature X. Similarly, the object is composed of the distribution of N topics z, and each topic is formed by the distribution of its appearance features A and geometric features G. Finally, the event node E is a categorical variable, representing different events. An event is made up of a certain distribution of scene S and object O. Hence, the model captures semantic relationships between the topics of the scene and those of the objects. For a query video, event recognition can be performed by identifying the event class with the highest probability, given the image features (X, A, and G) extracted from the query video: $E^* = \arg\max_E p(E|X, A, G)$.

Wang and Ji [79] introduced an augmented DBN to incorporate the scene, event object interaction, and the event temporal contexts into a standard DBNs for event recognition in surveillance videos. As shown in Fig. 5.89, event nodes E (both E_{n-1} and E_n) have K discrete values, where K is the number of different categories of events. The subscripts $n-1$ and n on E nodes stand for the events at two different times, where E_{n-1} stands for the previous event, and E_n stands for the current event. The link between E_{n-1} and E_n captures the temporal dependency, that is, the event temporal context. The S node stands for the scene. The link between S and E_n captures the causal influence of the scene context on the event. The O_n node stands for the contextual object for the current event clip.

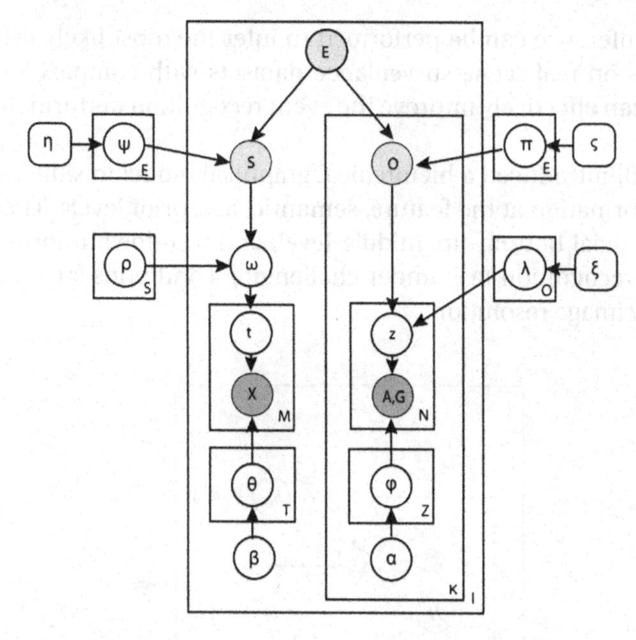

FIGURE 5.88 Hierarchical probabilistic model for classifying events by scene and object recognition [78], where E represents the event, S represents the event scene, and O represents the event objects.

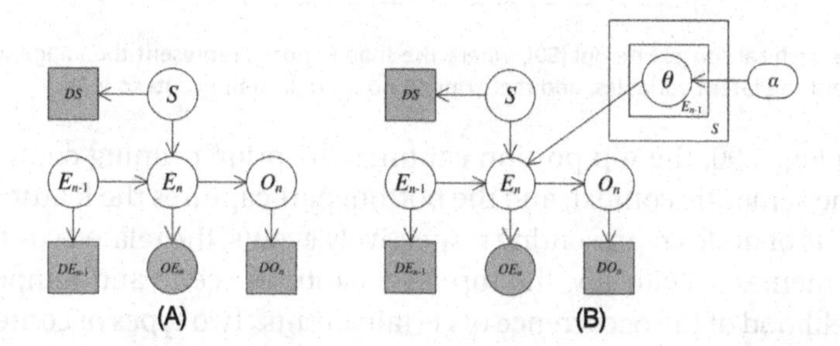

FIGURE 5.89 The augmented DBN for incorporating scene and semantic contexts for human event recognition [79]. (A) context model; and (B) hierarchical context model with hyperparameters.

The link between E_n and O_n captures the event–object interaction context. All four nodes E_n, E_{n-1}, S, and O_n have their corresponding measurements, represented respectively by OE_n, DE_{n-1}, DS, and DO_n. The circular OE_n node is a continuous vector representing the event observation for the current event generated by a conventional event classifier. The remaining three measurement nodes DE_{n-1}, DS, and DO_n are discrete nodes, resulting from the classifier detections of the corresponding contexts. Their model incorporates various contexts and target information simultaneously into one unified model. To account for intraclass variation for each event, they further extend their model to include hyperparameters (Fig. 5.89B) to specify the distribution of E_n node. The parameters of the context model are estimated using the MAP approach. During recognition, given the measurements, MAP inference can be performed to infer the most likely event E_n at current time. Experiments on real scene surveillance datasets with complex backgrounds show that the contexts can effectively improve the event recognition performance for real-world videos.

Wang and Ji [80] introduced a hierarchical graphical model to simultaneously capture the contextual information at the feature, semantic, and prior levels. These three levels of context provide crucial bottom-up, middle-level, and top-down information, which can benefit the event recognition task under challenging conditions such as large intraclass variations and low image resolution.

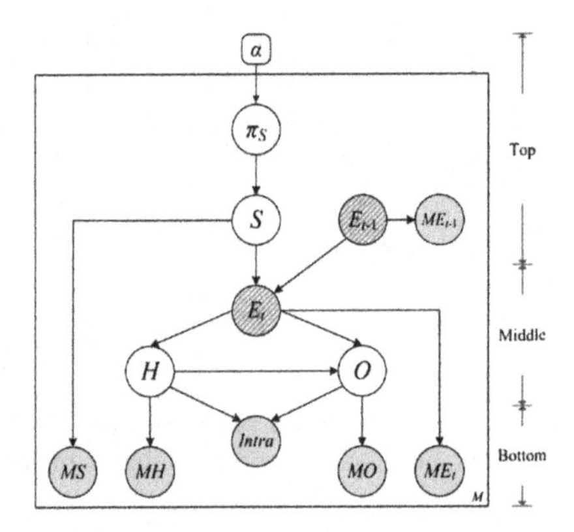

FIGURE 5.90 The hierarchical context model [80], where the shaded nodes represent the image measurements, the white nodes represent the latent variables, and the stripped nodes represent events to infer.

As shown in Fig. 5.90, the top portion captures the prior/priming context, the middle part captures the semantic context, and the bottom part captures the feature level context. Each part consists of nodes representing respectively events, the related contexts, and their image measurements. Specifically, the top level captures scene and temporal prior that predicts the likelihood of the occurrence of certain events. Two types of contextual priming are considered: the scene context and the dynamic context. The scene context provides an environmental context within which events occur. It provides the prior probability for an event to happen for a given scene. In the model, scene is captured by a discrete scene node S, which represents different possible scene types. The link from S to E_t captures the causal relation between the scene S and the event E_t at time t. In addition, parameters and hyperparameters are introduced to model the prior probability distribution of the scene node, allowing variability in scene contexts. The second contextual priming is the dynamic context. It provides temporal support as to what event is likely to happen, given the events that have already happened so far. The event at the current time is influenced by events at previous times. Dynamic context is captured by the E_{t-1} node. The link between E_{t-1} and E_t captures the temporal causal relation between E_t and E_{t-1}. Both nodes S and E_{t-1} provide top-down priming information for inference of the current event.

The semantic midlevel captures components of an event and their interactions. For this research, it captures human, object, and their interactions. Specifically, the middle level consists of an event node E_t and its components human and object, represented by discrete nodes H and O, respectively. As the exact human and object states are unknown, they are treated as latent variables, and their optimal states are learned from data during training. The interactions between H and O are captured by the links between them, which specify the probabilistic dependencies among their states. Finally, feature context modeling in the bottom level consists of image features that describe the event and its context. Various spatio-temporal image features are used as image features for the target event, including the human and object. To measure the event context at the feature level, several types of context features are extracted from the event neighborhoods. In addition, additional features are extracted to measure the interevent and intraevent spatio-temporal relations. The target features are represented by MH and MO, whereas the contextual features are represented by nodes ME and $Intra$. The hierarchical context model simultaneously exploits contexts at all three levels and systematically incorporates them into event recognition.

The hierarchical model is a directed PGM. As its structure is manually fixed, learning involves only learning the parameters, that is, the conditional probabilities for each node. As the model consists of latent variables, EM method can be employed to learn the model parameters. Furthermore, Wang and Ji employ a variational EM to address the complexity brought about by the continuous latent variable. Given the learned hierarchical model and a query video sequence, an MAP inference is performed to determine the most likely event. Again, variational inference is employed to approximately solve for the MAP inference. Experiments on benchmark datasets demonstrate the effectiveness of their method for event recognition under realistic challenges such as large intraclass variations and low image resolution.

References

[1] L. Zhang, Q. Ji, Image segmentation with a unified graphical model, IEEE Transactions on Pattern Analysis and Machine Intelligence 32 (8) (2009) 1406–1425.
[2] J. Jeong, T. Sung Yoon, J. Park, Towards a meaningful 3D map using a 3D lidar and a camera, Sensors 18 (08 2018) 2571.
[3] G. Gilboa, ROF denoising example [online], available: https://en.wikipedia.org/wiki/File:ROF_Denoising_Example.png, 2010.
[4] S.J. Prince, Computer Vision: Models, Learning, and Inference, Cambridge University Press, 2012.
[5] C. Bouman, B. Liu, Multiple resolution segmentation of textured images, IEEE Transactions Pattern Analysis and Machine Intelligence (1991) 99–113.
[6] P. Awasthi, A. Gagrani, B. Ravindran, Image modeling using tree structured conditional random fields, in: International Joint Conference on Artificial Intelligence, 2007, pp. 2060–2065.
[7] Y. Wang, Q. Ji, A dynamic conditional random field model for object segmentation in image sequences, in: IEEE Conference on Computer Vision and Pattern Recognition, vol. 1, 2005, pp. 264–270.
[8] S. Zheng, S. Jayasumana, B. Romera-Paredes, V. Vineet, Z. Su, D. Du, C. Huang, P.H. Torr, Conditional random fields as recurrent neural networks, in: Proceedings of the IEEE International Conference on Computer Vision, 2015, pp. 1529–1537.
[9] E.N. Mortensen, J. Jia, Real-time semi-automatic segmentation using a Bayesian network, IEEE Conference on Computer Vision and Pattern Recognition (2006) 1007–1014.
[10] X. Feng, C. Williams, S. Felderhof, Combining belief networks and neural networks for scene segmentation, IEEE Transactions on Pattern Analysis and Machine Intelligence 24 (4) (2002) 467–483.

[11] P.F. Felzenszwalb, D.P. Huttenlocher, Pictorial structures for object recognition, International Journal of Computer Vision 61 (1) (2005) 55–79.
[12] L. Fei-Fei, R. Fergus, P. Perona, Learning generative visual models from few training examples: an incremental Bayesian approach tested on 101 object categories, Computer Vision and Image Understanding (2007) 59–80.
[13] R. Fergus, P. Perona, A. Zisserman, A sparse object category model for efficient learning and exhaustive recognition, in: IEEE Conference on Computer Vision and Pattern Recognition, vol. 1, 2005, pp. 380–387.
[14] M.P. Kumar, P.H.S. Torr, A. Zisserman, Obj cut, in: IEEE Conference on Computer Vision and Pattern Recognition, 2005, pp. 18–25.
[15] E.B. Sudderth, A. Torralba, W.T. Freeman, A.S. Willsky, Learning hierarchical models of scenes, objects, and parts, in: International Conference on Computer Vision, vol. 2, 2005, pp. 1331–1338.
[16] K.P. Murphy, A. Torralba, W.T. Freeman, Using the forest to see the trees: a graphical model relating features, objects, and scenes, in: Advances in Neural Information Processing Systems, 2004, pp. 1499–1506.
[17] M. Szummer, Learning diagram parts with hidden random fields, in: Eighth International Conference on Document Analysis and Recognition, IEEE, 2005, pp. 1188–1193.
[18] A. Kapoor, J. Winn, Located hidden random fields: learning discriminative parts for object detection, in: European Conference on Computer Vision, 2006, pp. 302–315.
[19] Y. Zhang, Z. Tang, B. Wu, Q. Ji, H. Lu, A coupled hidden conditional random field model for simultaneous face clustering and naming in videos, IEEE Transactions on Image Processing 25 (12) (2016) 5780–5792.
[20] Y. Wu, Q. Ji, Facial landmark detection: a literature survey, International Journal of Computer Vision (2018).
[21] Y. Wu, Z. Wang, Q. Ji, Facial feature tracking under varying facial expressions and face poses based on restricted Boltzmann machines, in: Proceedings of the IEEE Conference on Computer Vision and Pattern Recognition, 2013, pp. 3452–3459.
[22] Y. Wu, Q. Ji, Discriminative deep face shape model for facial point detection, International Journal of Computer Vision 113 (1) (2015) 37–53.
[23] The CMU multi-pie face database [online], available: http://www.cs.cmu.edu/afs/cs/project/PIE/MultiPie/Multi-Pie/Home.html.
[24] Y. Wu, Z. Wang, Q. Ji, A hierarchical probabilistic model for facial feature detection, in: Proceedings of the IEEE Conference on Computer Vision and Pattern Recognition, 2014, pp. 1781–1788.
[25] L. Fei-Fei, P. Perona, A Bayesian hierarchical model for learning natural scene categories, in: 2005 IEEE Computer Society Conference on Computer Vision and Pattern Recognition, CVPR'05, vol. 2, 2005, pp. 524–531.
[26] B.C. Russell, A. Torralba, C. Liu, R. Fergus, W.T. Freeman, Object recognition by scene alignment, in: Advances in Neural Information Processing Systems, 2007.
[27] Y. Wang, S. Gong, Conditional random field for natural scene categorization, in: British Machine Vision Conference, 2007.
[28] A. Doucet, N. De Freitas, K. Murphy, S. Russell, Rao–Blackwellised particle filtering for dynamic Bayesian networks, in: Proceedings of the Sixteenth Conference on Uncertainty in Artificial Intelligence, Morgan Kaufmann Publishers Inc., 2000, pp. 176–183.
[29] L. Zhang, J. Chen, Z. Zeng, Q. Ji, 2D and 3D upper body tracking with one framework, in: 19th International Conference on Pattern Recognition, IEEE, 2008, pp. 1–4.
[30] V. Pavlovic, J.M. Rehg, T.-J. Cham, K.P. Murphy, A dynamic Bayesian network approach to figure tracking using learned dynamic models, in: International Conference on Computer Vision, 1999, pp. 94–101.
[31] L. Sigal, S. Bhatia, S. Roth, M. Black, M. Isard, Tracking loose-limbed people, in: IEEE Conference on Computer Vision and Pattern Recognition, 2004.
[32] D. Ramanan, D. Forsyth, Finding and tracking people from the bottom up, in: IEEE Conference on Computer Vision and Pattern Recognition, 2003.
[33] C. Sminchisescu, A. Kanaujia, Z. Li, D. Metaxas, Discriminative density propagation for 3D human motion estimation, in: IEEE Conference on Computer Vision and Pattern Recognition, 2005.

[34] M. Kim, V. Pavlovic, Conditional state-space model for discriminative motion estimation, in: International Conference on Computer Vision, 2007.
[35] M.F. Tappen, W.T. Freeman, Comparison of graph cuts with belief propagation for stereo, using identical MRF parameters, in: Proceedings of the International Conference on Computer Vision, 2003, p. 900.
[36] A. Saxena, S.H. Chung, A.Y. Ng, 3-D depth reconstruction from a single still image, International Journal of Computer Vision 76 (1) (2008) 53–69.
[37] A. Saxena, M. Sun, A.Y. Ng, Make3D: learning 3D scene structure from a single still image, IEEE Transactions on Pattern Analysis and Machine Intelligence 31 (5) (2009) 824–840.
[38] E. Delage, H. Lee, A.Y. Ng, A dynamic Bayesian network model for autonomous 3D reconstruction from a single indoor image, in: IEEE Computer Society Conference on Computer Vision and Pattern Recognition, vol. 2, 2006, pp. 2418–2428.
[39] Y. Tong, W. Liao, Q. Ji, Facial action unit recognition by exploiting their dynamic and semantic relationships, IEEE Transactions on Pattern Analysis and Machine Intelligence 29 (10) (2007).
[40] Z. Wang, Y. Li, S. Wang, Q. Ji, Capturing global semantic relationships for facial action unit recognition, in: Proceedings of the IEEE International Conference on Computer Vision, 2013, pp. 3304–3311.
[41] Y. Zhang, Q. Ji, Active and dynamic information fusion for facial expression understanding from image sequences, IEEE Transactions on Pattern Analysis and Machine Intelligence 27 (5) (2005) 699–714.
[42] Y. Li, S. Wang, Y. Zhao, Q. Ji, Simultaneous facial feature tracking and facial expression recognition, IEEE Transactions on Image Processing 22 (7) (2013) 2559–2573.
[43] J. Yamato, J. Ohya, K. Ishii, Recognizing human action in time-sequential images using hidden Markov model, in: IEEE Conference on Computer Vision and Pattern Recognition, 1992, pp. 379–385.
[44] M. Brand, N. Oliver, A. Pentland, Coupled hidden Markov models for complex action recognition, in: IEEE Computer Society Conference on Computer Vision and Pattern Recognition, 1997, pp. 994–999.
[45] D. Wu, L. Shao, Leveraging hierarchical parametric networks for skeletal joints based action segmentation and recognition, in: Proceedings of the IEEE Conference on Computer Vision and Pattern Recognition, 2014, pp. 724–731.
[46] T. Starner, J. Weaver, A. Pentland, Real-time American sign language recognition using desk and wearable computer based video, IEEE Transactions on Pattern Analysis and Machine Intelligence 20 (12) (1998) 1371–1375.
[47] A. Ramamoorthy, N. Vaswani, S. Chaudhury, S. Banerjee, Recognition of dynamic hand gestures, Pattern Recognition 36 (9) (2003) 2069–2081.
[48] C. Vogler, D. Metaxas, Parallel hidden Markov models for American sign language recognition, in: IEEE International Conference on Computer Vision, vol. 1, 1999, pp. 116–122.
[49] S. Nie, Q. Ji, Capturing global and local dynamics for human action recognition, in: International Conference on Pattern Recognition, ICPR, IEEE, 2014, pp. 1946–1951.
[50] G.W. Taylor, G.E. Hinton, S.T. Roweis, Modeling human motion using binary latent variables, in: Advances in Neural Information Processing Systems, 2007, pp. 1345–1352.
[51] S. Nie, Z. Wang, Q. Ji, A generative restricted Boltzmann machine based method for high-dimensional motion data modeling, Computer Vision and Image Understanding 136 (2015) 14–22.
[52] N. Oliver, B. Rosario, A. Pentland, A Bayesian computer vision system for modeling human interactions, IEEE Transactions on Pattern Analysis and Machine Intelligence (2000).
[53] P. Natarajan, R. Nevatia, Coupled hidden semi Markov models for activity recognition, in: IEEE Workshop on Motion and Video Computing, 2007, p. 10.
[54] Y. Luo, T.-D. Wu, J.-N. Hwang, Object-based analysis and interpretation of human motion in sports video sequences by dynamic Bayesian networks, Computer Vision and Image Understanding 92 (2) (2003) 196–216.
[55] X. Wang, Q. Ji, Learning dynamic Bayesian network discriminatively for human activity recognition, in: International Conference on Pattern Recognition, 2012, pp. 3553–3556.
[56] J. Wu, A. Osuntogun, T. Choudhury, M. Philipose, J. Regh, A scalable approach to activity recognition based on object use, in: International Conferences on Computer Vision, 2007.
[57] L. Zhang, Z. Zeng, Q. Ji, Probabilistic image modeling with an extended chain graph for human activity recognition and image segmentation, IEEE Transactions on Image Processing 20 (9) (2011) 2401–2413.

[58] F. Lv, R. Nevatia, Single view human action recognition using key pose matching and Viterbi path searching, in: IEEE Conference on Computer Vision and Pattern Recognition, 2007.
[59] D. Vail, M. Veloso, J. Lafferty, Conditional random fields for activity recognition, in: Proceedings of the 2007 Conference on Autonomous Agents and Multiagent Systems, 2007.
[60] L. Wang, D. Suter, Recognizing human activities from silhouettes: motion subspace and factorial discriminative graphical model, in: IEEE Conference on Computer Vision and Pattern Recognition, 2007.
[61] C. Sutton, A. McCallum, K. Rohanimanesh, Dynamic conditional random fields: factorized probabilistic models for labeling and segmenting sequence data, Journal of Machine Learning Research 8 (Mar 2007) 693–723.
[62] L.-P. Morency, A. Quattoni, T. Darrell, Latent-dynamic discriminative models for continuous gesture recognition, in: IEEE Conference on Computer Vision and Pattern Recognition, 2007, pp. 1–8.
[63] R. Filipovych, E. Ribeiro, Recognizing primitive interactions by exploring actor-object states, in: IEEE Conference on Computer Vision and Pattern Recognition, 2008.
[64] J. Niebles, H. Wang, F.-F. Li, Unsupervised learning of human action categories using spatio-temporal words, International Journal of Computer Vision (2008).
[65] S. Hongeng, R. Nevatia, Large-scale event detection using semi-hidden Markov models, in: International Conference on Computer Vision, 2003, pp. 1455–1462.
[66] D. Zhang, D. Perez, I. McCowan, Semi-supervised adapted HMMs for unusual event detection, in: IEEE Conference on Computer Vision and Pattern Recognition, 2005.
[67] T. Duong, H. Bui, D. Phung, Activity recognition and abnormality detection with the switching hidden semi-Markov model, in: IEEE Conference on Computer Vision and Pattern Recognition, 2005.
[68] Y. Du, F. Chen, W. Xu, W. Zhang, Activity recognition through multi-scale motion detail analysis, Neurocomputing 71 (16) (2008) 3561–3574.
[69] T. Xiang, S. Song, Beyond tracking: modelling activity and understanding behavior, International Journal of Computer Vision (2006).
[70] N.T. Nguyen, D.Q. Phung, S. Venkatesh, H. Bui, Learning and detecting activities from movement trajectories using the hierarchical hidden Markov model, in: IEEE Conference on Computer Vision and Pattern Recognition, vol. 2, 2005, pp. 955–960.
[71] B. Laxton, J. Lim, D. Kriegman, Leveraging temporal, contextual and ordering constraints for recognizing complex activities in video, in: IEEE Conference on Computer Vision and Pattern Recognition, 2007.
[72] I. Lillo, A. Soto, J. Carlos Niebles, Discriminative hierarchical modeling of spatio-temporally composable human activities, in: Proceedings of the IEEE Conference on Computer Vision and Pattern Recognition, 2014, pp. 812–819.
[73] Y. Zhang, Y. Zhang, E. Swears, N. Larios, Z. Wang, Q. Ji, Modeling temporal interactions with interval temporal Bayesian networks for complex activity recognition, IEEE Transactions on Pattern Analysis and Machine Intelligence 35 (10) (2013) 2468–2483.
[74] S. Hongeng, F. Bremond, R. Nevatia, Representation and optimal recognition of human activities, in: IEEE Conference on Computer Vision and Pattern Recognition, 2000.
[75] A. Gupta, L.S. Davis, Objects in action: an approach for combining action understanding and object perception, in: IEEE Conference on Computer Vision and Pattern Recognition, 2007, pp. 1–8.
[76] A. Gupta, A. Kembhavi, L.S. Davis, Observing human–object interactions: using spatial and functional compatibility for recognition, IEEE Transactions on Pattern Analysis and Machine Intelligence 31 (10) (2009) 1775–1789.
[77] B. Yao, L. Fei-Fei, Modeling mutual context of object and human pose in human–object interaction activities, in: IEEE Conference on Computer Vision and Pattern Recognition, 2010, pp. 17–24.
[78] L.-J. Li, L. Fei-Fei, What, where and who? Classifying events by scene and object recognition, in: International Conference on Computer Vision, 2007, pp. 1–8.
[79] X. Wang, Q. Ji, Context augmented dynamic Bayesian networks for event recognition, Pattern Recognition Letters 43 (2014) 62–70.
[80] X. Wang, Q. Ji, A hierarchical context model for event recognition in surveillance video, in: Proceedings of the IEEE Conference on Computer Vision and Pattern Recognition, 2014, pp. 2561–2568.
[81] S. Geman, D. Geman, Stochastic relaxation, Gibbs distributions, and the Bayesian restoration of images, IEEE Transactions on Pattern Analysis and Machine Intelligence 6 (6) (1984) 721–741.

[82] D. Melas, S. Wilson, Double Markov random fields and Bayesian image segmentation, IEEE Transactions on Signal Processing 50 (2002) 357–365.
[83] W. Pieczynski, A. Tebbache, Pairwise Markov random fields and segmentation of textured images, Machine Graphics and Vision 9 (3) (2000) 705–718.
[84] C. D'Elia, G. Poggi, G. Scarpa, A tree-structured Markov random field model for Bayesian image segmentation, IEEE Transactions on Image Processing 12 (10) (2003) 1259–1273.
[85] S. Geman, D. Geman, Stochastic relaxation, Gibbs distribution and the Bayesian restoration of images, IEEE Transactions on Pattern Analysis and Machine Intelligence 6 (6) (1984) 721–741.
[86] S.Z. Li, Markov Random Field Modeling in Image Analysis, Springer Science & Business Media, 2009.
[87] Y. Li, D.P. Huttenlocher, Sparse long-range random field and its application to image denoising, in: European Conference on Computer Vision, Springer, 2008, pp. 344–357.
[88] Z. Kato, M. Berthod, J. Zerubia, Multiscale Markov random field models for parallel image classification, in: International Conferences on Computer Vision, 1993, pp. 253–257.
[89] C. Bouman, M. Shapiro, A multiscale random field model for Bayesian image segmentation, IEEE Transactions on Image Processing 3 (2) (1994) 162–177.
[90] S. Todorovic, M. Nechyba, Dynamic trees for unsupervised segmentation and matching of image regions, IEEE Transactions on Pattern Analysis and Machine Intelligence 27 (11) (2005) 1762–1777.
[91] W. Irving, P. Fieguth, A. Willsky, An overlapping tree approach to multiscale stochastic modeling and estimation, IEEE Transactions on Image Processing 6 (11) (1997) 1517–1529.
[92] R. Szeliski, R. Zabih, D. Scharstein, O. Veksler, V. Kolmogorov, A. Agarwala, M. Tappen, C. Rother, A comparative study of energy minimization methods for Markov random fields with smoothness-based priors, IEEE Transactions on Pattern Analysis and Machine Intelligence 30 (6) (2008) 1068–1080.
[93] A. Fix, A. Gruber, E. Boros, R. Zabih, A graph cut algorithm for higher-order Markov random fields, in: IEEE International Conference on Computer Vision, ICCV, 2011, pp. 1020–1027.
[94] R. Dubes, A. Jain, S. Nadabar, C. Chen, MRF model-based algorithms for image segmentation, in: 10th International Conference on Pattern Recognition, vol. 1, 1990, pp. 808–814.
[95] J. Kappes, B. Andres, F. Hamprecht, C. Schnorr, S. Nowozin, D. Batra, S. Kim, B. Kausler, J. Lellmann, N. Komodakis, et al., A comparative study of modern inference techniques for discrete energy minimization problems, in: Proceedings of the IEEE Conference on Computer Vision and Pattern Recognition, 2013, pp. 1328–1335.
[96] S. Kumar, M. Hebert, Discriminative fields for modeling spatial dependencies in natural images, in: Advances in Neural Information Processing Systems, 2004, pp. 1531–1538.
[97] P. Krähenbühl, V. Koltun, Efficient inference in fully connected CRFs with Gaussian edge potentials, in: Advances in Neural Information Processing Systems, 2011, pp. 109–117.
[98] L.-C. Chen, G. Papandreou, I. Kokkinos, K. Murphy, A.L. Yuille, DeepLab: semantic image segmentation with deep convolutional nets, atrous convolution, and fully connected CRFs, arXiv preprint, arXiv:1606.00915, 2016.
[99] J. Shotton, J. Winn, C. Rother, A. Criminisi, TextonBoost: joint appearance, shape and context modeling for multi-class object recognition and segmentation, in: European Conference on Computer Vision, Springer, 2006, pp. 1–15.
[100] C. Sutton, A. McCallum, Piecewise training for undirected models, arXiv preprint, arXiv:1207.1409, 2012.
[101] Y.Y. Boykov, M.-P. Jolly, Interactive graph cuts for optimal boundary & region segmentation of objects in nd images, in: International Conference on Computer Vision, vol. 1, 2001, pp. 105–112.
[102] X. Ren, C.C. Fowlkes, J. Malik, Cue integration in figure/ground labeling, in: Advances in Neural Information Processing Systems 18, 2005.
[103] X. He, R.S. Zemel, D. Ray, Learning and incorporating top-down cues in image segmentation, in: European Conference on Computer Vision, 2006, pp. 338–351
.
[104] J. Reynolds, K. Murphy, Figure-ground segmentation using a hierarchical conditional random field, in: Proceedings of the Fourth Canadian Conference on Computer and Robot Vision, 2007.
[105] X. He, R.S. Zemel, M.Á. Carreira-Perpiñán, Multiscale conditional random fields for image labeling, in: IEEE Conference on Computer Vision and Pattern Recognition, vol. 2, 2004, p. II.

[106] P. Kohli, P.H. Torr, et al., Robust higher order potentials for enforcing label consistency, International Journal of Computer Vision 82 (3) (2009) 302–324.
[107] C. Russell, P. Kohli, P.H. Torr, et al., Associative hierarchical CRFs for object class image segmentation, in: International Conference on Computer Vision, 2009, pp. 739–746.
[108] V. Vineet, J. Warrell, P.H. Torr, Filter-based mean-field inference for random fields with higher-order terms and product label-spaces, International Journal of Computer Vision 110 (3) (2014) 290–307.
[109] V. Vineet, G. Sheasby, J. Warrell, P.H. Torr, Posefield: an efficient mean-field based method for joint estimation of human pose, segmentation, and depth, in: International Workshop on Energy Minimization Methods in Computer Vision and Pattern Recognition, Springer, 2013, pp. 180–194.
[110] P. Kohli, M.P. Kumar, P.H. Torr, P3 & beyond: solving energies with higher order cliques, in: IEEE Conference on Computer Vision and Pattern Recognition, 2007, pp. 1–8.
[111] B. Potetz, T.S. Lee, Efficient belief propagation for higher-order cliques using linear constraint nodes, Computer Vision and Image Understanding 112 (1) (2008) 39–54.
[112] T. Toyoda, O. Hasegawa, Random field model for integration of local information and global information, IEEE Transactions on Pattern Analysis and Machine Intelligence 30 (8) (2008) 1483–1489.
[113] N. Payet, S. Todorovic, $(RF)^2$ – random forest random field, in: Advances in Neural Information Processing Systems, 2010, pp. 1885–1893.
[114] P. Krähenbühl, V. Koltun, Parameter learning and convergent inference for dense random fields, in: International Conference on Machine Learning, 2013, pp. 513–521.
[115] V. Vineet, J. Warrell, P. Sturgess, P.H. Torr, Improved initialization and Gaussian mixture pairwise terms for dense random fields with mean-field inference, in: BMVC, 2012, pp. 1–11.
[116] P. Kohli, M.P. Kumar, P.H. Torr, Solving energies with higher order cliques, in: IEEE Conference on Computer Vision and Pattern Recognition, 2007.
[117] S. Paris, P. Kornprobst, J. Tumblin, F. Durand, et al., Bilateral filtering: theory and applications, Foundations and Trends® in Computer Graphics and Vision 4 (1) (2009) 1–73.
[118] A. Hosni, C. Rhemann, M. Bleyer, C. Rother, M. Gelautz, Fast cost-volume filtering for visual correspondence and beyond, IEEE Transactions on Pattern Analysis and Machine Intelligence 35 (2) (2013) 504–511.
[119] C. Harris, M. Stephens, A combined corner and edge detector, in: Alvey Vision Conference, vol. 15, no. 50, Manchester, UK, 1988, pp. 10–5244.
[120] P. Alvarado, A. Berner, S. Akyol, Combination of high-level cues in unsupervised single image segmentation using Bayesian belief networks, in: Proceedings of the International Conference on Imaging Science, Systems, and Technology, vol. 2, 2002, pp. 675–681.
[121] S. Todorovic, M. Nechyba, Interpretation of complex scenes using dynamic tree-structure Bayesian networks, Computer Vision and Image Understanding 106 (1) (2007) 71–84.
[122] A. Andreopoulos, J.K. Tsotsos, 50 years of object recognition: directions forward, Computer Vision and Image Understanding 117 (8) (2013) 827–891.
[123] J. Yang, Y.-G. Jiang, A.G. Hauptmann, C.-W. Ngo, Evaluating bag-of-visual-words representations in scene classification, in: Proceedings of the International Workshop on Workshop on Multimedia Information Retrieval, ACM, 2007, pp. 197–206.
[124] Y. Freund, R.E. Schapire, A decision-theoretic generalization of on-line learning and an application to boosting, in: Computational Learning Theory: Eurocolt, 1995, pp. 23–37.
[125] M.A. Fischler, R.A. Elschlager, The representation and matching of pictorial structures, IEEE Transactions on Computers 100 (1) (1973) 67–92.
[126] M.P. Kumar, P. Torr, A. Zisserman, Extending pictorial structures for object recognition, in: Proc. BMVC, 2004, p. 81.
[127] M.P.K.P. Torr, A. Zisserman, Learning layered pictorial structures from video, in: Indian Conference on Vision, Graphics and Image Processing, 2004.
[128] M. Burl, M. Weber, P. Perona, A probabilistic approach to object recognition using local photometry and global geometry, in: European Conference on Computer Vision, 1998, pp. 628–641.
[129] M. Weber, M. Welling, P. Perona, Unsupervised learning of models for recognition, in: European Conference on Computer Vision, 2000, pp. 18–32.
[130] R. Fergus, P. Perona, A. Zisserman, Object class recognition by unsupervised scale-invariant learning, in: IEEE Conference on Computer Vision and Pattern Recognition, 2003, pp. 264–271.

[131] H. Arora, N. Loeff, A. Sorokin, D. Forsyth, Efficient unsupervised learning for localization and detection in object categories, in: Neural Information Processing Systems, 2005.
[132] M.D. Gupta, S. Rajaram, N. Petrovic, T.S. Huang, Restoration and recognition in a loop, in: IEEE Conference on Computer Vision and Pattern Recognition, 2005.
[133] V. Kolmogorov, R. Zabih, What energy functions can be minimized via graph cuts?, IEEE Transactions on Pattern Analysis & Machine Intelligence (2) (2004) 147–159.
[134] S. Kumar, M. Hebert, Discriminative random fields, International Journal of Computer Vision 68 (2) (2006) 179–201.
[135] A. Quattoni, M. Collins, T. Darrell, Conditional random fields for object recognition, in: Neural Information Processing Systems, 2004.
[136] J. Winn, J. Shotton, The layout consistent random field for recognizing and segmenting partially occluded objects, in: IEEE Conference on Computer Vision and Pattern Recognition, vol. 1, 2006, pp. 37–44.
[137] D. Hoiem, C. Rother, J. Winn, 3D LayoutCRF for multi-view object class recognition and segmentation, in: IEEE Conference on Computer Vision and Pattern Recognition, 2007.
[138] M. Valstar, B. Martinez, X. Binefa, M. Pantic, Facial point detection using boosted regression and graph models, in: IEEE Conference on Computer Vision and Pattern Recognition, IEEE, 2010, pp. 2729–2736.
[139] B. Martinez, M.F. Valstar, X. Binefa, M. Pantic, Local evidence aggregation for regression-based facial point detection, IEEE Transactions on Pattern Analysis and Machine Intelligence 35 (5) (2013) 1149–1163.
[140] A. Rabinovich, A. Vedaldi, C. Galleguillos, E. Wiewiora, S. Belongie, Objects in context, in: International Conference on Computer Vision, 2007, pp. 1–8.
[141] C. Galleguillos, A. Rabinovich, S. Belongie, Object categorization using co-occurrence, location and appearance, in: IEEE Conference on Computer Vision and Pattern Recognition, 2008, pp. 1–8.
[142] P. Quelhas, F. Monay, J. Odobez, D. Gatica-Perez, T. Tuytelaars, L.V. Gool, Modeling scenes with local descriptors and latent aspects, in: International Conference on Computer Vision, 2005.
[143] A. Bosch, A. Zisserman, X. Munoz, Scene classification using a hybrid generative/discriminative approach, IEEE Transactions on Pattern Analysis and Machine Intelligence 30 (2008).
[144] E. Sudderth, A. Torralba, W. Freeman, A. Willsky, Describing visual scenes using transformed Dirichlet processes, in: Advances in Neural Information Processing Systems, vol. 19, 2006.
[145] J. Kivinen, E. Sudderth, M. Jordan, Learning multiscale representations of natural scenes using Dirichlet processes, in: International Conference on Computer Vision, 2007, pp. 1–8.
[146] V. Jain, A. Singhal, J. Luo, Selective hidden random fields: exploiting domain specific saliency for event classification, in: IEEE Conference on Computer Vision and Pattern Recognition, vol. 1, 2008.
[147] A. Gunawardana, M. Mahajan, A. Acero, J. Platt, Hidden conditional random fields for phone classification, in: Eurospeech, 2005.
[148] S. Wang, A. Quattoni, L. Morency, D. Demirdjian, T. Darrell, Hidden conditional random fields for gesture recognition, in: IEEE Conference on Computer Vision and Pattern Recognition, 2006.
[149] A. Singhal, J. Luo, W. Zhu, Probabilistic spatial context models for scene content understanding, in: IEEE Conference on Computer Vision and Pattern Recognition, vol. 1, 2003.
[150] L.-J. Li, F.-F. Li, What, where and who? Classifying event by scene and object recognition, in: International Conference on Computer Vision, 2007.
[151] K. Murphy, A. Torralba, W. Freeman, Using the forest to see the trees: a graphical model relating features, objects and scenes, in: Neural Information Processing Systems, vol. 15, 2003.
[152] Z. Tu, Auto-context and its application to high-level vision tasks, in: IEEE Conference on Computer Vision and Pattern Recognition, 2008.
[153] A. Yilmaz, O. Javed, M. Shah, Object tracking: a survey, ACM Computing Surveys 38 (4) (2006).
[154] D. Ross, J. Lim, R.-S. Lin, M.-H. Yang, Incremental learning for robust visual tracking, International Journal of Computer Vision (2007).
[155] Y. Bar-Shalom, X.-R. Li, Estimation and Tracking: Principles, Techniques, and Software, Artech House, Boston, 1993.
[156] A. Blake, M. Isard, The condensation algorithm-conditional density propagation and applications to visual tracking, in: Advances in Neural Information Processing Systems, 1997, pp. 361–367.

[157] J. Vermaak, N.D. Lawrence, P. Perez, Variational inference for visual tracking, in: IEEE Conference on Computer Vision and Pattern Recognition, 2003.
[158] R. van der Merwe, A. Doucet, J.F.G. de Freitas, E. Wan, The unscented particle filter, in: Neural Information Processing Systems, 2000.
[159] C. Andrieu, J.F.G. de Freitas, A. Doucet, Rao–Blackwellised particle filtering via data augmentation, in: Neural Information Processing Systems, 2001.
[160] G. Casella, C.P. Robert, Rao-Blackwellisation of sampling schemes, Biometrika 83 (1) (1996) 81–94.
[161] L. Taycher, D. Demirdjian, T. Darrell, G. Shakhnarovich, Conditional random people: tracking humans with CRFs and grid filters, in: IEEE Computer Society Conference on Computer Vision and Pattern Recognition, vol. 1, 2006, pp. 222–229.
[162] D. Ross, S. Osindero, R. Zemel, Combining discriminative features to infer complex trajectories, in: International Conference on Machine Learning, 2006.
[163] S. Ju, M. Black, Y. Yacoob, Cardboard people: a parameterized model of articulated motion, in: IEEE Conference on Automatic Face and Gesture Recognition, 1996.
[164] J. Deutscher, A. Blake, I. Reid, Articulated body motion capture by annealed particle filtering, in: IEEE Conference on Computer Vision and Pattern Recognition, 2000.
[165] H. Sidenbladh, M. Black, D. Fleet, Stochastic tracking of 3D human figures using 2D image motion, in: European Conference on Computer Vision, 2000.
[166] S.B.J. Malik, J. Puzicha, Shape matching and object recognition using shape contexts, IEEE Transactions on Pattern Analysis and Machine Intelligence 24 (4) (2002) 509–522.
[167] G. Mori, J. Malik, Estimating human body configurations using shape context matching, in: European Conference on Computer Vision, 2002.
[168] O. Cula, K. Dana, 3D texture recognition using bidirectional feature histograms, International Journal of Computer Vision 59 (1) (2004) 33–60.
[169] P. Viola, M. Jones, Rapid object detection using a boosted cascade of simple features, in: IEEE Conference on Computer Vision and Pattern Recognition, 2001.
[170] A.D. Jepson, D.J. Fleet, T.F. El-Maraghi, Robust online appearance models for visual tracking, IEEE Transactions on Pattern Analysis and Machine Intelligence 25 (10) (2001) 1296–1311.
[171] E. Sudderth, A. Ihler, W. Freeman, A. Willsky, Nonparametric belief propagation, in: IEEE Conference on Computer Vision and Pattern Recognition, 2003.
[172] M. Kim, V. Pavlovic, Discriminative learning of dynamical systems for motion tracking, in: IEEE Conference on Computer Vision and Pattern Recognition, 2007.
[173] A. Mccallum, D. Freitag, F. Pereira, Maximum entropy Markov models for information extraction and segmentation, in: International Conference on Machine Learning, 2000.
[174] T. Yu, Y. Wu, N.O. Krahnstoever, P.H. Tu, Distributed data association and filtering for multiple target tracking, in: IEEE Conference on Computer Vision and Pattern Recognition, 2008.
[175] Z. Khan, T. Balch, F. Dellaert, An MCMC-based particle filter for tracking multiple interacting targets, in: European Conference on Computer Vision, 2004.
[176] C. Rasmussen, G. Hager, Joint probabilistic techniques for tracking multi-part objects, in: IEEE Conference on Computer Vision and Pattern Recognition, 1998.
[177] Y. Ohta, T. Kanade, Stereo by intra- and inter-scanline search using dynamic programming, IEEE Transactions on Pattern Analysis and Machine Intelligence (2) (1985) 139–154.
[178] O. Veksler, Stereo correspondence by dynamic programming on a tree, in: IEEE Conference on Computer Vision and Pattern Recognition, vol. 2, 2005, pp. 384–390.
[179] J. Sun, N.-N. Zheng, H.-Y. Shum, Stereo matching using belief propagation, IEEE Transactions on Pattern Analysis and Machine Intelligence 25 (7) (2003) 787–800.
[180] D. Scharstein, R. Szeliski, A taxonomy and evaluation of dense two-frame stereo correspondence algorithms, International Journal of Computer Vision 47 (1–3) (2002) 7–42.
[181] J.J. Atick, P.A. Griffin, A.N. Redlich, Statistical approach to shape from shading: reconstruction of three-dimensional face surfaces from single two-dimensional images, Neural Computation 8 (6) (1996) 1321–1340.
[182] A. Saxena, M. Sun, A.Y. Ng, Learning 3-D scene structure from a single still image, in: International Conference on Computer Vision, IEEE, 2007, pp. 1–8.
[183] A. McCallum, C. Pal, G. Druck, X. Wang, Multi-conditional learning: generative/discriminative training for clustering and classification, in: AAAI Conference on Artificial Intelligence, 2006, pp. 433–439.

[184] B. Liu, S. Gould, D. Koller, Single image depth estimation from predicted semantic labels, in: IEEE Conference on Computer Vision and Pattern Recognition, 2010, pp. 1253–1260.
[185] P. Ekman, E.L. Rosenberg, What the Face Reveals: Basic and Applied Studies of Spontaneous Expression Using the Facial Action Coding System (FACS), Oxford University Press, USA, 1997.
[186] A. Bobick, J. Davis, The recognition of human movement using temporal template, IEEE Transactions on Pattern Analysis and Machine Intelligence (2001).
[187] T. Xiang, S. Song, Video behavior profiling for anomaly detection, IEEE Transactions on Pattern Analysis and Machine Intelligence (2008).
[188] G. Cheng, Y. Wan, A.N. Saudagar, K. Namuduri, B.P. Buckles, Advances in human action recognition: a survey, CoRR [online], available: arXiv:1501.05964, 2015.
[189] R. Poppe, A survey on vision-based human action recognition, Image and Vision Computing 28 (6) (2010) 976–990.
[190] M.Á. Mendoza, N.P. De La Blanca, Applying space state models in human action recognition: a comparative study, in: International Conference on Articulated Motion and Deformable Objects, Springer, 2008, pp. 53–62.
[191] S. Herath, M. Harandi, F. Porikli, Going deeper into action recognition: a survey, Image and Vision Computing 60 (2017) 4–21.
[192] H.-S. Yoon, J. Soh, Y.J. Bae, H.S. Yang, Hand gesture recognition using combined features of location, angle and velocity, Pattern Recognition 34 (7) (2001) 1491–1501.
[193] K. Tang, L. Fei-Fei, D. Koller, Learning latent temporal structure for complex event detection, in: IEEE Conference on Computer Vision and Pattern Recognition, 2012, pp. 1250–1257.
[194] Q. Shi, L. Cheng, L. Wang, A. Smola, Human action segmentation and recognition using discriminative semi-Markov models, International Journal of Computer Vision 93 (1) (2011) 22–32.
[195] V. Pavlovic, B.J. Frey, T.S. Huang, Time-series classification using mixed-state dynamic Bayesian networks, in: IEEE Conference on Computer Vision and Pattern Recognition, vol. 2, 1999, pp. 609–615.
[196] I. Alexiou, T. Xiang, S. Gong, Learning a joint discriminative-generative model for action recognition, in: International Conference on Systems, Signals and Image Processing, IWSSIP, 2015, pp. 1–4.
[197] J. Lafferty, A. McCallum, F. Pereira, Conditional random fields: probabilistic models for segmenting and labeling sequence data, in: Proceedings of the Eighteenth International Conference on Machine Learning, ICML, vol. 1, 2001, pp. 282–289.
[198] C. Sminchisescu, A. Kanaujia, D. Metaxas, Conditional models for contextual human motion recognition, Computer Vision and Image Understanding 104 (2) (2006) 210–220.
[199] A. Quattoni, S. Wang, L.-P. Morency, M. Collins, T. Darrell, Hidden conditional random fields, IEEE Transactions on Pattern Analysis and Machine Intelligence 29 (10) (2007).
[200] Y. Wang, G. Mori, Hidden part models for human action recognition: probabilistic versus max margin, IEEE Transactions on Pattern Analysis and Machine Intelligence 33 (7) (2011) 1310–1323.
[201] C. Sun, R. Nevatia, Active: activity concept transitions in video event classification, in: Proceedings of the IEEE International Conference on Computer Vision, 2013, pp. 913–920.
[202] Summary of TRECVID datasets – NIST [online], available: https://trecvid.nist.gov/past.data.table.html, 2017.
[203] Y. Ivanov, A. Bobick, Recognition of visual activities and interactions by stochastic parsing, IEEE Transactions on Pattern Analysis and Machine Intelligence (2000).
[204] T. Truyen, D. Phung, H. Bui, AdaBoost.MRF: boosted Markov random forests and application to multilevel activity recognition, in: IEEE Conference on Computer Vision and Pattern Recognition, 2006.
[205] A.L. Yuille, A. Rangarajan, The concave-convex procedure (CCCP), in: Advances in Neural Information Processing Systems, 2002, pp. 1033–1040.
[206] X. Wang, X. Ma, W. Grimson, Unsupervised activity perception by hierarchical Bayesian models, in: IEEE Conference on Computer Vision and Pattern Recognition, 2007.

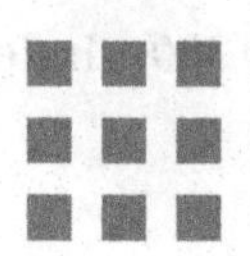

Index

Printed in the United States
by Baker & Taylor Publisher Services

Printed in the United States
by Baker & Taylor Publisher Services